ROWAN UNIVERSITY
CAMPBELL LIBRARY
201 MULLICA HILL RD.
GLASSBORO, NJ 08028-1701

DIGITAL TERRAIN MODELING
Principles and Methodology

DIGITAL TERRAIN MODELING
Principles and Methodology

Dr. Zhilin Li
Professor in Geo-Informatics
Department of Land Surveying and Geo-Informatics
The Hong Kong Polytechnic University

Dr. Qing Zhu
Professor in GIS
State Key Laboratory for Information Engineering in
Surveying, Mapping and Remote Sensing (LIESMARS)
Wuhan University

Dr. Christopher Gold
Professor, EU Marie-Curie Chair
School of Computing
University of Glamorgan

CRC PRESS

Boca Raton   London   New York   Washington, D.C.

Library of Congress Cataloging-in-Publication Data

Li, Zhilin, 1960–
  Digital terrain modeling: principles and methodology /
    Zhilin Li, Qing Zhu, and Chris Gold.
  p. cm.
  Includes bibliographical references and index.
  ISBN 0-415-32462-9
  1. Digital mapping–Methodology. I. Zhu, Qing, 1966– II. Gold, Chris, 1944– III. Title.

526–dc22

2004054578

This book contains information obtained from authentic and highly regarded sources. Reprinted material is quoted with permission, and sources are indicated. A wide variety of references are listed. Reasonable efforts have been made to publish reliable data and information, but the author and the publisher cannot assume responsibility for the validity of all materials or for the consequences of their use.

Neither this book nor any part may be reproduced or transmitted in any form or by any means, electronic or mechanical, including photocopying, microfilming, and recording, or by any information storage or retrieval system, without prior permission in writing from the publisher.

The consent of CRC Press does not extend to copying for general distribution, for promotion, for creating new works, or for resale. Specific permission must be obtained in writing from CRC Press for such copying.

Direct all inquiries to CRC Press, 2000 N.W. Corporate Blvd., Boca Raton, Florida 33431.

Trademark Notice: Product or corporate names may be trademarks or registered trademarks, and are used only for identification and explanation, without intent to infringe.

Visit the CRC Press Web site at www.crcpress.com

© 2005 by CRC Press

No claim to original U.S. Government works
International Standard Book Number 0-415-32462-9
Library of Congress Card Number
Printed in the United States of America 1 2 3 4 5 6 7 8 9 0
Printed on acid-free paper

Contents

Preface   xv

1 Introduction   1
  1.1 Representation of Digital Terrain Surfaces   1
    1.1.1 Representation of Terrain Surfaces   1
    1.1.2 Representation of Digital Terrain Surfaces   4
  1.2 Digital Terrain Models   4
    1.2.1 The Concept of Model and Mathematical Models   4
    1.2.2 The Terrain Model and the Digital Terrain Model   6
    1.2.3 Digital Elevation Models and Digital Terrain Models   7
  1.3 Digital Terrain Modeling   9
    1.3.1 The Process of Digital Terrain Modeling   9
    1.3.2 Development of Digital Terrain Modeling   9
  1.4 Relationships Between Digital Terrain Modeling and Other Disciplines   11

2 Terrain Descriptors and Sampling Strategies   13
  2.1 General (Qualitative) Terrain Descriptors   13
  2.2 Numeric Terrain Descriptors   14
    2.2.1 Frequency Spectrum   14
    2.2.2 Fractal Dimension   15
    2.2.3 Curvature   16
    2.2.4 Covariance and Auto-Correlation   17
    2.2.5 Semivariogram   17
  2.3 Terrain Roughness Vector: Slope, Relief, and Wavelength   18
    2.3.1 Slope, Relief, and Wavelength as a Roughness Vector   18
    2.3.2 The Adequacy of the Terrain Roughness Vector for DTM Purposes   19
    2.3.3 Estimation of Slope   20
  2.4 Theoretical Basis for Surface Sampling   21
    2.4.1 Theoretical Background for Sampling   21
    2.4.2 Sampling from Different Points of View   22

	2.5	Sampling Strategy for Data Acquisition	24
	2.5.1	Selective Sampling: Very Important Points plus Other Points	24
	2.5.2	Sampling with One Dimension Fixed: Contouring and Profiling	25
	2.5.3	Sampling with Two Dimensions Fixed: Regular Grid and Progressive Sampling	25
	2.5.4	Composite Sampling: An Integrated Strategy	26
	2.6	Attributes of Sampled Source Data	26
	2.6.1	Distribution of Sampled Source Data	26
	2.6.2	Density of Sampled Source Data	28
	2.6.3	Accuracy of Sampled Source Data	28

3 Techniques for Acquisition of DTM Source Data — 31

- 3.1 Data Sources for Digital Terrain Modeling — 31
  - 3.1.1 The Terrain Surface as a Data Source — 31
  - 3.1.2 Aerial and Space Images — 32
  - 3.1.3 Existing Topographic Maps — 34
- 3.2 Photogrammetry — 35
  - 3.2.1 The Development of Photogrammetry — 35
  - 3.2.2 Basic Principles of Photogrammetry — 36
- 3.3 Radargrammetry and SAR Interferometry — 39
  - 3.3.1 The Principle of Synthetic Aperture Radar Imaging — 40
  - 3.3.2 Principles of Interferometric SAR — 43
  - 3.3.3 Principles of Radargrammetry — 48
- 3.4 Airborne Laser Scanning (LIDAR) — 50
  - 3.4.1 Basic Principle of Airborne Laser Scanning — 53
  - 3.4.2 From Laser Point Cloud to DTM — 55
- 3.5 Cartographic Digitization — 56
  - 3.5.1 Line-Following Digitization — 56
  - 3.5.2 Raster Scanning — 57
- 3.6 GPS for Direct Data Acquisition — 58
  - 3.6.1 The Operation of GPS — 58
  - 3.6.2 The Principles of GPS Measurement — 60
  - 3.6.3 The Principles of Traditional Surveying Techniques — 61
- 3.7 A Comparison between DTM Data from Different Sources — 62

4 Digital Terrain Surface Modeling — 65

- 4.1 Basic Concepts of Surface Modeling — 65
  - 4.1.1 Interpolation and Surface Modeling — 65
  - 4.1.2 Surface Modeling and DTM Networks — 66
  - 4.1.3 Surface Modeling Function: General Polynomial — 66
- 4.2 Approaches for Digital Terrain Surface Modeling — 67
  - 4.2.1 Surface Modeling Approaches: A Classification — 68
  - 4.2.2 Point-Based Surface Modeling — 68

		4.2.3	Triangle-Based Surface Modeling	69
		4.2.4	Grid-Based Surface Modeling	70
		4.2.5	Hybrid Surface Modeling	71
	4.3	The Continuity of DTM Surfaces		72
		4.3.1	The Characteristics of DTM Surfaces: A Classification	72
		4.3.2	Discontinuous DTM Surfaces	72
		4.3.3	Continuous DTM Surfaces	73
		4.3.4	Smooth DTM Surfaces	74
	4.4	Triangular Network Formation for Surface Modeling		75
		4.4.1	Triangular Regular Network Formation from Regularly Distributed Data	75
		4.4.2	Triangular Irregular Network Formation from Regularly Distributed Data	77
		4.4.3	Triangular Irregular Network Formation from Irregularly Distributed Data	79
		4.4.4	Triangular Irregular Network Formation from Specially Distributed Data	80
	4.5	Grid Network Formation for Surface Modeling		80
		4.5.1	Coarser Grid Network Formation from Finer Grid Data: Resampling	81
		4.5.2	Grid Network Formation from Randomly Distributed Data	82
		4.5.3	Grid Network Formation from Contour Data	83
5	**Generation of Triangular Irregular Networks**			**87**
	5.1	Triangular Irregular Network Formation: Principles		87
		5.1.1	Approaches for Triangular Irregular Network Formation	87
		5.1.2	Principles of Triangular Irregular Network Formation	88
	5.2	Vector-Based Static Delaunay Triangulation		90
		5.2.1	Selection of a Starting Point for Delaunay Triangulation	90
		5.2.2	Searching for a Point to Form a New Triangle	92
		5.2.3	The Process of Delaunay Triangulation	93
	5.3	Vector-Based Dynamic Delaunay Triangulation		94
		5.3.1	The Principle of Bowyer–Watson Algorithm for Dynamic Triangulation	94
		5.3.2	Walk-Through Algorithm for Locating the Triangle Containing a Point	95
		5.3.3	Numerical Criterion for Edge Swapping	97
		5.3.4	Removal of a Point from the Delaunay Triangulation	98
	5.4	Constrained Delaunay Triangulation		99
		5.4.1	Constraints for Delaunay Triangulation: The Issue and Solutions	99
		5.4.2	Delaunay Triangulation with Constraints	101

5.5	Triangulation from Contour Data with Skeletonization		102
	5.5.1	Extraction of Skeleton Lines from Contour Map	103
	5.5.2	Height Estimation for Skeleton Points	104
	5.5.3	Triangulation from Contour Data with Skeletons	106
5.6	Delaunay Triangulations via Voronoi Diagrams		107
	5.6.1	Derivation of Delaunay Triangulations from Voronoi Diagrams	108
	5.6.2	Vector-Based Algorithms for the Generation of Voronoi Diagram	108
	5.6.3	Raster-Based Algorithms for the Generation of Voronoi Diagram	111

6 Interpolation Techniques for Terrain Surface Modeling — 115

6.1	Interpolation Techniques: An Overview		115
6.2	Area-Based Exact Fitting of Linear Surfaces		117
	6.2.1	Simple Linear Interpolation	117
	6.2.2	Bilinear Interpolation	117
6.3	Area-Based Exact Fitting of Curved Surface		119
	6.3.1	Bicubic Spline Interpolation	119
	6.3.2	Multi-Surface Interpolation (Hardy Method)	120
6.4	Area-Based Best Fitting of Surfaces		123
	6.4.1	Least-Squares Fitting of a Local Surface	123
	6.4.2	Least-Squares Fitting of Finite Elements	126
6.5	Point-Based Moving Averaging		127
	6.5.1	The Principle of Point-Based Moving Averaging	127
	6.5.2	Searching for Neighbor Points	128
	6.5.3	Determination of Weighting Functions	129
6.6	Point-Based Moving Surfaces		130
	6.6.1	Principles of Moving Surfaces	131
	6.6.2	Selection of Points	131

7 Quality Control in Terrain Data Acquisition — 133

7.1	Quality Control: Concepts and Strategy		133
	7.1.1	A Simple Strategy for Quality Control in Digital Terrain Modeling	133
	7.1.2	Sources of Error in DTM Source (Raw) Data	134
	7.1.3	Types of Error in DTM Source Data	134
7.2	On-Line Quality Control in Photogrammetric Data Acquisition		135
	7.2.1	Superimposition of Contours Back to the Stereo Model	135
	7.2.2	Zero Stereo Model from Orthoimages	135
	7.2.3	Trend Surface Analysis	136
	7.2.4	Three-Dimensional Perspective View for Visual Inspection	136

7.3	Filtering of the Random Errors of the Original Data		136
	7.3.1	The Effect of Random Noise on the Quality of DTM Data	137
	7.3.2	Low-Pass Filter for Noise Filtering	139
	7.3.3	Improvement of DTM Data Quality by Filtering	140
	7.3.4	Discussion: When to Apply a Low-Pass Filtering	141
7.4	Detection of Gross Errors in Grid Data Based on Slope Information		142
	7.4.1	Gross Error Detection Using Slope Information: An Introduction	143
	7.4.2	General Principle of Gross Error Detection Based on an Adaptive Threshold	143
	7.4.3	Computation of an Adaptive Threshold	145
	7.4.4	Detection of Gross Error and Correction of a Point	146
	7.4.5	A Practical Example	147
7.5	Detection of Isolated Gross Errors in Irregularly Distributed Data		147
	7.5.1	Three Approaches for Developing Algorithms for Gross Error Detection	148
	7.5.2	General Principle Based on the Pointwise Algorithm	149
	7.5.3	Range of Neighbors (Size of Window)	149
	7.5.4	Calculating the Threshold Value and Suspecting a Point	150
	7.5.5	A Practical Example	150
7.6	Detection of a Cluster of Gross Errors in Irregularly Distributed Data		151
	7.6.1	Gross Errors in Cluster: The Issue	151
	7.6.2	The Algorithm for Detecting Gross Errors in Clusters	153
	7.6.3	A Practical Example	154
7.7	Detection of Gross Errors Based on Topologic Relations of Contours		155
	7.7.1	Gross Errors in Contour Data: An Example	155
	7.7.2	Topological Relations of Contours for Gross Error Detection	156

8 Accuracy of Digital Terrain Models — 159

8.1	DTM Accuracy Assessment: An Overview		159
	8.1.1	Approaches for DTM Accuracy Assessment	159
	8.1.2	Distributions of DTM Errors	160
	8.1.3	Measures for DTM Accuracy	161
	8.1.4	Factors Affecting DTM Accuracy	163
8.2	Design Considerations for Experimental Tests on DTM Accuracy		165
	8.2.1	Strategies for Experimental Tests	165
	8.2.2	Requirements for Checkpoints in Experimental Tests	166
8.3	Empirical Models for the Accuracy of the DTM Derived from Grid Data		170
	8.3.1	Three ISPRS Test Data Sets	170
	8.3.2	Empirical Models for the Relationship between DTM Accuracy and Sampling Intervals	170

		8.3.3	Empirical Models for DTM Accuracy Improvement with the Addition of Feature Data	172
	8.4	\multicolumn{2}{l	}{Theoretical Models of DTM Accuracy Based on Slope and Sampling Interval}	173

    8.3.3 Empirical Models for DTM Accuracy Improvement with the Addition of Feature Data 172
8.4 Theoretical Models of DTM Accuracy Based on Slope and Sampling Interval 173
    8.4.1 Theoretical Models for DTM Accuracy: An Overview 174
    8.4.2 Propagation of Errors from DTM Source Data to the DTM Surface 178
    8.4.3 Accuracy Loss Due to Linear Representation of Terrain Surface 180
    8.4.4 Mathematical Models of the Accuracy of DTMs Linearly Constructed from Grid Data 186
8.5 Empirical Model for the Relationship between Grid and Contour Intervals 188
    8.5.1 Empirical Model for the Accuracy of DTMs Constructed from Contour Data 188
    8.5.2 Empirical Model for the Relationship between Contour and Grid Intervals 189

9 Multi-Scale Representations of Digital Terrain Models 191
9.1 Multi-Scale Representations of DTM: An Overview 191
    9.1.1 Scale as an Important Issue in Digital Terrain Modeling 191
    9.1.2 Transformation in Scale: An Irreversible Process in Geographical Space 192
    9.1.3 Scale, Resolution, and Simplification of Representations 194
    9.1.4 Approaches for Multi-Scale Representations 195
9.2 Hierarchical Representation of DTM at Discrete Scales 196
    9.2.1 Pyramidal Structure for Hierarchical Representation 196
    9.2.2 Quadtree Structure for Hierarchical Representation 198
9.3 Metric Multi-Scale Representation of DTM at Continuous Scales: Generalization 200
    9.3.1 Requirements for Metric Multi-Scale Representation of DTM 200
    9.3.2 A Natural Principle for DTM Generalization 200
    9.3.3 DTM Generalization Based on the Natural Principle 202
9.4 Visual Multi-Scale Representation of DTM at Continuous Scales: View-Dependent LOD 205
    9.4.1 Principles for View-Dependent LOD 205
    9.4.2 Typical Algorithms for View-Dependent LOD for DTM Data 207
9.5 Multi-Scale DTM at a National Level 208
    9.5.1 Multi-Scale DTM in China 209
    9.5.2 Multi-Scale DTM in the United States 209

10 Management of DTM Data 211
10.1 Strategies for management of DTM data 211
    10.1.1 Strategy for Making DTM Data Management Operational 211
    10.1.2 Strategy for Using Databases for DTM Data Management 212

	10.2	Management of DTM Data with Files	213
		10.2.1 File Structure for Grid DTM	213
		10.2.2 File Structure for TIN DTM	214
		10.2.3 File Structure for Additional Terrain Feature Data	216
	10.3	Management of DTM Data with Spatial Databases	217
		10.3.1 Organization of Tables for Grid DTM Data	218
		10.3.2 Organization of Tables for TIN DTM Data	221
		10.3.3 Organization of Tables for Additional Terrain Feature Data	223
		10.3.4 Organization of Tables for Metadata	225
	10.4	Compression of DTM Data	226
		10.4.1 Concepts and Approaches for DTM Data Compression	226
		10.4.2 Huffman Coding	227
		10.4.3 Differencing Followed by Coding	228
	10.5	Standards for DTM Data Format	229
		10.5.1 Concepts and Principles of DTM Data Standards	230
		10.5.2 Standards for DTM Data Exchange of the United States	231
		10.5.3 Standards for DTM Data Exchange of China	231
11	**Contouring from Digital Terrain Models**		**233**
	11.1	Approaches for Contouring from DTM	233
	11.2	Vector-Based Contouring from Grid DTM	233
		11.2.1 Searching for Contour Points	234
		11.2.2 Interpolation of Contour Points	235
		11.2.3 Tracing Contour Lines	236
		11.2.4 Smoothing Contour Lines	238
	11.3	Raster-Based Contouring from Grid DTM	238
		11.3.1 Binary and Edge Contouring	239
		11.3.2 Gray-Tone Contouring	241
	11.4	Vector-Based Contouring from Triangulated DTM	241
	11.5	Stereo Contouring from Grid DTM	243
		11.5.1 The Principle of Stereo Contouring	243
		11.5.2 Generation of Stereomate for Contour Map	245
12	**Visualization of Digital Terrain Models**		**247**
	12.1	Visualization of Digital Terrain Models: An Overview	247
		12.1.1 Variables for Visualization	247
		12.1.2 Approaches for the Visualization of DTM Data	250
	12.2	Image-Based 2-D DTM Visualization	250
		12.2.1 Slope Shading and Hill Shading	251
		12.2.2 Height-Based Coloring	252
	12.3	Rendering Technique for Three-Dimensional DTM Visualization	253
		12.3.1 Basic Principles of Rendering	253
		12.3.2 Graphic Transformations	254
		12.3.3 Visible Surfaces Identification	256
		12.3.4 The Selection of an Illumination Model	257
		12.3.5 Gray Value Assignment for Graphics Generation	259

	12.4	Texture Mapping for Virtual Landscape Generation	260
		12.4.1 Mapping Texture onto DTM Surfaces	260
		12.4.2 Mapping Other Attributes onto DTM Surfaces	262
	12.5	Animation Techniques for DTM Visualization	262
		12.5.1 Principles of Animation	263
		12.5.2 Seamless Pan-View on DTM in a Large Area	264
		12.5.3 "Fly-Through" and "Walk-Through" for DTM Visualization	266
13	**Interpretation of Digital Terrain Models**		**267**
	13.1	DTM Interpretation: An Overview	267
	13.2	Geometric Terrain Parameters	267
		13.2.1 Surface and Projection Areas	268
		13.2.2 Volume	270
	13.3	Morphological Terrain Parameters	271
		13.3.1 Slope and Aspect	271
		13.3.2 Plan and Profile Curvatures	274
		13.3.3 Rate of Change in Slope and Aspect	275
		13.3.4 Roughness Parameters	275
	13.4	Hydrological Terrain Parameters	276
		13.4.1 Flow Direction	276
		13.4.2 Flow Accumulation and Flow Line	278
		13.4.3 Drainage Network and Catchments	279
		13.4.4 Multiple Direction Flow Modeling: A Discussion	280
	13.5	Visibility Terrain Parameters	281
		13.5.1 Line-of-Sight: Point-to-Point Visibility	282
		13.5.2 Viewshed: Point-to-Area Visibility	283
14	**Applications of Digital Terrain Models**		**285**
	14.1	Applications in Civil Engineering	285
		14.1.1 Highway and Railway Design	285
		14.1.2 Water Conservancy	286
	14.2	Applications in Remote Sensing and Mapping	288
		14.2.1 Orthoimage Generation	288
		14.2.2 Remote Sensing Image Analysis	290
	14.3	Applications in Military Engineering	290
		14.3.1 Flight Simulation	290
		14.3.2 Virtual Battlefield	291
	14.4	Applications in Resources and Environment	291
		14.4.1 Wind Field Models for Environmental Study	291
		14.4.2 Sunlight Model for Climatology	292
		14.4.3 Flood Simulation	292
		14.4.4 Agriculture Management	293
	14.5	Marine Navigation	293
	14.6	Other Applications	295

15	**Beyond Digital Terrain Modeling**	297
	15.1 Digital Terrain Modeling with Complex Construction	297
	15.1.1 Manual Addition of Constructions on Terrain Surface	297
	15.1.2 Semiautomated Modification of the Terrain Surface	298
	15.2 Digital Terrain Modeling on the Sphere	300
	15.2.1 Generation of TIN and Voronoi Diagram on Sphere	300
	15.2.2 Voronoi Diagram for Modeling Changes in Sea Level on Sphere	301
	15.3 Three-Dimensional Volumetric Modeling	302
Epilogue		305
References		307
Index		319

Preface

Terrain models have always appealed to military personnel, planners, landscape architects, civil engineers, as well as other experts in various earth sciences. Originally, terrain models were physical models, made of rubber, plastic, clay, sand, etc. Since the later 1950s, the computer has been introduced into this area and the modeling of terrain surface has since then been carried out numerically or digitally, leading to the current discipline — digital terrain modeling.

Digital terrain modeling is a process to obtain desirable models of the land surface. Such models have found wide applications, since its origin in the late 1950s, in various disciplines such as mapping, remote sensing, civil engineering, mining engineering, geology, geomorphology, military engineering, land planning, and communications. Therefore, digital terrain modeling has become a discipline receiving increasing attention.

It is encouraging that more literature is now available in this discipline. After 30 years of development, the first book in this area, entitled *Terrain Modelling in Surveying and Civil Engineering*, was published by Whittles Publishing in 1990, which was edited by Prof. G. Petrie of Glasgow University together with his former student Tom Kennie. This book has been serving as the text book in this area since its publication. On the other hand, as one could imagine, some of the materials in this book have become outdated during another 10 years of rapid development. A revision of this book was desirable. This became difficult after the retirement of Prof. Petrie and Tom Kennie's leaving of the academic community.

Therefore, Zhilin Li, as a former Ph.D. student of Prof. G. Petrie at Glasgow University, felt obliged to do something. He talked to Qing Zhu of Wuhan University and decided to write a book. In 2000, a book entitled *Digital Elevation Model* was written in Chinese and published by the then Wuhan Technical University of Surveying and Mapping Press (now Wuhan University Press). This book was largely based on some of the materials from the Ph.D. thesis of Zhilin Li (1990) and the research work of both Zhilin and Qing, thus some traditional topics such as contouring and interpolation are either very simplified or completely neglected. This book has been well received in China and is widely used as a textbook for postgraduate students in geo-information. As a result, Zhilin and Qing were presented an "Excellent Textbook Award" (second prize) by the Ministry of Education of China in 2002.

However, the omission of some traditional topics made it deficient as a textbook and there was an urgent need for a revision of this book. At that critical moment, Chris Gold joined the Hong Kong Polytechnic University in 2000 and became a colleague of Zhilin. This presented Zhilin and Qing with a golden opportunity to cooperate with Chris not only to revise the book but also to produce an English edition. Chris happily accepted an offer to be one of the coauthors as he has been working in terrain modeling using triangulation and Voronoi diagrams for nearly 30 years and had a lot of materials to be included. As a result, the current English edition is produced, which is indeed more a rewritten book than a revised version.

This book contains 15 chapters. Apart from the introduction, Chapters 2 and 3 are about sampling and data acquisition. Chapters 4 to 6 are about the theories, methods, and algorithms for digital terrain modeling. Chapters 7 and 8 are on quality control and accuracy of digital terrain modeling. Chapters 9 to 12 are about presentation of DTMs, in databases, in contour form and in other forms of computer graphics. Chapters 13 and 14 are about interpretation and applications. Chapter 15 discusses some extensions of digital terrain models for specific problems, to present an opinion on where the research in this area will lead. Chapters 9, 11, and 15 are newly added to make the original edition more complete. There are major revisions in all other chapters.

As the authors of this book, we are pleased to present you with this volume. However, we must do justice to the many who have contributed to the various earlier versions. We appreciate Prof. G. Petrie's assistance to Zhilin while writing his Ph.D. dissertation. We would like to express our thanks to Valerie Gold (Chris's wife) for editing the language; to Prof. D. Li of Wuhan University for his encouragement of the writing of this book; to a number of our students for producing some of the diagrams; and to the publisher for making this volume available to you. We hope you like it. Last but not the least, we would also like to thank Lingyun Liu, Yijun Zhang, and Valerie Gold (i.e., our wives) for their support.

Z. Li, Q. Zhu, and C. Gold

CHAPTER 1

Introduction

1.1 REPRESENTATION OF DIGITAL TERRAIN SURFACES

People live on Earth and learn to cope with its terrain. Civil engineers design and construct buildings on it; geologists try to study its underlying construction; geomorphologists are interested in its shape and the processes by which the landscape was formed; and topographic scientists are concerned with measuring and describing its surface and presenting it in different ways, for example, using maps, orthoimages, perspective views, etc. Despite these differences in emphasis and interest, these specialists have a common interest, that is, they wish the surface of the terrain to be represented conveniently and with a certain accuracy.

1.1.1 Representation of Terrain Surfaces

People have tried every means to represent phenomena on the terrain that they have been familiar with since ancient times, and painting may be the oldest representation. A painting offers some general information (e.g., shape and color) about the terrain which it depicts; however, the metric quality (or accuracy) is extremely low and, thus, it cannot be used for engineering purposes.

Another ancient but effective terrain representation is maps, which are still widely used today. Maps have played as important a role in the development of society as language. Indeed, maps have been used to represent the environments during the history of civilization.

In ancient times, semi-symbolic and semi-pictorial descriptions were used to depict the actual three-dimensional (3-D) terrain surface. Again, the metric quality (or accuracy) was very low. Modern maps employ a well-designed symbol system and a well-established mathematical basis for representation so that they possess three major characteristics:

1. measurability warranted by the mathematical rules
2. overview provided by generalization
3. intuition by symbolization.

A contoured topographic map is perhaps the most familiar way of representing terrain. On a topographic map, all features present on the terrain are projected orthogonally onto a 2-D horizontal datum. Detail is then reduced in scale and represented by lines and symbols. Terrain height and morphological information are represented by contour lines. The use of such maps can be traced back to the 18th century. It is believed by many that the contour map is one of the most important inventions in the history of mapping due to its convenience and intuition to perceive. Figure 1.1 is an example of the contour map.

Essentially, a map is a scientific generalization and abstraction of features on the terrain. Typically, and perhaps most importantly, topographic maps make use of 2-D representation for 3-D reality. There is always a gulf between the 2-D representation and the 3-D reality. Because of this gulf, cartographers have been devoting themselves to the 3-D representation of terrain topography for years. Scenography, hachuring, shading and hypermetric tints (color layers) have been traditionally used on topographic maps; however, only shading is still widely in use because it can be easily generated by computers. Figure 1.2 is an example of a topographic map with shading.

Compared to various line drawings, images have some advantages: for instance, they are more detailed and easier to understand. Therefore, as soon as photography was invented, it was used extensively to record the colorful world we live in. Since 1849,

Figure 1.1 Contour map of a small island.

INTRODUCTION

photographs, and later aerial photographs, have been used for terrain representation. However, in an aerial photograph, one dimension of the 3-D surface, the height, is essentially absent, so that a single aerial photograph cannot be used to derive information about the true heights of ground points. The rectified aerial images can be used as a plan in some sense. However, 3-D surfaces can be reconstructed by using a pair of aerial photographs with a certain percentage of overlap (i.e., 60% normally). This technique is called photogrammetry.

Satellite images have been used to complement aerial photography since the 1970s. Many satellite systems take overlapping images of the terrain so that these images can also be used to construct 3-D models. SPOT and, more recently, IKONOS are two examples. Figure 1.3 is an example of IKONOS satellite images. However, the resolution of satellite images is still not compatible with aerial images.

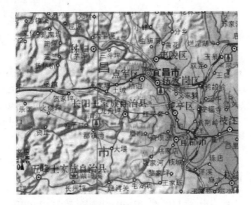

Figure 1.2 A topographic map with shading.

Figure 1.3 An IKONOS image of Hong Kong with 4 m resolution. The color plate can be viewed at http://www.crcpress.com/e_products/downloads/download.asp?cat_no=TF1732.

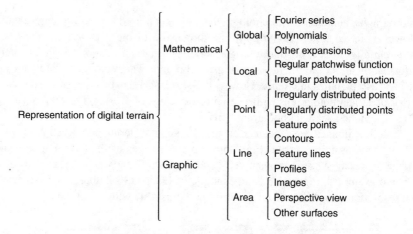

Figure 1.4 A classification scheme of representation of digital terrain surfaces.

Terrain can also be represented by a perspective view. The process of representing a surface in this way includes projecting it onto a plane and removing those lines that are not visible from the point of projection. One such product is the so-called *block diagram* and another is the perspective contour diagram. For easy production, a digital model of the terrain surface is essential.

1.1.2 Representation of Digital Terrain Surfaces

Since the middle of the 20th century, various digital terrain representation techniques have been developed with the development of computing technology, modern mathematics, and computer graphics. Nowadays, the use of the computer has become a significant landmark in the information era. Indeed, computers have become an important means for the representation of digital terrain surface.

Digital terrain surfaces can be represented mathematically and graphically. Fourier series and polynomials are common mathematic representations. Regular grid, irregular grid, contouring and the sectional diagram are common graphic representations. Figure 1.4 illustrates these.

1.2 DIGITAL TERRAIN MODELS

In representing the terrain surface, the digital terrain model (DTM) is one of the most important concepts. This section will discuss this concept, starting from the general model.

1.2.1 The Concept of Model and Mathematical Models

A "*Model* is an object or a concept which is used to represent something else. It is reality scaled down and converted to a form which we can comprehend" (Meyer 1985).

A model may have a few specific purposes such as prediction and control, etc., in which case, the model only needs to have just enough significant detail to satisfy these purposes. The model may be used to represent the original situation (system or phenomenon) or it may be used to represent some proposed or predicted situation.

Thus, the word *model* usually means a representation and in many situations it is used to describe the system at hand. Consequently, there are strong differences of opinion as to the appropriate use of the word *model*. For example, it may be applied to a photogrammetric replication of a piece of the terrain surface which has been photographed or it may suggest a perspective view of the piece of terrain.

Generally speaking, there are three types of models:

1. conceptual
2. physical
3. mathematical.

The *conceptual model* is the model borne in a person's mind about a situation or an object based on his knowledge or experience. Often this particular type of model forms the primary stage of modeling and will be followed later by a physical or mathematical model. However, if the situation or object is too difficult to represent in any other way, then the model will remain conceptual.

A physical model is usually an analog model. An example of this kind of model would be a terrain model made of rubber, plastic, or clay. A stereo model of terrain based on optical or mechanical projection principles, which is widely used in photogrammetry, would also fall into this category. A physical model is usually smaller than the real object in geosciences.

A mathematical model represents a situation, object, or phenomenon in mathematical terms. In other words, a mathematical model is a model whose components are mathematical concepts, such as constants, variables, functions, equations, inequalities, etc.

Mathematical models may be divided into two types (Saaty and Alexander 1981):

1. quantitative models, based on a number system
2. qualitative models, based on set theory, etc., and not reducible to numbers.

Also, a problem may be either deterministic or subject to changes and therefore probabilistic. Therefore, mathematical models may also be classified into

1. functional models, which are those intended to solve deterministic problems
2. stochastic models, which are those used to solve probabilistic problems.

One very important question about mathematical models is "what kind of benefit can one have by using mathematical models" or "why should we make use of mathematical models?" Saaty and Alexander (1981) give the following reasons:

1. Models permit abstraction based on logical formation using a convenient language expressed in a shorthand notation, thus enabling one to better visualize the main

elements of a problem while at the same time satisfying communication, decreasing ambiguity, and improving the chance of agreement on the results.
2. They allow one to keep track of a line of thought, focusing attention on the important parts of the problem.
3. They help one to generalize or apply the results of solving problems on the other areas.
4. They provide an opportunity to consider all the possibilities, to evaluate alternatives, and to eliminate the impossible ones.
5. They are tools for understanding the real world and discovering natural laws.

Next, comes the question "what kind of mathematical models should be used?" This is related to the problem of how to judge the *goodness* or value of a mathematical model. Meyer (1985) provides six criteria for the evaluation of mathematical models:

1. *accuracy*: the output of the model is correct or very nearly correct
2. *descriptive realism*: based on correct assumptions
3. *precision*: the prediction of the model is definite numbers, functions, or geometric figures
4. *robustness*: relative immunity to errors in the input data
5. *generality*: applicability to a wide variety of situations
6. *fruitfulness*: the conclusions are useful, or inspiring and pointing the way to other good models.

To this list, Li (1990) has added one more criterion, namely, simplicity: the smallest possible number of parameters are used in the model.

This is based on the fact that complicated models are not always needed even though a phenomenon may be complicated and is also in accordance with the *principle of parsimony* (Cryer 1986).

1.2.2 The Terrain Model and the Digital Terrain Model

Terrain models have always appealed to military personnel, planners, landscape architects, civil engineers, as well as other experts in various earth sciences. Originally, terrain models were physical models, made of rubber, plastic, clay, sand, etc. For example, during the Second World War, many models were made by the American Navy and reproduced in rubber (Baffisfore 1957). In the recent Folklands War in 1982, the British forces in the field used sand and clay models extensively to plan military operations.

The introduction of mathematical, numerical, and digital techniques to terrain modeling owes much to the activities of photogrammetrists working in the field of civil engineering. In the 1950s, photogrammetry had begun to be used widely to collect data for highway design. Roberts (1957) first proposed the use of the digital computer with photogrammetry as a new tool for acquiring data for planning and design in highway engineering. Miller and Laflamme (1958) of Massachussetts Institute of Technology (MIT) described the development in detail. They selected and measured from stereo models the 3-D coordinates of the terrain points along designed roads and formed digital profiles in the computer to assist road design. They also introduced

INTRODUCTION

the concept of the digital terrain model. The definition given by them is as follows:

> The digital terrain model (DTM) is simply a statistical representation of the continuous surface of the ground by a large number of selected points with known X, Y, Z coordinates in an arbitrary coordinate field.

Compared to traditional analog representation, a DTM has the following specific features:

1. *A variety of representation forms*: In digital form, various forms of representations can be easily produced, such as topographic maps, vertical and cross sections, and 3-D animation.
2. *No accuracy loss of data over time*: As time goes by, paper maps may be deformed, but the DTM can keep its precision owing to the use of digital medium.
3. *Greater feasibility of automation and real-time processing*: In digital form, data integration and updating are more flexible than in analog form.
4. *Easier multi-scale representation*: DTM can be arranged in different resolutions, corresponding to representations at different scales.

1.2.3 Digital Elevation Models and Digital Terrain Models

In a sense, the DTM was defined as a digital (numerical) representation of the terrain. Since Miller and Laflamme (1958) coined the original term, other alternatives have been brought into use. These include digital elevation models (DEMs), digital height models (DHMs), digital ground models (DGMs), as well as digital terrain elevation models (DTEMs). These terms originated from different countries. DEM was widely used in America; DHM came from Germany; DGM was used in the United Kingdom; and DTEM was introduced and used by USGS and DMA (Defense Mapping Agency) (Petrie and Kennie 1987).

In practice, these terms (DTM, DEM, DHM, and DTEM) are often assumed to be synonymous and indeed this is often the case. But sometimes they actually refer to different products. That is, there may be slight differences between these terms. Li (1990) has made a comparative analysis of these differences as follows:

1. *Ground*: "the solid surface of the earth"; "a solid base or foundation"; "a surface of the earth"; "bottom of the sea"; etc.
2. *Height*: "measurement from base to top"; "elevation above the ground or recognized level, especially that of the sea"; "distance upwards"; etc.
3. *Elevation*: "height above a given level, especially that of sea"; "height above the horizon"; etc.
4. *Terrain*: "tract of country considered with regarded to its natural features, etc."; "an extent of ground, region, territory"; etc.

From these definitions, some of the differences between DGM, DHM, DEM, and DTM begin to manifest themselves. So, a DGM more or less has the meaning of "a digital model of a solid surface." In contrast to the use of *ground*, the terms *height* and *elevation* emphasize the "measurement from a datum to the top" of an object. They do not necessarily refer to the altitude of the terrain surface, but in practice, this is the aspect that is emphasized in the use of these terms. The meaning of "terrain" is more complex and embracing. It may contain the concept of "height" (or "elevation"), but also attempts to include other geographical elements and natural features. Therefore, the term DTM tends to have a wider meaning than DHM or DEM and will attempt to incorporate specific terrain features such as rivers, ridge lines, break lines, etc. into the model (Li 1990).

Indeed, the term *terrain* means different things to specialists in different areas and so does the term DTM. Surveyors study DTM from the viewpoint of terrain representation and are especially interested in the topography of the terrain and objects in the terrain. The ideal DTM in their mind could be the new generation of topographic maps, of course, in digital form.

Specialists in other geosciences combine the non-topographic information with topographic information to construct the DTM according to their own specific needs. For example, at the very beginning, Miller and Laflamme (1958) intended to add geotechnical information to regular grid nodes of the strip area for computer-assisted highway design. Generally, a DTM could contain the following four groups of (topographic and nontopographic) information as follows:

1. Landforms, such as elevation, slope, slope form, and the other more complicated geomorphological features that are used to depict the relief of the terrain.
2. Terrain features, such as hydrographic features (i.e., rivers, lakes, coast lines), transportation networks (i.e., roads, railways, paths), settlements, boundaries, etc.
3. Natural resources and environments, such as soil, vegetation, geology, climate, etc.
4. Socioeconomic data, such as the population distribution in an area, industry and agriculture and capital income, etc.

From the discussion above, the definition of the DTM may be generalized as: A DTM is an ordered set of sampled data points that represent the spatial distribution of various types of information on the terrain. The mathematical expression could be something like:

$$K_P = f(u_P, v_P), \qquad K = 1, 2, 3, \ldots, m, \qquad P = 1, 2, 3, \ldots, n \qquad (1.1)$$

where K_P is one attribute value of the kth type of terrain feature at the location of point P (which can be a single point, but is usually a small area centered by P); u_P, v_P is the 2-D coordinate pair of point P; m ($m \geq 1$) is the total number of terrain information types; and n is the total number of sampled points. For example, suppose soil type is categorized as ith type of terrain information, then the DTM of this component is expressed as

$$I_P = f_i(u_P, v_P), \qquad P = 1, 2, 3, \ldots, n. \qquad (1.2)$$

A DTM is a digital representation of the spatial distribution of one or more types of terrain information and is represented by 2-D locations plus a mathematical representation of terrain information. It is commonly regarded as a 2.5-D representation of the terrain information in 3-D geographical space.

In Equation (1.1), when $m = 1$ and the terrain information is height, then the result is the mathematical expression of DEM. Obviously, DEM is a subset of DTM and the most fundamental component of DTM. However, DEM usually refers to the elevation data organized in the form of a matrix. In fact, other terms such as DGM, DHM, and DTEM have all been superseded by DEM to refer to terrain models with elevation information only. In the context of this book, we are interested in terrain information much more than just the elevation, although the socioeconomic information and resources and environmental information are not considered. Therefore, the term DTM will be used throughout this book.

1.3 DIGITAL TERRAIN MODELING

1.3.1 The Process of Digital Terrain Modeling

The process for the construction of a DTM surface is called digital terrain modeling. It is also a process of mathematical modeling. In such a process, points are sampled from the terrain to be modeled with a certain observation accuracy, density, and distribution; the terrain surface is then represented by the set of sample points. If attributes on locations on the digital surface other than the sample points need to be obtained, interpolation is then applied by forming a DTM surface from the sampled data points. Other attributes could be the height value, slope and aspect, and so on.

Figure 1.5 of Li (1990) describes the whole process of digital terrain modeling. It can be seen clearly that there are six different stages, in each of which one or more actions are needed to move to the next one. A total of 12 actions (tasks) are listed in the figure although actually, a specific DTM project may need only some of them. In fact, some actions are omitted in this book, such as feasibility study, project planning and design, contracting and shipment. In other words, this book deals mainly with the theoretical and methodological aspects of digital terrain modeling. The chapters are organized following the data flow shown in Figure 1.5.

1.3.2 Development of Digital Terrain Modeling

In the late 1950s, Miller and Laflamme (1958) introduced DTM into civil engineering. They also made use of DTM to monitor the changes in Earth's surface (e.g., subsidence and erosion). Furthermore, they suggested automated data acquisition by scanning stereo pairs of aerial photographs.

Since the 1960s, DTM has been an important research area for the International Society for Photogrammetry and Remote Sensing, as photogrammetrists are usually DTM producers. In the 1960s and early 1970s, the main research was on surface modeling and contouring from DEM. At this stage, many interpolation methods were proposed such as different types of moving averages (Schuts 1976), HIFI (height interpolation by finite element) (Ebner et al. 1980), projective interpolation, and even Kriging. Many triangulation methods have been proposed (e.g., McCullagh and Ross 1980; Gannapathy and Dennehy 1982; Christensen 1987). For contouring, threading and smoothing methods were studied (e.g., Yeoli 1977; Elfick 1979). It has been gradually recognized that sampling interval is the single critical factor. From the 1970s focus has shifted to quality control and sampling strategies. Both experimental

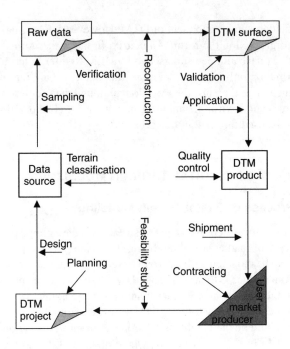

Figure 1.5 The process of digital terrain modeling (Li 1990).

studies and theoretical analysis have been conducted to produce mathematical models for the prediction of DTM accuracy (e.g., Makarovic 1972; Kubik and Botman 1976; Ackermann 1979; Frederiksen 1980; Li 1993b). The progressive sampling proposed by Makarovic (1973, 1979) is a typical example of sampling strategies used in photogrammetry. Determination of optimum sampling intervals has also been tried (Frederiksen et al. 1986; Balce 1987; Li 1990) and it relies heavily on the reliability of mathematical models for predicting DTM accuracy (e.g., Torlegard et al. 1986; Li 1992a, 1993a,b, 1994). From the late 1980s, large-scale production came into practice (e.g., Toomey 1988).

Analytical plotters are the most widely used machines for DTM data acquisition. The invention of the analytical plotter is attributed to Helava (1958). The concept was first used in AP1 and AP2 in the early 1960s. In the late 1980s, image-matching techniques (Heleva and Chapelle 1972; Masry 1974; Keating and Wolf 1976; Sarjakoski 1981) were developed in photogrammetry and automated data acquisition has been made possible since then.

In the 1990s, with the development of geographical information systems (GIS), DTM has become an important part of a national geospatial data infrastructure. DTM is used more and more in geospatial information science and technology. Indeed, DTM has found wide application in all geosciences and engineering, such as

1. planning and design of civil, road, and mine engineering
2. 3-D animation for military purposes, landscape design, and urban planning
3. analysis of catchments and hydraulic simulation

INTRODUCTION

4. analysis of visibility between objects on the terrain surface
5. terrain analysis and volume computation
6. geomorphological and soil erosion analysis
7. remote sensing image interpretation and processing
8. various types of geographical analysis
9. others.

1.4 RELATIONSHIPS BETWEEN DIGITAL TERRAIN MODELING AND OTHER DISCIPLINES

To discuss the relationships between digital terrain modeling and other disciplines, it is necessary to examine who are involved in the business. As discussed previously, the early development of digital terrain modeling involved photogrammetrists and civil engineers. Scientists in computational geometry and applied mathematics are involved in the development of modeling algorithms, and scientists in computing technology are involved in data management and system development. Nowadays, specialists from various geo-disciplines are involved in the applications of DTMs. Therefore, digital terrain modeling comprises four major components, that is, data acquisition, modeling, data management, and application development. However, they are not in a linear connection. For example, photogrammetry is a tool for data acquisition for terrain modeling; however, DTM is also applied to photogrammetry for ortho-rectification of aerial photographs and satellite images. Therefore, the inter-relationships are like those shown in Figure 1.6.

1. In "data acquisition," photogrammetry, surveying (including global positioning system [GPS] surveying), remote sensing, and cartography (mainly digitization of contour maps) are the main disciplines.
2. In "computation and modeling," photogrammetry, surveying, cartography, geography, computational geometry, computer graphics, and image processing are the main disciplines.

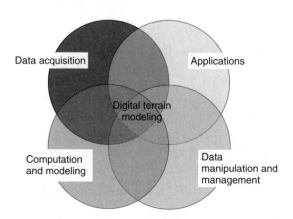

Figure 1.6 Relationships between digital terrain modeling and other disciplines.

3. In "data management and manipulation," spatial database technique, data coding and compression techniques, data structuring, and computer graphics, are the main disciplines.
4. In "applications," all geosciences are involved, including surveying, photogrammetry, cartography, remote sensing, geography, geomorphology, civil engineering, mining engineering, geological engineering, landscape design, urban planning, environmental management, resources management, facility management, and so on.

Indeed, DTM has also found wide application in military engineering (such as flight simulation, battle simulation, tank route planning, missile and airplane navigation, etc.).

Apart from these applications in science, technology, and engineering, DTM has also found wide use in computer games. That is, DTM in involved in our daily life.

CHAPTER 2

Terrain Descriptors and Sampling Strategies

To model a piece of terrain surface, first a set of data points needs to be acquired from the surface. Indeed, data acquisition is the primary (and perhaps the single most important) stage in digital terrain modeling. For this, two stages are distinguished, that is, sampling and measurement. Sampling refers to the selection of the location while measurement determines the coordinates of the location. Sampling will be discussed in this chapter while measurement methods will be discussed in the next chapter. Three important issues related to acquired DTM source (or raw) data are density, accuracy, and distribution. The accuracy is related to measurements. The optimum density and distribution are closely related to the characteristics of the terrain surface. For example, if a terrain is a plane, then three points on any location will be sufficient. This is not a realistic assumption and, therefore, an analysis of the terrain surface precedes the discussion of sampling strategies in this chapter.

2.1 GENERAL (QUALITATIVE) TERRAIN DESCRIPTORS

In general, two basic types of descriptors may be distinguished:

1. *qualitative descriptors*, which are expressed in general terms, so that they are referred to as *general descriptors*
2. *quantitative descriptors*, which are those specified by *numeric descriptors*.

In this section, a brief discussion of general descriptors is given and numeric descriptors are described in the next section. As discussed in Chapter 1, different groups of people are concerned with different attributes of the terrain surface. Therefore, a variety of general descriptors can be found based on these different interests. However, some of them are irrelevant to the concern of digital terrain modeling. Indeed, those that indicate the roughness and the coverage of terrain surface are more important in the context of terrain surface modeling. The following are some of these descriptors:

1. *Descriptors based on terrain surface cover*: Vegetation, water, desert, dry soil, snow, artificial or man-made features (e.g., roads, buildings, airports, etc.), and so on.
2. *Descriptors based on genesis of landforms*: Two such forms have been distinguished (Demek 1972), each of which has its own special characteristics —
    - *endogenetic forms*: formed by internal forces, including neotectonic forms, volcanic forms, and those forms resulting from deposition of hot springs
    - *exogenetic forms*: formed by external forces, including denudation forms, fluvial forms, karst forms, glacial forms, marine forms, and so on.
3. *Descriptors based on physiography*: Generalized regions according to the structure and characteristics of its landforms, each of which is kept as homogenous as possible and has dominant characteristics, for example, high mountains, high plateau, mountains, low mountains, hills, plateau, etc.
4. *Descriptors based on other classifications*.

Those descriptors are so broad that they can only provide the user with some very general information about a particular landscape and thus they can only be used for general planning but not for project design. To design a particular project, more precise numeric descriptors are essential.

2.2 NUMERIC TERRAIN DESCRIPTORS

The complexity of a terrain surface may be described by the concepts of roughness and irregularity and characterized by different numerical parameters.

2.2.1 Frequency Spectrum

A surface can be transformed from the space domain to the frequency domain by means of a Fourier transformation. The terrain surface in its frequency domain is characterized by the frequency spectrum. The estimation of such a spectrum from equally spaced discrete (profile) data has been discussed by Frederiksen et al. (1978). The spectrum can be approximated by the following expression:

$$S(F) = E \times F^a \tag{2.1}$$

where F denotes the frequency at which the spectrum magnitude is $S(F)$ and E and a are constants (i.e., characteristic parameters), which are two statistics expressing the complexity of the terrain surface (or profiles) over all of the area. Thus, they can be considered as parameters to provide more detailed information about the terrain surface, although still general in some sense.

Different values for E and a can be obtained from different types of terrain surfaces. According to the study carried out by Frederiksen (1981), if the parameter a is greater than 2, the landscape is hilly with a smooth surface, and if the value of

a is smaller than 2, it indicates a flat landscape with a rough surface since the surface contains large variations with high frequency (short wavelength). The value of a provides us with general topographic information.

2.2.2 Fractal Dimension

Fractal dimension is another statistical parameter which can be used to characterize the complexity of a curve or a surface. The discussion will start with the concept of effective dimension.

It is well known that in Euclidean geometry, a curve has a dimension of 1 and a surface has a dimension of 2 regardless of its complexity. However, in reality, a very irregular curve is much longer than a straight line between the same points, and a complex surface has a much larger area than a plane over the same area. In the extreme, if a line is so irregular that it fills a plane fully, then it becomes a plane, thus having a dimension of 2. Similarly, a surface could have a dimension of 3.

In fractal geometry, which was introduced by Mandelbrot (1981), the dimensionality of an object is defined by necessity (i.e., practical need), leading to the so-called effective dimension. This can be explained by taking the example of the shape of the Earth's surface when viewed from different distances.

1. If it is viewed from an infinite distance, the Earth appears as a point, thus having a dimension of 0.
2. If it is viewed from a position on the Moon, it appears to be a small ball, thus having a dimension of 3.
3. If the viewer comes nearer, for example, to a distance above the Earth's surface of about 830 km (the altitude of the SPOT satellite's orbit), the height information is extractable but not in detail. Thus, in general terms, the observer can see a mainly smooth surface with a dimension of nearly 2.
4. If the Earth's surface is viewed on the ground, then the roughness of the surface can be seen clearly, thus the effective dimension of the surface should be greater than 2.

In fractal geometry, the effective dimension could be a fraction, leading to the jargon *fractal dimension* or *fractal*. For example, the fractal dimension of a curve changes between 1 and 2, and that of a surface between 2 and 3. The fractal dimension is calculated as follows:

$$L = C \times r^{1-D} \tag{2.2a}$$

where r is the scale of measurement (a principal unit), L is the length of measurement, C is a constant, and D is interpreted as the fractal dimension of the curve line. When measuring a fractal dimension of curve surface, r becomes the principal unit of surface used for measurement and the resultant area is A instead of L; the expression becomes

$$A = C \times r^{2-D} \tag{2.2b}$$

Figure 2.1 shows an example of Koch line with a fractal dimension of 1.26. The process of curve generation is as follows: (a) draw a line with its length as a unit;

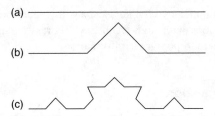

Figure 2.1 A complex Koch line having a fractal dimension of 1.26. (a) A line with unit. (b) Divided into three line segments and mid-segment split into two. (c) Process repeated.

Figure 2.2 Relationship between curvatures and complexity: the curvatures of the left two lines are 0 as the radius is infinite while the line on the right side has large curvatures as the radiuses are small.

(b) divide the line into three segments; (c) the middle segment will be replaced by two polylines with length equal to $\frac{1}{3}$ unit. The same procedure is repeatedly applied to all line segments. As a result, the line will become more and more complex, resulting in a fractal dimension of 1.26.

From the discussion above, it can be concluded that a fractal dimension approaching 3 indicates a very complex and probably rough surface, while a simple (near planar) surface has a fractal dimension value near 2.

2.2.3 Curvature

The terrain surface can be synthesized by combing terrain form elements, defined as relief unit of homogenous plan and profile curvatures (see Chapter 13 for more details). Suppose a profile can be expressed as $y = f(x)$, then the curvature at position x can be computed as follows:

$$c = \frac{d^2y/dx^2}{[1 + (dy/dx)^2]^{3/2}} \qquad (2.3)$$

In this formula, curvature c is inversely proportional to the radius of the curve (R), that is, a large curvature is associated with a small radius (Figure 2.2). Thus, intuitively, it can be seen that the larger the curvature, the rougher is the surface. Therefore, curvatures can also be used as a measure for the roughness of the terrain. This criterion has already been used for terrain analysis (e.g., Dikau 1989).

This is a comparatively useful method for planning DTM sampling strategies. However, a rather large volume of data (that of a DTM) needs to be available to allow the curvature values to be derived — which leads to a chicken-and-egg situation at the stage.

2.2.4 Covariance and Auto-Correlation

The degree of similarity between pairs of surface points can be described by a correlation function. This may take many forms like covariance or an auto-correlation function. The auto-correlation function is described as follows:

$$R(d) = \frac{\text{Cov}(d)}{V} \tag{2.4}$$

where $R(d)$ is the correlation coefficient of all the points with horizontal interval d, $\text{Cov}(d)$ is the covariance of all the points with horizontal interval d, and V is the variance calculated from all the (N) points. The mathematical functions are as follows:

$$V = \frac{\sum_{i=1}^{N}(Z_i - M)^2}{N-1} \tag{2.5}$$

$$\text{Cov}(d) = \frac{\sum_{i=1}^{N}(Z_i - M)(Z_{i+d} - M)}{N-1} \tag{2.6}$$

where Z_i is the height of point i, Z_{i+d} is the elevation of the point with an interval of d from point i, M is the average height value of all the points, and N is the total number of points.

When the value of d changes, $\text{Cov}(d)$ and $R(d)$ will also change because the height difference of two points with different d values is different. Covariance and auto-correlation values can be plotted against the distance between pairs of data points. Figure 2.3 is an example of auto-correlations varying with d. In general, if the value of d increases, the values of $\text{Cov}(d)$ and $R(d)$ will decrease. The curve is usually described (Kubik and Botman 1976) by the exponential function:

$$\text{Cov}(d) = V \times e^{-2d/c} \tag{2.7}$$

and the Gaussian model:

$$\text{Cov}(d) = V \times e^{-2d^2/c^2} \tag{2.8}$$

where c is the parameter indicating the correlation distance at which the value of covariance approaches 0. Therefore, the smaller the value of c, the less similar are the surface points.

The value of similarity is also an indicator of the complexity of the terrain surface. The relationship between them is that the smaller the similarity over the same given distance, the more complex is the terrain surface.

2.2.5 Semivariogram

The variogram is another parameter used to describe the similarity of a DTM surface, similar to (auto-)covariance. The expression for its computation is as follows:

$$2\gamma(d) = \frac{\sum_{i=1}^{N}(Z_i - Z_{i+d})^2}{N} \tag{2.9}$$

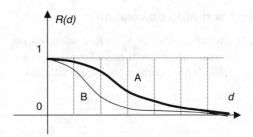

Figure 2.3 Two auto-correlation functions, whose values decrease with an increase in distance between points from 1 to 0.

where $\gamma(d)$ is called the semivariogram. Similar to covariance, the value of $\gamma(d)$ will vary with distance. But the change in direction is opposite to the case of covariance. That is, $\gamma(d)$ will increase with an increase in the value of d. The values of $\gamma(d)$ can also be plotted against d, resulting a curved line. Such a curve can be approximated by an exponential function as follows:

$$\gamma(d) = A \times d^b \tag{2.10}$$

where A and b are two constants, i.e. the two parameters for the description of terrain roughness. A larger b indicates a smother terrain surface. When b is approaching zero, the terrain is very rough. Some examples of semivariograms are given in Figure 8.6.

Indeed, Frederiksen et al. (1983, 1986) used the semivariogram to describe terrain roughness in digital terrain modeling. They also tried to connect this variable to the covariance used by Kubik and Botman (1976).

2.3 TERRAIN ROUGHNESS VECTOR: SLOPE, RELIEF, AND WAVELENGTH

The numerical descriptors discussed in Section 2.2 are essentially statistical. They are computed from a sample of terrain points from the project area. Usually, some profiles are used as the sample and then a parameter is calculated from these profiles. However, there are some problems associated with this approach. One of these is that the parameters calculated from the selected profiles can be different from those derived from the whole surface. If one tries to compute these for the whole surface, then a sample from the whole surface is necessary. In this case, the original purpose of having a terrain descriptor for project planning and design is lost. For these reasons, Li (1990) recommended slope and wavelength as the main descriptors for DTM purposes.

2.3.1 Slope, Relief, and Wavelength as a Roughness Vector

The parameters for roughness or complexity of a terrain surface used in geomorphology have also been reviewed by Mark (1975). It was found that roughness cannot be completely defined by any single parameter, but must be a *roughness vector* or a set of parameters.

TERRAIN DESCRIPTORS AND SAMPLING STRATEGIES

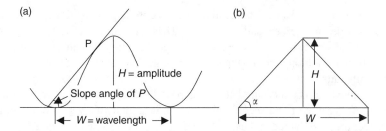

Figure 2.4 The relationship between slope, wavelength, and relief: (a) their full relationship and (b) simplified diagram.

In this set of parameters, *relief* is used to describe the vertical dimension (or amplitude of the topography), while the terms *grain* and *texture* (the longest and shortest significant wavelengths) are used to describe the horizontal variations (in terms of the frequency of change). The parameters for these two dimensions are connected by slope. Thus, relief, wavelength, and slope are the roughness parameters. The relationship between them can be illustrated in Figure 2.4. It can clearly be seen that the slope angle at a point on the *wave* varies from position to position. The following mathematical equation may be used as an approximate expression of their relationship (for a more rigorous definition, see Chapter 13):

$$\tan \alpha = \frac{H}{W/2} = \frac{2H}{W} \qquad (2.11)$$

where α denotes the average value of the slope angle, H is the local relief value (or the amplitude), and W is the so-called wavelength. It is clear that if two of them are known, then the third can be computed from Equation (2.11). For the reasons to be discussed in the next section, slope and wavelength together are recommended as the terrain roughness vector for DTM purposes.

2.3.2 The Adequacy of the Terrain Roughness Vector for DTM Purposes

From both the theoretical and the practical points of view, slope, altitude, and wavelength are the important parameters for terrain description.

In geomorphology, Evan (1981) states

> a useful description of the landform at any point is given by altitude and the surface derivatives, i.e. slope and convexity (curvature)... Slope is defined by a plane tangent to the surface at a given point and is completely specified by the two components: gradient (vertical component) and aspect (plane component)... Gradient is essentially the first vertical derivative of the altitude surface while aspect is the first horizontal derivative.

Further, land surface properties are specified by convexity (positive and negative convexity — concavity). These are the changes in gradient at a point (in profile) and the aspect (in the plane tangential to the contour passing through the point). In other words, they are second derivatives. These five attributes (altitude, gradient, aspect, profile convexity, and plane convexity) are the main elements used to describe terrain surfaces. Among them, slope, comprising of both gradient and aspect, is the fundamental attribute.

Gradient should be measured at the steepest direction. However, when taking the gradient of a profile or in a specific direction, it is actually the vector of the gradient and aspect that is obtained and used. Therefore, the term slope or slope angle is used in this context to refer to the gradient in any specific direction.

The importance of slope has also been realized by others. As quoted by Evans (1972), Strahler (1956) pointed out that "slope is perhaps the most important aspect of surface form, since surfaces may be formed completely from slope angles ...." Slope is the first derivative of altitude on the terrain surface. It shows the rate of change in height of the terrain over distance.

From the practical point of view, using slope (and relief) as the main terrain descriptor for DTM purposes can be justified for the following reasons:

1. Traditionally, slope has been recognized as very important and used in surveying and mapping. For example, map specifications for contours are given in terms of slope angle all over the world.
2. In the determination of vertical contour intervals (CIs) for topographic maps, slope and relief (height range) are the two main parameters considered. For example, Table 2.1 is a classification system adopted by the Chinese State Bureau of Surveying and Mapping (SBSM) in its specifications for 1:50,000 topographic maps.
3. In DTM practice, many researchers (e.g., Ackermann 1979; Ley 1986; Li 1990, 1993b) have noted the high correlation between DTM errors and the mean slope angle of the region.

2.3.3 Estimation of Slope

To use slope together with wavelength or relief to describe terrain, two problems related to the estimation of its values need to be considered, that is, *availability* and *variability*.

By *availability* we mean that slope values should be available or estimated before sampling takes place, to assist in the determination of sampling intervals. If a DTM exists in an area, then the slope values for DTM points can be computed and the average can be used as the representative (Zhu et al. 1999). Otherwise, slope may be estimated from a stereo model formed by a pair of aerial photographs with overlap (see Chapter 3) or from contour maps. The method proposed by Wentworth (1930) is still widely used to estimate the average slope of an area from the contour maps.

Table 2.1 Terrain Classification by Means of Slope and Relief

Terrain Type	CI (m)	Slope (°)	Relief (Height Range) (m)
Plain	10 (5)	<2	<80
Upland	10	2–6	80–300
Hill	20	6–25	300–600
Mountain	20	>25	>600

TERRAIN DESCRIPTORS AND SAMPLING STRATEGIES

The average slope value (α) of a homogeneous are can be estimated as follows:

$$\alpha = \arctan\left(\Delta H \times \frac{\Sigma L}{A}\right) \tag{2.12}$$

where ΔH is the contour interval, ΣL is the total length of contours in the area and A is the size of the area. If there is no contour map for such an area, then the slope may be estimated from an aerial photograph. Some of the methods that are available for measurement of slope from aerial photographs have been reviewed by Turner (1997).

By *variability* we mean that slope values may vary from place to place so that the slope estimate that is representative for one area may not be suitable for another. In this case, average values may be used as suggested by Ley (1986). If slope varies too greatly in an area, then the area should be divided into smaller parts for slope estimation. Different sampling strategies could be applied to each area.

2.4 THEORETICAL BASIS FOR SURFACE SAMPLING

After estimating slope and relief (height range), the wavelengths of terrain variation can be computed. These parameters are used to determine the sampling strategy and intervals for data acquisition. First, some theories related to surface sampling are discussed.

2.4.1 Theoretical Background for Sampling

From the theoretical point of view, a point on the terrain surface is 0-D, thus without size, while a terrain surface comprises an infinite number of points. If full information about the geometry of a terrain surface is required, it is necessary to measure an infinite number of points. This means that it is impossible to obtain full information about the terrain surface. However, in practice, a point measured on a surface represents the height over an area of a certain size; therefore, it is possible to use a set of finite points to represent the surface. Indeed, in most cases, full or complete information about the terrain surface is not required for a specific DTM project, so it is necessary only to measure enough data points to represent the surface to the required degree of accuracy and fidelity.

The problem a DTM specialist is concerned with is how to adequately represent the terrain surface by a limited number of elevation points, that is, what sampling interval to use with a known surface (or profile). The fundamental sampling theorem that is being widely used in mathematics, statistics, engineering, and other related disciplines can be used as the theoretical basis. The sampling theorem can be stated as follows:

> If a function $g(x)$ is sampled at an interval of d, then the variations at frequencies higher than $1/(2d)$ cannot be reconstructed from the sampled data.

That is, when sampling takes two samples (i.e., points) from each period of waves with the highest frequency in the function $g(x)$, the original $g(x)$ can be completely reconstructed with the sampled data. In the case of terrain modeling, if a terrain profile is long enough to be representative of the local terrain, it can then be represented by the

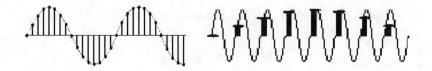

Figure 2.5 The relationship between the least sampling interval and the highest functional frequency. Left: sampling interval is less than half the functional frequency so that full reconstruction is possible; right: sampling interval is larger than half the functional frequency so that information about the function is lost.

sum of its sine and cosine waves. If it is assumed that the number of terms in this sum is finite, there is, therefore, a maximum frequency value, F, for this set of sinusoidal. According to the sampling theorem, the terrain profile can be completely reconstructed if the sampling interval along the profile is smaller than $1/(2F)$ (see Figure 2.5, left). Therefore, extending this idea to surfaces, the sampling theorem can also be used to determine the sampling interval between profiles to obtain adequate information about a terrain surface. In contrast, if a terrain profile is sampled at an interval of d, then the terrain information with a wavelength less than $2d$ will be completely lost (Figure 2.5, right). Therefore, as Peucker (1972) has pointed out, "a given regular grid of sampling points can depict only those variations of the data with wave lengths of twice the sampling interval or more."

2.4.2 Sampling from Different Points of View

Points on a terrain surface can be viewed in various ways from the differing viewpoints inherent in subjects such as statistics, geometry, topographic, science, etc. Therefore, different sampling methods can be designed and evaluated according to each of these different viewpoints as follows (Li 1990):

1. statistics-based sampling
2. geometry-based sampling
3. feature-based sampling.

From the statistical point of view, a terrain surface is a population (called a sample space) and the sampling can be carried out either randomly or systematically. The population can then be studied by the sampled data. In random sampling, any sampled point is selected by a chance mechanism with known chance of selection. The chance of selection may differ from point to point. If the chance is equal for all sampled points, it is referred to as simple random sampling. In systematic sampling, the points are selected in a specially designed way, each with a chance of 100% probability of being selected. Other possible sampling strategies are stratified sampling and cluster sampling. However, they are not suitable for terrain modeling and thus are omitted here.

From the geometric point of view, a terrain surface can be represented by different geometric patterns, either regular or irregular in nature. The regular pattern can be subdivided into 1-D or 2-D patterns. If sampling is conducted with a regular pattern

TERRAIN DESCRIPTORS AND SAMPLING STRATEGIES

that is only regular in one dimension, then the corresponding method is referred to as profiling (or contouring). A 2-D regular pattern could be a square or a regular grid, or a series of contiguous equilateral triangles, hexagons, or other regularly shaped geometric figures.

From the viewpoint of features, a terrain surface is composed of a finite number of points, and the information content of these points may vary with their positions. Therefore, surface points are classified into two groups, one of which comprises *feature-specific* (F-S) (or *surface-specific*) *points* (and lines) while the other comprises *random points*. An F-S point is a *local extrema* point on the terrain surface, such as peaks, pits, and passes. These points may not only present their own elevation values but also provide more topographic information to their surroundings. Peaks are the summits of mountains and hills, so they have a set of points of lower height around them. By contrast, pits are the bottoms of valleys (holes), so they have a set of greater height values around them. That is, F-S points are more important because they not only contain the coordinate information about themselves, but also implicitly represent some information about their surroundings. Thus, F-S points represent surface features with higher or more significant information content than the average points. The lines connecting certain types of F-S points are referred to as *feature-specific lines*, such as *ridge lines*, *course lines* (rivers, valleys, ravines, etc.), *break lines*, and so on. Figure 2.6 shows the F-S points and lines. Ridge lines are the lines connecting pairs of points such that the points on them are local maxima (see Figure 2.7). Similarly, course lines are linking pairs or strings of points so that the points defined by them are local minima.

The crossing points of these two types of lines are referred to as passes. They are, therefore, the points that, at the same time, can be a maxima elevation in one direction and a minima in the other direction.

From the morphological point of view, a terrain surface is characterized completely by its slope angles. Therefore, the importance of F-S points comes from the fact that at these points, slope changes not only in direction but also in sign and magnitude. For example, at peaks, it changes from positive to negative and at pits, it changes from negative to positive. There are also two other types of points where the slope changes its vertical angle but not its sign. They are *convex* and *concave points*.

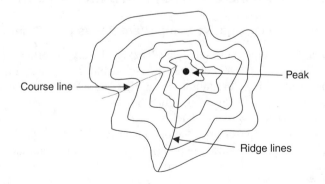

Figure 2.6 Terrain feature points and lines.

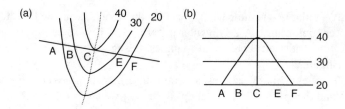

Figure 2.7 Points (e.g., C) on a ridge line being local maxima.

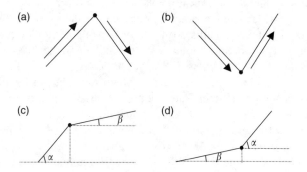

Figure 2.8 Slope changes at F-S points (peaks, pits, and convex and concave points). (a) Peak ($+ \Rightarrow -$). (b) Pits ($- \Rightarrow +$). (c) Convex point ($\alpha \neq \beta$). (d) Concave point ($\alpha \neq \beta$).

If a slope is viewed as an up–down transition, the slope change is from gentle to steep at a convex point and from steep to gentle at a concave point. Figure 2.8 shows such points. The convex and concave points are also invariably F-S points, connected to become linear features. If there is a special case where the slope change is very sudden, then these linear features are referred to as *break lines*.

2.5 SAMPLING STRATEGY FOR DATA ACQUISITION

2.5.1 Selective Sampling: Very Important Points plus Other Points

Selective sampling mimics field surveying. That is, all very important points (VIPs) discussed in Section 2.4.2 are selected, thereby ensuring that data are reasonably comprehensive in coverage. In addition, some others are selected to make the sampled data have a certain density. This method has the distinct advantage that fewer points can represent the surface with high fidelity.

However, in sampling using the photogrammetric method (see Chapter 3), this is not an efficient way of selecting data points because it requires substantial interpretation of the stereo model (i.e., reconstructed terrain surface from a pair of aerial photographs) by a trained operator. In practice, no automated procedure can be implemented on the basis of this strategy. So, it is not popular in certain mapping organizations (e.g., military survey organizations) where speed of data acquisition is of prime importance.

2.5.2 Sampling with One Dimension Fixed: Contouring and Profiling

In analog photogrammetry, stereo models are constructed from a pair of aerial photographs and direct measurement of contours from the reconstructed stereo model is the most common practice. The height value is fixed for each contour and float marks (one for the left photograph and the other for the right photograph, both of which should coincide if they are just on the surface of the stereo model) are moved on the surface of the stereo model, which is realized by a combination of movement in X and Y directions, driven by two mechanical wheels.

The term *contouring* means that the data sampling is along contours. This is exactly the same as the traditional contour measurement on the stereo model. The only difference is that in the DTM data sampling, all points on the contour lines are recorded in digital form and point recording could be selective along a contour line.

In contouring, the height value in Z dimension, is fixed when measuring a contour line. On the other hand, if the fixed dimension is X, then the movement of floating marks on the stereo model surface is on the YZ plane. The result is a profile on the YZ plane. The process to obtain a profile in digital form is called *profiling*. Of course, profiling could be in any direction apart from the XZ and YZ planes.

2.5.3 Sampling with Two Dimensions Fixed: Regular Grid and Progressive Sampling

As the name implies, *regular grid sampling* ensures that the data points are obtained in the form of a regular grid. This can be achieved by setting the fixed intervals in both X and Y directions to form the plane grid. Then, all points on the grid nodes are measured.

But in terms of sampling, a heavy redundancy of data is required to ensure that all slope discontinuities are detected or that changes in the topography are represented in an adequate manner.

To solve the problem of data redundancy in regular grid sampling, Makarovic (1973) designed a modified strategy, which he called *progressive sampling*. In this procedure, the sampling is carried out in a grid pattern whose interval changes progressively from coarse to fine over an area.

The procedure is as follows. First, a set of grid points is measured at a low density, then the elevation values at these data points are analyzed by an on-line computer. In turn, the computer generates the locations of new points to be sampled in the next run. The procedure is repeated until some prior criteria are satisfied.

For such criteria, Makarovic (1973) proposed initially to use the second differences of elevation values computed along both rows and columns of the measured (sampled) coarse grid. Several additional or alternative criteria have also been proposed later (Makarovic 1975), such as the so-called random-variation, parabolic, distance, and contour criteria. Of course, other criteria may also be used as the basis of the sampling strategy for a particular type of terrain.

Progressive sampling can solve part of the redundancy problem that is inherent in regular grid sampling, but still there are shortcomings, as Makarovic (1979) noted:

1. The sampled data points exhibit a high degree of redundancy in the proximity of abrupt changes in the terrain surface.
2. Pertinent features may be lost in the first run with its wide (coarse) spacing. These cannot be recovered by the following sampling runs.
3. The tracking path is rather long, which decreases efficiency.

2.5.4 Composite Sampling: An Integrated Strategy

The idea of progressive sampling sounds great. Indeed, it was implemented by some photogrammetric systems such as the analytical plotter. However, in practice, it was not widely implemented due to the reasons mentioned in the previous section.

A more natural line of thinking is to combine a regular grid sampling with selective sampling, because the former is very efficient in measurement and the latter is very effective in surface representation. Such a combination is referred to as *composite sampling*. In this way, abrupt changes — specific features on the terrain such as ridges, break lines, etc. — are sampled selectively. And the values and F-S points — peaks, passes and hollows — are added to the regular grid-sampled data.

Indeed, there are two types of composite sampling. The first one is mentioned already, and the second one is a combination of selective and progressive sampling. It has proved in practice that the use of composite sampling may solve many problems encountered in regular grid sampling and progressive sampling.

2.6 ATTRIBUTES OF SAMPLED SOURCE DATA

In the context of digital terrain modeling, sampling is the process of selecting those points that have to be measured in certain positions. The operation can be characterized by two parameters, that is, *distribution* and *density*. Measurement is to determine the X, Y coordinates of a point and is concerned with *accuracy*. Sampling can take place before or after measurement. Sampling after measurement is to select points from a set of measured data points, usually with great density. Therefore, accuracy can also be included in the attribute set for the sampled data, called DTM source data, raw data, or simply source data.

2.6.1 Distribution of Sampled Source Data

The distribution of sampled data is usually specified by the terms of *location* and *pattern*. The location is defined in terms of two positional coordinates, that is, longitude and latitude in a geographical coordinate system or easting and northing in a grid coordinate system. Regarding *pattern*, a variety of these are available for selection, such as regular or rectangular grids. These patterns can be classified in different ways. Figure 2.9 shows one such classification.

TERRAIN DESCRIPTORS AND SAMPLING STRATEGIES

Regular 2-D data are produced by means of regular grid or progressive sampling. The resulting pattern could be a rectangular grid, a square grid, or a hierarchical (or progressive) structure of these two. The square grid is most commonly used. The hierarchical structured data, sampled by means of progressive sampling, can be decomposed into a normal square grid.

Data that are regular in one dimension are produced by sampling with one dimension fixed (X, Y, or Z). That is, such a pattern is generated by using contouring or profiling.

There are other special regular patterns, for instance, equilateral triangles and hexagons, etc. However, it seems that these structures are not as widely used as profiled or regular grid data.

As has been discussed before, data patterns can be divided into two categories, that is, regular and irregular patterns. *Regular patterns* have been discussed above. *Irregular patterns* may generally be classified into three groups, that is, *random*, *cluster*, and *string data*. By random data we mean that the measured points are located randomly, that is, not in any specific form. By clustered data we mean that the measured points are clustered, which is often the case in geology. String data are not located in a regular pattern, yet they follow certain features (such as break lines).

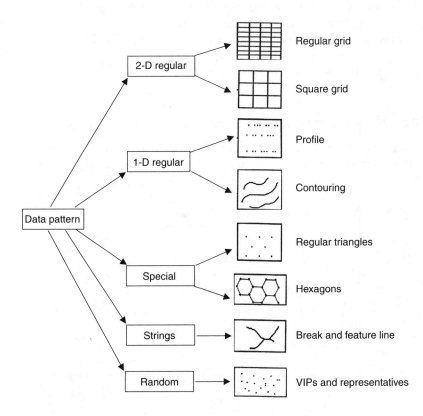

Figure 2.9 Patterns of sampled data points.

The data sets that are sampled along rivers, break lines, or feature lines all belong to this pattern. Actually, it is not an independent pattern, but rather a supplemental one that is F-S. For example, the pattern of the data resulting from composite sampling is usually a combination of string data with regular grid data.

2.6.2 Density of Sampled Source Data

Density is another attribute of sampled data. It can be specified by measures like the distance between two points, the number of points per unit area, the cutoff frequency (Nyquist), and so on.

The distance between two sampled points is usually referred to as the sampling interval (or distance or spacing). If the sampling interval varies with position, then an average value can be used. This measure is specified by a number with a unit, for instance, 20 m. Another measure that could be used in terrain modeling practice is the number of points per unit area, for example, 500 points per square kilometer.

If the sampling interval is transformed from space domain to frequency domain, then the cutoff frequency (the maximum frequency that the sampled data represent) can be obtained. From another point of view, the required maximum frequency can also be used as a measure of data density because the sampling interval can also be obtained from it (the value of maximum frequency). Figure 2.10 sketches the frequency of a curve. The frequency at point B can be considered as the cutoff. In fact, the swing of point A is already near 0 and the value at point A may also be regarded as cutoff frequency in some sense.

2.6.3 Accuracy of Sampled Source Data

The accuracy of sampled data largely depends on the methods used for measurement, such as the mode of measurement, instruments used, and technique adopted.

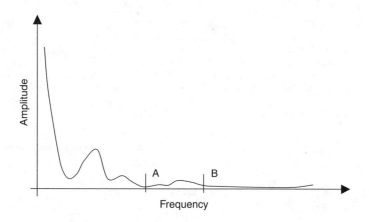

Figure 2.10 Cutoff frequency: the swing approaching 0.

TERRAIN DESCRIPTORS AND SAMPLING STRATEGIES

Technique means the field survey, photogrammetry, or map digitization. Generally speaking, data acquired by field survey are usually the most accurate and data acquired by map digitization are less accurate. Of course, there are always exceptions. For example, if the instruments used for field surveying are of very low accuracy but the existing maps are at large scale and digitized by a very accurate instrument, then the data digitized from maps may be more accurate than those acquired by field survey. Therefore, there are conditions to the above general statement, that is, whether the techniques are compatible in terms of scale.

By instrument we mean the type of instrument, which in turn implies potential accuracy limitation. Highly accurate results can be obtained only when the instruments used for measurement are of high quality.

Mode of measurement refers to either static or dynamic mode. Dynamic mode means that measurement is carried out dynamically. In the field survey using GPS, the GPS receiver is in motion, either carried by a surveyor or in a vehicle. In photogrammetry, the measurement is carried out when the float marks are still in motion. In digitization, points are recorded while the cursor is in motion. In dynamic mode, the data acquired are usually of much lower accuracy.

There will be more discussion on data measurement and the accuracy of measured data using different techniques in the next chapter.

CHAPTER 3

Techniques for Acquisition of DTM Source Data

In Chapter 2, sampling strategies were discussed, on the selection of points on the terrain (or reconstructed stereo model) surface. In this chapter, the techniques used for actual measurement of such selected positions are presented.

3.1 DATA SOURCES FOR DIGITAL TERRAIN MODELING

Data sources means the materials from which data can be acquired for terrain modeling and *DTM source data* means data acquired from data sources of digital terrain modeling. Such data can be measured by different techniques:

1. field surveying by using total station theodolite and GPS for direct measurement from terrain surfaces
2. photogrammetry by using stereo pairs of aerial (or space) images and photogrammetric instruments
3. cartographic digitization by using existing topographic maps and digitizers.

3.1.1 The Terrain Surface as a Data Source

The continents occupy about 150 million km^2, accounting for 29.2% of the Earth's surface. Relief varies from place to place, ranging from a few meters in flat areas to a few thousand meters in mountainous areas. The highest peak of the Earth is about 8,884 m at Mt Everest. Most oceans are kilometers deep while some trenches in the Pacific plunge in excess of 10,000 m. In this book, terrain means the continental part of the Earth's surface.

The Earth's surface is covered by natural and cultural features, apart from water. Vegetation, snow/ice, and desert are the major natural features. Indeed, in the polar regions and some high mountainous areas, terrain surfaces are covered by ice and snow all the time. Settlements and transportation networks are the major cultural features.

For terrain surfaces with different types of coverage, different measurement techniques may be used because some techniques may be less suitable for some areas. For example, it is not easy to directly measure the terrain surface in highly mountainous areas. For this, photogrammetric techniques using aerial or space images are more suitable.

3.1.2 Aerial and Space Images

Aerial images are the most effective way to produce and update topographic maps. It has been estimated that about 85% of all topographic maps have been produced by photogrammetric techniques using aerial photographs. Aerial photographs are also the most valuable data source for large-scale production of high-quality DTM.

Such photographs are taken by metric cameras mounted on aerial planes. Figure 3.1(a) is an example of an aerial camera. The cameras are of such high metric quality that image distortions due to imperfections of camera lens are very small. Four fiducial marks are on the four corners (see Figure 3.2) or sides of each photograph and are used to precisely determine the center (principal point) of the photograph. The standard size of aerial photographs is 23 cm × 23 cm.

Aerial photographs can be classified into different types based on different criteria:

Color: Color (true or false) and monochromatic photographs.
Attitude of photography: Vertical (i.e., main optical axis vertical), titled ($\leq 3°$), and oblique ($>3°$) photographs. Commonly used aerial photographs are titled photographs.
Angular field of view: Normal, wide-angle and super wide-angle photography (see Table 3.1). In practice, over 80% of modern aerial photographs belong to the wide-angle category.

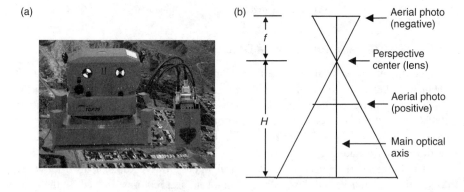

Figure 3.1 Aerial camera and aerial photography. (a) An aerial camera. (Courtesy of Zeiss.) (b) Geometry of aerial photography.

TECHNIQUES FOR ACQUISITION OF DTM SOURCE DATA

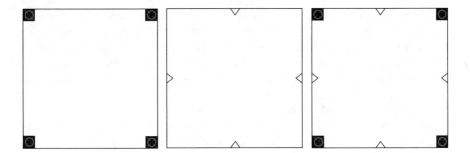

Figure 3.2 Different types of fiducial marks.

Table 3.1 Types of Aerial Photographs Based on Angular Field of View

Type	Super-Wide Angle	Wide Angle	Normal Angle
Focal length	≈85 mm	≈150 mm	≈310 mm
Angular field of view	≈120°	≈90°	≈60°

The principle of photography is described by the following mathematical formula:

$$\frac{1}{u} + \frac{1}{v} = \frac{1}{f} \tag{3.1}$$

where u is the distance between the object and the lens, v is the distance between the image plane and the lens, and f is the focal length of the lens. In the case of aerial photography, the value of u is large, about a few thousand meters. Therefore, $1/u$ approaches 0 and v approaches f. That is, the image is formed at a plane very close to the focal plane. Figure 3.1(b) illustrates the geometry of aerial photography. The ratio f/H determines the scale of the aerial photograph, where H is the flying height of the airplane (thus the camera):

$$\frac{1}{S} = \frac{f}{H} \tag{3.2}$$

Traditionally, aerial photographs are in analog form and the images are recorded on films. If images in digital form are required, then a scanning process is applied. Experimental studies show that a pixel size as large as 30 μm is sufficient to retain the geometric quality of analog images. On the other hand, aerial images can also be directly recorded by an electronic device to form digital images, using a CCD (charge-coupled device) camera. However, the optical principle of imaging is the same as analog photography.

There is another type of aerial image obtained by airborne scanners. However, they are not widely used for acquisition of data for digital terrain modeling. On the other hand, scanned space images, particularly those from SPOT satellite system, are widely used for the generation of small-scale DTM over large areas. However, with high-resolution images such as IKONOS 1-m resolution images, space images will find more applications in DTM generation.

These images are all obtained by passive systems, where the sensors record the electromagnetic radiations reflected by the terrain surface and objects on the terrain surface. It is also possible to use active systems, which send off electromagnetic waves, and then to receive the waves reflected by terrain surfaces and objects on the terrain surface. Radar is such a system. As radar images are a potential source for medium- and small-scale DTM over large areas, the use of them for DTM data acquisition will be discussed later at some length although they are still not widely used.

3.1.3 Existing Topographic Maps

Every country has topographic maps and these may be used as another main data source for digital terrain modeling. In many developing countries, these data sources may be poor due to the lack of topographic map coverage or the poor quality of the height and contour information contained in the map. However, in most developed countries and even some developing countries like China, most of the terrain is covered by good-quality topographic maps containing contours. Therefore, these form a rich source of data for digital terrain modeling provided that the limitations of extracting height data from contour maps are kept in mind.

The largest scale of topographic maps that cover the whole country with contour lines is usually referred to as the basic map scale. This may also vary from country to country. For example, the basic map scales for China, United Kingdom, and United States are 1:50,000, 1:10,000, and 1:24,000, respectively. This indicates the best quality of DTM that can be obtained from existing contour maps. There are usually some other topographic maps at scales smaller than the basic map scale. Of course, such smaller-scale topographic maps have a higher degree of generalization and thus lower accuracy. Table 3.2 shows the characteristics of such maps.

One important concern with topographic maps is the quality of the data contained in them, especially the metric quality, which is then specified in terms of accuracy. The fidelity of the terrain representation given by a contour map is largely determined by the density of contour lines and the accuracy of the contour lines themselves.

Table 3.2 Topographic Maps at Different Scales (Konecny et al. 1979)

Topographic Map	Scale	Characteristics
Large- to medium-scale maps	>1:10,000	Representation true to plan
Medium- to small-scale maps	1:20,000–1:75,000	Representation similar to plan
General topographic map	<1:100,000	High degree of generalization or signature representation

TECHNIQUES FOR ACQUISITION OF DTM SOURCE DATA 35

Table 3.3 Map Scales and Commonly Used Contour Intervals (Konecny et al. 1979)

Scale of the Topographic Map	Interval between Contour Lines (m)
1:200,000	25–100
1:100,000	10–40
1:50,000	10–20
1:25,000	5–20
1:10,000	2–10

Table 3.4 Map Scales and Commonly Used Contour Intervals

Country	Scale	Height Accuracy (m)
Germany	1:5,000	$0.4 + 3 \times \tan\alpha$
Switzerland	1:10,000	$1.0 + 3 \times \tan\alpha$
Britain	1:10,000/1:10,560	$\sqrt{1.8^2 + (3 \times \tan\alpha)^2}$
Italy		$1.8 + 12.5 \times \tan\alpha$
Norway		$2.5 + 7.5 \times \tan\alpha$
Switzerland		$1.0 + 7.5 \times \tan\alpha$
Israel	1:25,000	$1.5 + 5.0 \times \tan\alpha$
Germany		$0.8 + 5.0 \times \tan\alpha$
Finland		$1.5 + 3.0 \times \tan\alpha$
The Netherlands		$0.3 + 4.0 \times \tan\alpha$
Switzerland	1:50,000	$1.5 + 10 \times \tan\alpha$
United States		$1.8 + 15 \times \tan\alpha$

One important measure of contour density is the vertical contour interval, or simply contour interval (CI). The commonly used contour intervals for different map scales are shown in Table 3.3.

The accuracy requirements of a contour map are given by the map specifications. Examples of the specifications for the accuracy of contours for different map scales used in different countries are given in Table 3.4 (Imhof 1965; Konecny et al. 1979), α is the slope angle. In general, it is expected that the height accuracy of any point interpolated from contour lines will be about 1/2 to 1/3 of the CI.

3.2 PHOTOGRAMMETRY

3.2.1 The Development of Photogrammetry

The word photogrammetry comes from the Greek words *photos* (meaning *light*), *gramma* (meaning that which is drawn or written), and *metron* (meaning *to measure*). It originally signified "measuring graphically by means of light" (Whitmore and Thompson 1966).

Table 3.5   The Characteristics of the Four Stages of Photogrammetry (Li et al. 1993)

Components and Parameters	Stages of Development in Photogrammetry			
	Analog	Numerical	Analytical	Digital
Input component	Analog	Analog	Analog	Digit
Model component	Analog	Analog	Analytical	Analytical
Output component	Analog	Digit	Digit	Digit
Degree of "hardness"	3	2	1	0
Degree of flexibility	0	1	2	3

The development of photogrammetry can be traced back to the middle of the 19th century.

> In 1849, A. Laussedat, an officer in the Engineering Corps of the French Army, embarked on a determined effort to prove that photography could be used with advantage in the preparation of topographic maps . . . In 1858, Laussedat experimented with a glass-plate camera in the air, first supported by a string of kites. Laussedat also made a few maps with the aid of a ballon. (Whitmore and Thompson 1966)

With his pioneering work, Laussedat is regarded by many as the "father of photogrammetry."

In early times, maps were made by graphic methods. The credit for the development of measurement instruments goes to two members of the Geographical Institute of Vienna — A. von Hubl and E. von Orel, who developed the stereocomparator and stereoautograph. It is also said that a stereocomparator was developed independently by Zeiss in 1901. In the early stages, these were all optical instruments. Later, optical–mechanical and mechanical projections were adopted to improve the accuracy of measurement. In the late 1950s, the computer was introduced in photogrammetry. The first attempt was to record the output digitally, resulting in numerical photogrammetry, then optical–mechanical projections were replaced by the computational model, resulting in analytical photogrammetry (Helava 1958). From the early 1980s, images in digital form were in use, resulting in digital or softcopy photogrammetry (Sarjakoski 1981).

In summary, photogrammetry has undergone four stages of development, that is, analog, numerical, analytical, and digital photogrammetry. The characteristics of these four stages are given in Table 3.5. Some examples of such instruments are shown in Figure 3.3.

3.2.2   Basic Principles of Photogrammetry

The fundamental principle of photogrammetry is to make use of a pair of stereo images (or simply stereo pair) to reconstruct the original shape of 3-D objects, that is, to form the stereo model, and then to measure the 3-D coordinates of the objects on the stereo model. *Stereo pair* refers to two images of the same scene photographed at two slightly different places so that they have a certain degree of overlap. Figure 3.4 is an example of such a pair. Actually, only in the overlapping area can one reconstruct the 3-D models (see Figure 3.5).

TECHNIQUES FOR ACQUISITION OF DTM SOURCE DATA

Figure 3.3 Some examples of photogrammetric instruments (a) Optical plotter (photo courtesy of Bruce King), (b) Optical-mechanical plotter (photo courtesy of Bruce King), (c) Analytical plotter, (d) Digital photogrammetric system (courtesy of 3D Mapper).

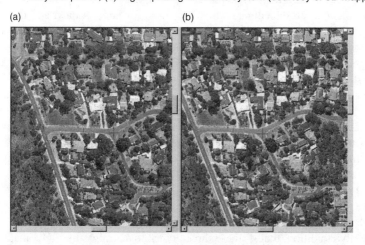

Figure 3.4 A pair of stereo images with 60% overlap, partially displayed on screen (courtesy of 3D Mapper).

In aerial photography, there is generally a 60% overlap degree in the flight direction and 30% between the flight strips. Each photograph is characterized by six orientation elements, three angular elements (one for each of X, Y, and Z axes) and three translations (X, Y, and Z coordinates in a coordinate system, usually geodetic coordinate system). Any two images with overlap can be used to generate a stereo

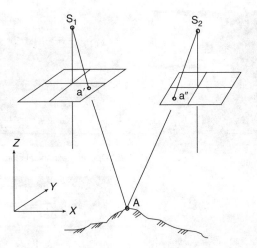

Figure 3.5 A stereo model is formed by projecting image points from a stereo pair.

model. With space images, the percentage of overlap is not that standardized but as long as overlaps exist, they can be used to reconstruct stereo models. However, for scanned images, each strip must have six orientation elements to be determined. Here, aerial photographs are used as an illustration, as they are more used widely for DTM data acquisition.

Imagine that the left and right photographs of a stereo pair are put in two projectors that are identical to the camera which was used for photography, and the positions and orientations of these two projectors are restored to the same situations as when the camera took the two photographs. Then, the light rays projected from the two photographs through the two projectors will intersect in the air to form a 3-D model (i.e., a stereo model) of the objects on the photographs. However, the scale of the stereo model will certainly not be 1:1. Practically, the model can be reduced to a manageable scale by reducing the length of the base line (i.e., the distance between the two projectors). In this way, the operator can measure 3-D points on the stereo model. This is the basic principle of analog photogrammetry and is shown in Figure 3.5. In this figure, S_1 and S_2 are the projection centers, a' and a'' are the two image points on the left and right images, respectively. The light rays from $S_1 a'$ and $S_2 a''$ intersect at point A which is on the stereo model.

The relationship between an image point, the corresponding ground point, and the projection center (camera) is described by an analytical function, called the colinearity condition, that is, these three points on a straight line. The mathematical expression is as follows:

$$x = -f \frac{a_1(X_A - X_S) + b_1(Y_A - Y_S) + c_1(Z_A - Z_S)}{a_3(X_A - X_S) + b_3(Y_A - Y_S) + c_3(Z_A - Z_S)}$$
$$y = -f \frac{a_2(X_A - X_S) + b_2(Y_A - Y_S) + c_2(Z_A - Z_S)}{a_3(X_A - X_S) + b_3(Y_A - Y_S) + c_3(Z_A - Z_S)}$$
(3.3)

where X, Y, Z is a geodesic coordinate system; $S-xy$ is a photocoordinate system; x, y is the pair of image coordinates; A is point on the ground; S is the perspective center of the camera; X_S, Y_S, Z_S is the set of ground coordinates of projection center S in the geodetic coordinate system; X_A, Y_A, Z_A is the set of ground coordinates of point A in the geodetic coordinate system; f is the distance from S to the photo, that is, the focal length of the camera; and a_i, b_i, and c_i ($i = 1, 2, 3$) are the functions of the three angular orientation elements (i.e., ϕ, ω, κ) as follows:

$$a_1 = \cos\phi \cos\kappa + \sin\phi \sin\omega \sin\kappa$$
$$b_1 = \cos\phi \sin\kappa + \sin\phi \sin\omega \cos\kappa$$
$$c_1 = \sin\phi \cos\omega$$
$$a_2 = -\cos\omega \sin\kappa$$
$$b_2 = \cos\omega \cos\kappa \qquad (3.4)$$
$$c_2 = \sin\omega$$
$$a_3 = \sin\phi \cos\kappa + \cos\phi \sin\omega \sin\kappa$$
$$b_3 = \sin\phi \sin\kappa - \cos\phi \sin\omega \cos\kappa$$
$$c_3 = \cos\phi \cos\omega$$

If the six orientation elements for each photograph are known, then when the coordinates of the image points a′, a″ are measured, the ground coordinates of A, (i.e., X_A, Y_A, Z_A) can be computed from Equation (3.3). The six orientation elements can be determined by mounting GPS receivers on the airplane or by measuring a few control points (both on the ground and on images) and using Equation (3.3).

In analytical photogrammetry, the measurement of image coordinates is still carried out by the operator. However, in digital photogrammetry, images are in digital form and thus the coordinates of a point are determined by row and column numbers. When given an image point on the left image, the system will search the corresponding point on the right image (called conjugate point) automatically by a procedure called *image matching*. Then, ground coordinates can be computed accordingly. Such an automated system is called a *Digital Photogrammetric Workstation* (DPW). Figure 3.3(d) is an example of such a system.

To use DPW, images must be in digital form already. If not, a scanning process needs to be applied to convert images from analog to digital form. However, a very high-quality photogrammetric scanner is required to avoid distortion. A pixel size of about 20 μm is usually used because the experimental tests shows that there is no significant difference between the images scanned with 15 and 30 μm.

3.3 RADARGRAMMETRY AND SAR INTERFEROMETRY

In practice, synthetic aperture radar (SAR), is widely used to acquire images. Images acquired by SAR are very sensitive to terrain variation. This is the basis for three types of techniques, that is, radargrammetry, interferometry, and radarclinometry (Polidori 1991). Radargrammetry acquires DTM data through the measurement of parallax while SAR interferometry acquires DTM data through the determination of phase shifts between two echoes. Radarclinometry acquires DTM data through shape from shading. Radarclinometry makes use of a single image and the height information is not accurate enough for DTM. Therefore, it is omitted in this section.

3.3.1 The Principle of Synthetic Aperture Radar Imaging

SAR is a microwave imaging radar developed in the 1960s to improve the resolution of traditional (real aperture) radar based on the principle of Doppler frequency shift.

Imaging radar is an active sensor — providing its own illumination in the form of microwaves. It receives and records echos reflected by the target, and then maps the intensity of the echo into a grey scale to form an image. Unlike optical and infrared imaging sensors, imaging radar is able to take clear pictures day and night under all weather conditions.

Figure 3.6 shows the geometry of the imaging radar often employed for Earth observation. The radar is onboard a flying platform such as an airplane or a satellite. It transmits a cone-shaped microwave beam (pulses) (1 to 1000 GHz) to the ground continuously with a side-looking angle θ_0 in the direction perpendicular to the flying track (azimuth direction). Each time, the energy sent by the imaging radar forms a radar footprint on the ground. This area may be regarded as consisting of many small cells. The echo backscattered from each ground cell within the footprint is received and recorded as a pixel in the image plane according to the slant range between the antenna and the ground cell (as shown in Figure 3.7). During the flying mission, the area swept by the radar footprint forms a swath of the ground, thus a radar image of the swath is obtained (Curlander and Mcdonough 1991; Chen et al. 2000).

The angular fields in the flying direction (ω_h) and the cross-track direction (ω_v) are related to the width (w) and the length (L) of the radar antenna of the radar, respectively, as shown in Equation (3.5). The Swath W_G can be approximated by Equation (3.6).

$$\omega_v = \frac{\lambda}{w}$$
$$\omega_h = \frac{\lambda}{L} \tag{3.5}$$

$$W_G \approx \frac{\lambda R_m}{w \cos \eta} \tag{3.6}$$

where λ is the wavelength of the microwave used by the radar system; R_m is the slant range from the center of the antenna to the center of the footprint; and η is the incident angle of radar beam pulses.

The minimum distance between two distinguishable objects is called the resolution of the radar image, which is the most important measure of radar image quality. Apparently, the smaller this value, the higher the resolution. The resolution of a radar

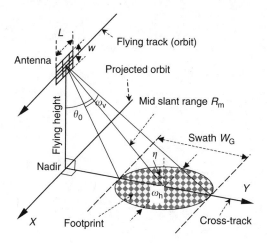

Figure 3.6 Radar imaging geometry.

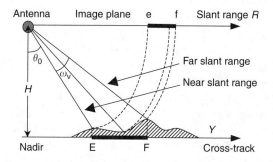

Figure 3.7 Projection of radar image.

image for Earth observation is defined by the azimuth resolution in the flying direction (Δx) and by the slant range resolution in the slant rage direction (ΔR) or the ground range resolution in the cross-track direction (Δy), as shown in Figure 3.8. According to the electromagnetic (EM) wave theory, the azimuth resolution is:

$$\Delta x = \frac{R\lambda}{L} \tag{3.7}$$

where R is the slant range, λ is the wavelength of the microwave, and L is the length of the aperture of the radar antenna. Here, Δx is the width of the footprint, as shown in Figure 3.8. The slant range and ground range resolutions are:

$$\Delta R = \frac{c\tau_p}{2} \tag{3.8}$$

$$\Delta y = \frac{\Delta R}{\sin \theta_i} = \frac{c\tau_p}{2 \sin \theta_i} \tag{3.9}$$

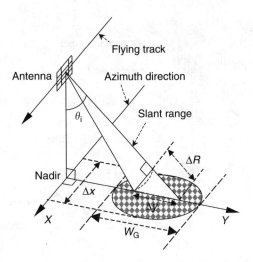

Figure 3.8 Resolution of radar images.

where c is the speed of light; τ_p is the pulse duration; and θ_i is the side-looking angle.

Equations (3.7) to (3.9) show that the slant range resolution (or ground range resolution) is characterized only by the property of the microwave and the look angle and they have nothing to do with the position and size of the antenna. However, the azimuth resolution (Δx) is dominantly determined by the position and size of the antenna. If a C-band microwave ($\lambda = 5.66$ cm) real aperture radar onboard the satellite (ERS-1/2) is employed to take images with an azimuth resolution of 10 m from 785 km away, the required length of its aperture is longer than 3 km. It seems impossible for any flying platform to carry such a long antenna. In other words, the azimuth resolution of radar images is too low for many applications.

To improve the resolution of radar images, SAR was developed in the 1960s. It is based on the principle of the Doppler frequency shift caused by the relative movement between the antenna and the target (Fritsch and Spiller 1999). Figure 3.9 shows the imaging geometry of synthetic aperture radar while it is being used to take a side-look image of the ground.

Assuming that a real aperture imaging radar with aperture length L moves from a to b, then to c, the slant range from any point, for example, target O, to the antenna varies from R_a to R_b, then to R_c. $R_a > R_b$, and $R_b < R_c$, which means that at first the antenna is flying nearer and nearer to the point object until the slant range becomes the shortest R_b, then it gets further away. The variation of slant range R will cause the frequency shift of the echo backscattered from target O, varying from an increase to a decrease. By precisely measuring the phase delay of the received echoes, tracing its frequency shift, and then synthesizing the corresponding echoes, the azimuth resolution can be sharpened, as the area of the intersection of the three footprints shown in Figure 3.9. Compared to the azimuth resolution of the full footprint width described earlier, the

TECHNIQUES FOR ACQUISITION OF DTM SOURCE DATA

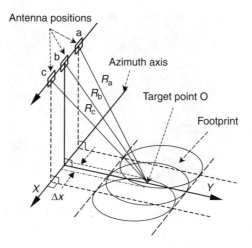

Figure 3.9 Imaging geometry of SAR.

azimuth resolution (Δx) of the SAR is much improved (Curlander and Mcdonough 1991), that is,

$$\Delta x = \frac{L}{2} \tag{3.10}$$

Indeed, it means that the azimuth resolution (Δx) of an SAR is only determined by the length of the real aperture of an antenna, independent of the slant range R and the wavelength λ. As a result, it is possible to acquire images with 5-m azimuth resolution by an SAR with a 10-m real aperture length onboard ERS-1/2.

Combined with some advanced range compressing techniques, an SAR whether on an aircraft or on a space platform can take high-quality images (with high resolution in both azimuth direction X and slant range direction R) day and night under all weather conditions. After processing, each pixel of the SAR image contains not only the grey value (i.e., amplitude image) but also the phase value related to the radar slant range. These two components can be expressed by a complex number. Therefore, the SAR image can also be called a radar complex image. Figure 3.10 shows an example of an amplitude image and Figure 3.11 illustrates the plane coordinate system of the SAR image and the complex number expression of the pixel. It is the use of phase information that makes interferometric SAR (InSAR) technologically special.

3.3.2 Principles of Interferometric SAR

SAR images (amplitude images) have been widely used for reconnaissance and environmental monitoring in remote sensing. In such cases, the phase component recorded simultaneously by the SAR has been overlooked for a long time. In 1974, Graham first reported that a pair of SAR images of the same area taken at slightly different positions can be used to form an interferogram and the phase differences recorded in the interferogram can be used to derive a topographic map of the Earth's surface

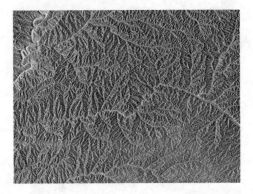

Figure 3.10 An example of the SAR image of Yan'an (C-band, by ERS-1 on August 9, 1998).

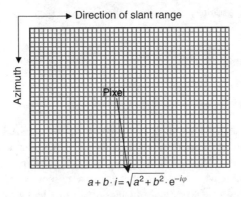

Figure 3.11 Complex number table of pixels.

(Graham 1974). This technology is called InSAR or SAR interferometry. As InSAR is new, discussions will be more detailed.

At present, InSAR is a signal processing technique rather than an instrument. It derives height information by using the interferogram, $\phi(x, r)$, which records the phase differences between two complex radar images of the same area taken by two SARs on board the same platform or by a single SAR revisited, as shown in Figure 3.12 and Figure 3.13, where B and α are the baseline and baseline orientation angles with respect to the horizon, respectively. Let $\hat{S}_1(x, r)$ be the complex image taken at position A_1 with its phase component $\Phi_1(x, r)$ and $\hat{S}_2(x, r)$ taken at position A_2 with its phase component $\Phi_2(x, r)$. According to radiowave propagation theory, the phase delay measured by an antenna is directly proportional to the slant range from the antenna to a target point, that is,

$$\Phi = \frac{2\pi R}{\lambda} \tag{3.11}$$

TECHNIQUES FOR ACQUISITION OF DTM SOURCE DATA

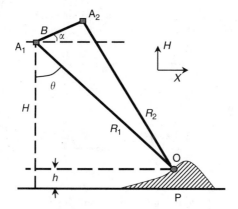

Figure 3.12 Geometry of SAR images for heighting.

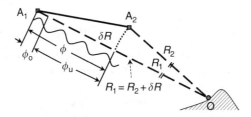

Figure 3.13 Different types of phase differences (ϕ, ϕ_u, and ϕ_o).

By subtracting $\Phi_1(x, r)$ from $\Phi_2(x, r)$, the differences form an interferogram $\phi(x, r)$ (see more detailed discussion later).

$$\phi = \Delta\Phi = \Phi_2 - \Phi_1 = \frac{2 \times \pi \times Q \times \delta R}{\lambda} \qquad (3.12)$$

where $Q = 1$ when the two antennas are mounted on the same flying platform, one transmitting wave but both receiving echoes simultaneously to form one-pass interferometry like TOPSAR (Zebker and Villasenor 1992); otherwise, $Q = 2$. That is, if the two SAR complex images are acquired at two different places by the same radar, then $Q = 2$.

From $\phi(x, r)$, the slant range difference (δR) between R_1 (the distance from a target point O to A_1) and R_2 (the distance from O to A_2) can be calculated by the following formula:

$$\delta R = R_1 - R_2 = \frac{\lambda}{2\pi Q} \phi \qquad (3.13)$$

where λ is the wavelength. As λ is in centimeters, the slant range difference is measured in centimeters.

When $B \ll R_1$, the difference between two slant ranges can be approximated by the baseline component in the slant direction (i.e., the so-called parallel baseline). Mathematically,

$$\delta R \approx B_{\parallel} = B \sin(\theta - \alpha) \qquad (3.14)$$

where θ is the side-looking angle. From Figure 3.12, it is not difficult to obtain the following relationship:

$$\sin(\theta - \alpha) = \frac{R_1^2 + B^2 - R_2^2}{2BR_1} = \frac{R_1^2 + B - (R_1 + \delta R)^2}{2BR_1} \qquad (3.15)$$

After the side-looking angle is determined by Equations (3.12) to (3.14), the height h of the point O can be derived from the following equations:

$$\theta = \sin^{-1}\left(\frac{\lambda \phi}{4\pi B}\right) + \alpha \qquad (3.16)$$

$$h = H - R_1 \times \cos\theta \qquad (3.17)$$

where α is the angle of the baseline with respect to the horizontal line, H is the flying height (from radar antenna to reference tatum) and h is the height of the point O (from O to the same reference datum).

From the previous discussions it can be seen that the key issues of heighting with InSAR are (a) the precise computation of the phase difference and (b) the precise estimation of the baseline. Of course, there are other processes involved. Figure 3.14 shows the whole process for DTM data acquisition by InSAR. As the baseline can be determined by GPS data on board, the following discussion concentrates on the computation of phase differences.

First, two SAR complex images are used, one referred to as the master image and the other as the slave image. These two images may have different orientations because the antennas may have slightly different attitudes at different times. Therefore, they need to be transformed to the same coordinate system and resampled into pixels with the same size in terms of ground distance so that they could match each other. These two processes can be performed simultaneously and the whole process is called co-registration. Commonly, polynomials are used as the mathematical function for such a transformation and bilinear interpolation is used for resampling.

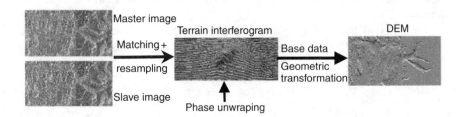

Figure 3.14 The process of DTM data acquisition by InSAR.

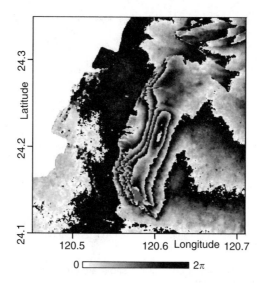

Figure 3.15 An example of InSAR interferogram (western coastline area of Taiwan, generated by use of a pair of ERS-1/2 Tandem SAR image data: ERS-1: 1996.3.15 and ERS-2: 1996.3.16).

The next step is to solve the coefficients (unknowns) of the polynomials. In doing so, some points (called tie points) on both images are selected as control points (i.e., with known x, y coordinates). The normal practice is to select some well-defined feature points such as road intersections. If such points are not available, then a grid with fixed x, y intervals is set, superimposed onto the master image, and these nodes are selected as tie points. The corresponding points on the slave image are found by using image-matching technique.

After images are co-registered, the phase image can be used to produce the interferogram (see Figure 3.15). The value of each pixel in the interferogram is in fact the phase difference of the conjugate pixels. It is computed by multiplying the two conjugate complex numbers, for example, $\hat{S}_1(i)$ and $\hat{S}_2(i)$ as follows:

$$G = \sum_{i}^{N} \hat{S}_1(i) \cdot \hat{S}_2^*(i) \tag{3.18}$$

and

$$\phi_o = \tan^{-1}[G] \in [-\pi, \pi) \tag{3.19}$$

where "$*$" represents the conjugate complex numbers, and N is the total number of pixels of the moving window, which means that a moving average is applied for the reduction of phase noise.

However, this is not the whole story. Actually, the difference, δR, in slant range from the ground point P to the two antennas at A_1 and A_2 corresponds to a number

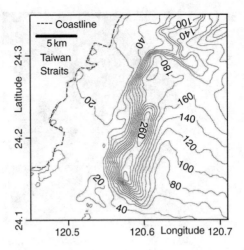

Figure 3.16 Contour diagram of DTM of the same area as shown in Figure 3.15 (produced from DTM generated by InSAR).

of whole waves (ϕ_u) plus a residual ϕ_o, that is,

$$\phi = \phi_u + \phi_o = \phi_o + 2\pi k \tag{3.20}$$

where k is the integral number of microwave cycles. However, the value of k cannot be determined. This is a cycle ambiguity problem, solved by a process called phase unwrapping, which makes use of information about the phase differences in neighboring pixels. This topic will not be discussed further, but interested readers may refer to the article by Goldstein et al. (1988). Figure 3.15 shows the interferogram of an area along the western coastline of Taiwan. The ground resolution is about 20 m × 20 m. The interferogram fringes look similar to contour lines, which actually reflect the undulation of the Earth's surface (Figure 13.16).

In fact, apart from terrain variations, phase information also includes several other types of information, that is, atmospheric effect and other noise. These are not desired components in the generation of interferograms and should be removed beforehand. More information about such processes could be found in a paper by Zebker et al. (1997).

3.3.3 Principles of Radargrammetry

Similar to photogrammetry, radargrammetry forms a stereo model for 3-D measurement. The difference is that in radargrammetry, two SAR images collected with the unique side-looking geometry (as shown in Figure 3.6) are used to form the stereo model. Only the intensity information of SAR images is used for radargrammetric measurement, unlike InSAR which works principally with interferometric phase information. The 3-D reconstruction is still performed in a way similar to

TECHNIQUES FOR ACQUISITION OF DTM SOURCE DATA

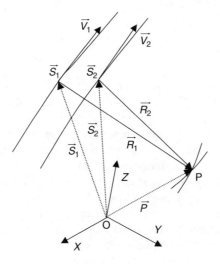

Figure 3.17 Stereo configuration of radargrammetry.

photogrammetry, relying on the following key issues:

1. determining the sensor–object stereo model
2. searching for corresponding pixels from two overlapping SAR images using image-matching techniques
3. determining 3-D coordinates by solving the intersection problem.

Figure 3.17 shows the general stereo configuration of radargrammetry, in which two SAR images are acquired with different radar look angles along two different flight paths (airplane tracks or satellite orbits). To satisfy the requirement of stereoscopy, a sufficient overlap between two SAR images is guaranteed.

Suppose O–XYZ is a geodetic coordinate system, then some rigorous formulae can be derived for radar stereo computation. As seen in Figure 3.17, there is a plane formed by the two-sensor positions (S_1, S_2) and the object position (P). This implies that the object position is determined by the intersection of two radar rays with different look angles, leading to a coplanarity condition expressed by

$$\vec{S}_1 + \vec{R}_1 - \vec{S}_2 - \vec{R}_2 = 0 \tag{3.21}$$

where $\vec{S}_1$ and $\vec{S}_2$ denote the 3-D position vectors of sensors 1 and 2, respectively, while $\vec{R}_1$ and $\vec{R}_2$ denote the sensor–object vectors of two radar rays. The above conditions can be interpreted as the intersection of range spheres and Doppler cones (Leberl 1990), and thus we have two range equations and two Doppler equations given as follows:

Range equations:

$$|\vec{P} - \vec{S}_1| = |\vec{R}_1| = R_1 \tag{3.22a}$$

$$|\vec{P} - \vec{S}_2| = |\vec{R}_2| = R_2 \tag{3.22b}$$

Doppler equations:

$$\vec{V}_1 \cdot (\vec{P} - \vec{S}_1) = 0 \quad (3.22c)$$

$$\vec{V}_2 \cdot (\vec{P} - \vec{S}_2) = 0 \quad (3.22d)$$

where $\vec{V}_1$ and $\vec{V}_2$ denote the 3-D velocity vectors of sensors 1 and 2, respectively. Equations (3.22c) and (3.22d) represent the general case of zero-Doppler projection (see Leberl 1990 for non-zero-Doppler projection).

In essence, these four equations represent the stereo model of radargrammetry. Before commencing stereo measurements, some parameters in the model should be solved or refined. In particular, the positions and velocities of the flying sensors should be determined, and each component of the vectors is generally modeled as a function (e.g., polynomial) of imaging time. Although the track or orbit data from differential GPS (DGPS) or orbit determination techniques may provide the input for such modeling, their accuracies are not always sufficient for accurate 3-D reconstruction. Therefore, based on the least-squares approach, the refinement of the stereo model using several ground control points (GCPs) can be performed to improve the accuracy of the parameters (Toutin 2000).

Existing studies indicate that a larger intersection angle between two SAR images results in a larger parallax and an equivalently higher geometric sensitivity to ground elevation, but makes image matching difficult due to a larger radiometric dissimilarity caused by different radar illumination directions (Leberl et al. 1986a, b; Toutin and Gray 2000). Therefore, a careful selection of intersection angle is needed to balance between geometric sensitivity and radiometric similarity.

Since the launch of the Canadian RADARSAT satellite in 1995, most experimental studies on radargrammetry have been carried out using radar images acquired in multi-modes. These experiments with different stereo configurations showed inconsistent accuracy of about 20 to 70 m in elevation results (Toutin 2000, 2002). Indeed, it has been found that the accuracy of DTM by radargrammetry is affected by the following factors:

1. terrain features such as topographic slopes
2. geographical conditions and geometric distortions in relation to radar looking angles
3. intersection angles.

Figure 3.18 shows an example of a DTM generated from a pair of ERS-1 SAR images acquired along two adjacent descending orbits over Hong Kong. There is only a 30% overlap (i.e., around 30 km) and the intersection angle is about 4.5°.

3.4 AIRBORNE LASER SCANNING (LIDAR)

The use of lasers as remote sensing instruments has an established history going back 30 years. Through the 1960s and 1970s various experiments demonstrated the power of using lasers in remote sensing including lunar laser ranging, satellite laser ranging, atmospheric monitoring, and oceanographic studies (Flood 2001). Due to advancements in reliability and resolution over the past decades, the airborne laser scanning (ALS) system is becoming an important operational tool in remote sensing,

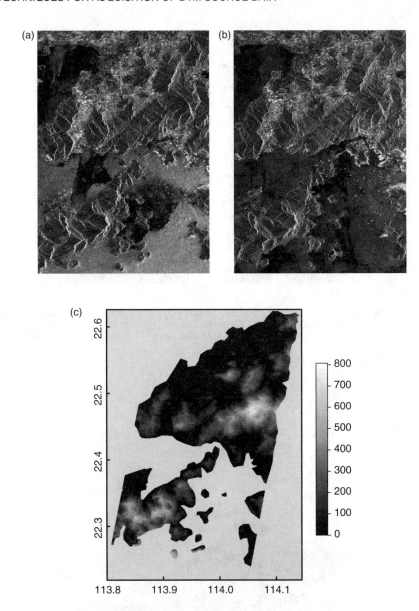

Figure 3.18 DTM generated from an ERS-1 SAR stereo pair over Hong Kong by radargrammetry: (a) ERS-1 SAR image on March 2, 1996; (b) ERS-1 SAR image on March 18, 1996; and (c) DTM generated using two SAR images as shown in (a) and (b).

photogrammetry, surveying, and mapping (Ackermann 1996). The ALS system, usually called airborne LIDAR (Light Detection And Ranging) in the commercial sector, is an active remote system. The usefulness of ALS systems has been demonstrated by a number of applications where traditional photogrammetric methods fail or become too expensive, for example, the acquisition of terrain elevation data over areas with dense vegetation (Kraus and Pfeifer 1998), acquisition of 3-D city data

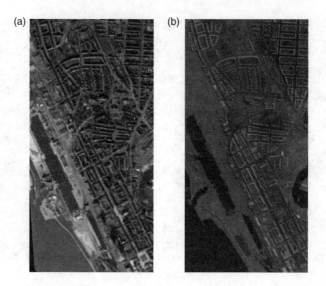

Figure 3.19 An example of 3-D city model acquired by LIDAR (Courtesy of GeoLas Consulting) (a) Aerial photograph. (b) 3D model acquired by laser scanning; both acquired by the LiteMapper system (www.LiteMapper.com).

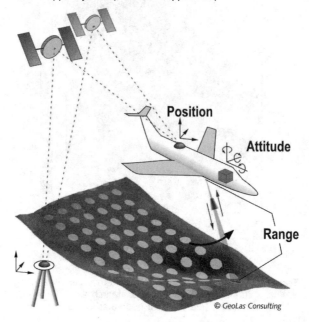

Figure 3.20 Principle of airborne laser scanning (Courtesy of GeoLas Consulting).

(Haala et al. 1998), or the surveying and modeling of power lines. An overview of resources on existing ALS systems has been produced by Baltsavias (1999a,b). An example of a 3-D model acquired by LIDAR is given in Figure 3.19.

3.4.1 Basic Principle of Airborne Laser Scanning

ALS is a complex integrated system, consisting of a laser range finder (LRF), a computer system to control the on-line data acquisition, a storage medium, a scanner, and a GPS/INS system for determining the position and orientation of the system. The basic scanning principle is illustrated in Figure 3.20.

As LIDAR is an active system, it sends off electromagnetic energy and records the energy scattered back from the terrain surface and the objects on the terrain surface. It is the type of materials hit by the pulses which determines the intensity of the returning signals. The wavelength of the laser lies in, or just above, the visual range of the electromagnetic spectrum, that is, in the range of 1040 to 1060 nm. The formulae (Baltsavias 1999a,b) governing the height determination by laser ranging will be given in this section. The formulae presented here relate mostly to pulse lasers. When they refer to continuous-wave lasers (CW lasers), it will be explicitly mentioned. For the sake of simplicity, it is assumed that

1. the roll and pitch angles are 0
2. the system scans along a plane perpendicular to the flight direction, with scanning lines equidistant
3. the terrain is flat (unless mentioned otherwise)
4. the area scanned consists of n overlapping parallel strips of equal length
5. the flying speed and height are constant.

3.4.1.1 Range and Range Resolution

Pulse laser:

$$R = c\frac{t}{2} \qquad \Delta R = c\frac{\Delta t}{2} \tag{3.23}$$

where R is the range distance between sensor and object (m); ΔR is the range resolution (cm); t is the time interval between sending and receiving a pulse (or echo) (ns); c is the speed of light, $\approx$300,000 km/s; and Δt is the resolution of time measurement (ns).

Time t is measured by a time interval counter relative to a specific point on the pulse, for example, the leading edge (i.e., the rising side of the pulse). Since the leading edge is not well defined (no rectangular pulses), time is measured for a point on the leading edge, where the signal voltage has reached a predetermined threshold. This is accomplished with a threshold trigger circuit to start and stop the time counting.

For CW lasers, the range and range resolution are as follows:

CW laser:

$$R = \frac{1}{4\pi}\frac{c}{f}\varphi \qquad \Delta R = \frac{1}{4\pi}\frac{c}{f}\Delta\varphi \tag{3.24}$$

where f is the frequency (Hz); φ is the phase (for CW lasers) (rad); and $\Delta\varphi$ is the phase resolution (for CW lasers) (rad).

3.4.1.2 Maximum Unambiguous Range

For pulse lasers, the maximum unambiguous range depends on two major factors:

1. the maximum range (number of bits) of the time interval counter
2. the pulse rate.

To avoid confusion in the time interval counter, it is usually required that no pulse be transmitted until the echo of the previous pulse has been received. For example, for a pulse rate of 25 kHz, the maximum unambiguous range is 6 km.

In practice, these two factors have never had any effect on the maximum range (and flying height). In contrast, the maximum range is limited by other factors such as the intensity of the laser power, the divergence of the laser beam, the transmission rate of the atmosphere, the reflectance rate of the target, the sensitivity of the detector, and the influence of flying height and attitude errors on the 3-D positional accuracy. For CW laser, the maximum unambiguous range is

CW laser:

$$R_{max} = \frac{\lambda_{long}}{2} \tag{3.25}$$

where λ_{long} refers to the long wavelength corresponding to the low frequency of a CW laser.

For example, a CW laser employs two frequencies, 1 and 10 MHz. The low frequency corresponds to a wavelength of 300 m, then the maximum unambiguous range is 150 m. This does not imply that the flying height over ground must be limited to 150 m. In fact, the flying height can be increased by making use of other supplementary information. If all other conditions are kept constant, the maximum range is typically proportional to the square root of reflectivity and of intensity of the laser.

Accuracy of laser ranging is

$$\sigma_R \sim \frac{1}{\sqrt{S/N}} \tag{3.26}$$

where σ_R is the ranging precision (m) and S/N is the signal-to-noise ratio.

For CW lasers, the accuracy of the ranging is proportional to the square root of the signal bandwidth (measurement rate), as the latter is inversely proportional to the average number of cycles required for one measurement. That is,

for pulse laser:

$$\sigma_{R_{pulse}} \sim \frac{c}{2} t_{rise} \frac{\sqrt{B_{pulse}}}{PR_{peak}}$$

for CW laser:

$$\sigma_{R_{cw}} \sim \frac{\lambda_{short}}{4\pi} \frac{\sqrt{B_{cw}}}{PR_{av}}$$

where $\sigma_{R_{pulse}}$ is the pulse laser ranging precision (m); t_{rise} is the rise (from 10% to 90% of its maximum value) time of the pulse (ns); B_{pulse} is the noise input bandwidth

TECHNIQUES FOR ACQUISITION OF DTM SOURCE DATA

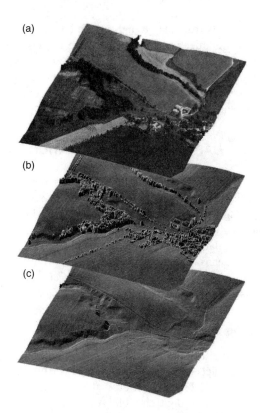

Figure 3.21 DTM obtained from DSM using filtering (Courtesy of GeoLas Consulting) (Top: Aerial photo; middle: scanned data; Bottom: DSM from laser data).

(pulse lasers) (Hz); $P_{R_{peak}}$ is the peak power (applies only to pulse lasers) (W); $\sigma_{R_{cw}}$ is the CW laser ranging precision (m); B_{cw} is the noise input bandwidth (CW lasers) (Hz); $P_{R_{av}}$ is the average power (applies to pulse and CW lasers) (W); and λ_{short} is the short wavelength corresponding to the high frequency of a CW laser.

3.4.2 From Laser Point Cloud to DTM

The ALS system produces data that can be characterized as sub-randomly distributed 3-D point clouds. The processing of ALS data often aims at either the removal of unwanted measurements (in the form of either erroneous measurements or objects) or the modeling of data for a given specific model (e.g., a DTM) as a subset of a measured digital surface model (DSM).

In the process of acquiring ALS data, the following steps are involved, that is, filtering, classification, and modeling. Filtering refers to the removal of unwanted measurements to find a ground surface from a mixture of ground and vegetation measurements. The unwanted measurements can, depending on applications, be characterized as noise, outliers, or gross errors. Classification means to find a specific geometric or statistic structure, such as buildings or vegetation. Generalization of classified objects is referred to as modeling. Figure 3.21 shows a DTM obtained from DSM using filtering.

The separation of objects from the ground surface (Axelsson 1999) is a general process common to most applications, if not all. Once objects are separated from the ground surface, height variation of the terrain surface is obtained.

The flying height of most existing ALS systems is in the range of 20 to 6000 m (typically 200 to 300 m) and the height accuracy is in the range of 10 to 60 cm (typically 15 to 20 cm) while planimetric accuracy is 0.1 to 3 m (typically 0.3 to 1 m).

3.5 CARTOGRAPHIC DIGITIZATION

There are basically two cartographic digitization techniques, that is, vector-based line following or raster-based scanning. Digitization can be done either manually or by automated devices, giving four possible solutions as shown in Figure 3.22.

Manual line following is the most widely used method. Semiautomated devices for line following are available but they are very expensive and thus not popular. Manual raster scanning means to superimpose a regular grid onto the map and then record whether contours pass through these grid cells. This is not a practical method. Automated scanning is also a popular method for cartographic digitization.

3.5.1 Line-Following Digitization

In manual line-following digitization, either a mechanically based digitization system or a solid-state digitizing tablet can be used — nowadays usually the latter, an example of which is shown in Figure 3.23. In either case, the digitization is done manually. The map is carefully put onto the digitizer table. A cursor with cross-hairs is used to trace the contour lines by hand and to record the coordinates. Manual line-following digitization can be done in two ways, that is, either stream or point mode.

With point mode digitization, each time the operator presses a button, the x, y coordinates of the cursor's position will be recorded. Therefore, each time, a decision needs to be made by the operator on which point is to be measured. Usually, the measurement is carried out in stationary (i.e., static) mode to give the best accuracy. The main advantage of point mode manual digitization is that the operator controls the selection of points to reduce data volume. However, this is a tedious process.

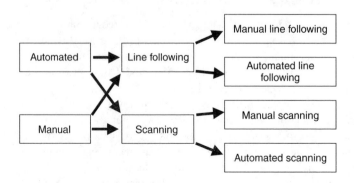

Figure 3.22 Cartographic digitization methods.

TECHNIQUES FOR ACQUISITION OF DTM SOURCE DATA

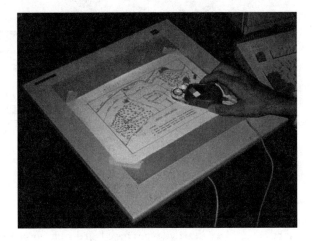

Figure 3.23 An example of tablet digitizer.

Stream mode means that the tracing/measurement process is carried out dynamically and is thus less accurate. In this mode, while the cursor is moving along the contour lines, point coordinates are recorded, either on a time or on a distance basis. The disadvantages of stream tracing is that the operator does not need to do the line following. However, data redundancy is a big problem. Another concern with this mode is the fidelity of the results to the original line because data points will be recorded at certain intervals irrespective of whether the cursor is following the line well or is quite deviated from the line, which often happens at the turns of curves.

To overcome the problems encountered in manual line-following digitization, automated line following devices such as Laser-Scan's Fastrak and Lasertrak systems have been developed to remove the need for manual movement of the cursor during measurement. However, an operator is still required to supervise the system and to execute various operations such as the initial positioning of the device on contours; guiding the device through areas of closely packed contours and cliffs; inserting contour elevation values, etc. Given the level of manual interventions required, the method is usually referred to as semiautomated digitization. Unfortunately, all semiautomated line-following digitizers are very expensive and beyond the means of many small mapping organizations. Therefore, this method will not be discussed further in this section.

Data obtained by digitization are in the digitizer coordinate system and must be transformed into a geodetic coordinate system. The normal practice is to digitize a few map grid notes as control points and then apply an affine transform to convert digitized data points into the geodetic system employed for the map.

3.5.2 Raster Scanning

Raster scanners make fully automated digitization possible. In raster scanning, each line scan is divided into resolution units, for example, 25 μm × 25 μm, and for each unit, the scan provides a return as to whether or not a contour line is present.

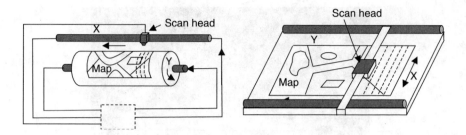

Figure 3.24 Illustration of drum (left) and flat-bed (right) scanners.

Each response is recorded the same way, for example, 0 if nothing is present and 1 if there is a line. Or the data can also be recorded as a grey image.

Scanners can be categorized into two types based on the platform, that is, flat-bed and drum scanners. Figure 3.24 illustrates these scanners. Based on the arrangement of detectors, they can be classified as point, line, and area scanners.

If a drum scanner is used, the map to be scanned is wrapped on the drum. With the rotation of the drum as moves in the y direction and the scanner-head's moves in the x direction, the whole map can be scanned. The working principle of the flat-bed scanner is similar. In this case, the moves in the y direction are achieved by movement along two rails.

Scanners for cartographic scanning are of high resolution. As a result, the amount of data is huge and there is a great deal of redundancy. Then vectorization follows, which can be manual or automated. Manual vectorization means to display the scanned map on a screen and then to carry out line-following digitization on screen. Automated vectorization is done completed with algorithms.

3.6 GPS FOR DIRECT DATA ACQUISITION

GPS are a popular technique for direct measurement of the Earth's surface. They are replacing the traditional theodolites and total stations. In this section, the basic principle of traditional ground surveying is also briefly discussed for comparison.

3.6.1 The Operation of GPS

The GPS has three parts, that is, the space segment, ground control segment, and user segment (see Figure 3.25).

Space segment refers to the GPS satellite constellation, which consists of 24 satellites and some spares. Figure 3.26 shows such a configuration. The satellites are about 20,000 km above the Earth's surface and they continuously broadcast measurement signals and navigation messages to GPS users.

The control segment consists of ground stations that ensure the satellites are working properly. It consists of one master control station (MCS) located at Schriever (formerly Falcon) AFB in Colorado, five monitor stations (Hawaii, Kwajalein,

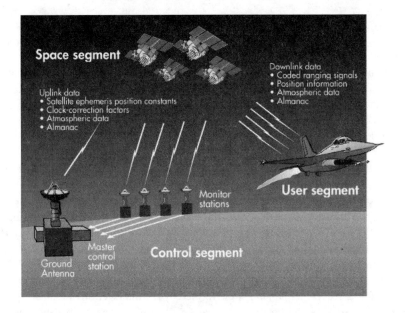

Figure 3.25 GPS segments (Reprinted from The Aerospace Corporation 2003 with permission).

Figure 3.26 The GPS satellite constellation (Reprinted from Garmin 2003 with permission).

Ascension Island, Diego Garcia, and Colorado Springs), and three ground antennas (Ascension Island, Diego Garcia, and Kwajalein). The locations of the monitor stations are precisely determined and each has a GPS receiver to receive signals from all satellites in view so as to monitor them continuously. The master control station gathers tracking and other related information collected at the monitor stations so that satellite orbits can be predicted and corrected and the health status of the satellites determined. Such information forms part of the navigation message, which is then sent to the transmission stations to be uploaded to all satellites. Such messages are continuously broadcasted by satellites and GPS receivers can receive them anytime.

The user segment consists of receivers, used by users for different applications. The receivers can be hand-held, mounted in cars, installed on aircraft, ships, tanks, submarines, cars, and trucks, or put on a tripod, which is the case in field surveying. A GPS receiver consists of hardware and software for receiving signals from satellites, and for decoding, storing, and processing these signals.

3.6.2 The Principles of GPS Measurement

The basic principle of GPS-based measurement is range intersection. To determine an unknown position in 3-D space, three distances from three known points are needed. Figure 3.27 illustrates this principle. If one knows that the point (say P) to be determined is d away from a known point, say A, one can only know that P is located somewhere on the sphere which is centred at A and has a radius of d. With two distances from two known points, two such spheres can be formed and one can then know that P is somewhere on the circle formed by the intersection of the two spheres. With three distances, the position of P can be determined exactly because the three spheres intersect at two points, one of which will obviously be wrong.

In the case of GPS, the positions of satellites are all known at any time because the orbit of each satellite is monitored and controlled by ground stations. To determine the location of a GPS receiver, one needs to measure at least three distances from the receiver to three GPS satellites. Therefore, to GPS users, the key issue is to measure distances from the receiver to the satellites.

As one can imagine, it is impossible to measure these distances using traditional equipment. The solution is to make electromagnetic waves that travel at the speed of light. Suppose a satellite sends a signal to a GPS receiver and it takes t sec for the GPS receiver to receive the signal, then the distance between the GPS receiver and the satellite is

$$D = c \times t \qquad (3.27)$$

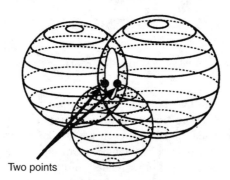

Figure 3.27 The positioning principle of GPS (Reprinted from McElroy 1992 with permission).

TECHNIQUES FOR ACQUISITION OF DTM SOURCE DATA

where D is the distance, t is the travelling time, and c is the velocity of light, that is, 299,792,458 m/sec.

Now the issue becomes how to determine the time required for the signal to travel from satellite to the receiver. The current practice is to have a clock onboard the satellite and a clock in the receiver. Then, Equation (3.27) should be rewritten as follows

$$D = c \times (T_s - T_r) \tag{3.28}$$

where T_s represents the time when the satellite transmits the signal and T_r represents the time when the signal reaches the GPS receiver.

From Equation (3.28), it can be noted that a tiny error in either clock will cause significant error in the distance computed. This requires that the clocks both onboard the satellites and in the receiver be extremely accurate. This would certainly make GPS receivers very expensive and thus make GPS application less popular. The solution to this problem is to assume an error between the clock in the GPS receiver and the clocks onboard the satellites but the error is constant to this particular GPS receiver because satellites are equipped with extremely accurate atomic clocks and satellite clock errors can be corrected by applying the correction parameters contained in the navigation message. Assuming this, one could then treat such an error as an unknown. With such an extra unknown (in addition to the X, Y, Z coordinates of the receiver), one needs to measure another distance in addition to the three required for position determination. That is, a total of four satellites need to be observed to determine the position of a point.

To make the measurement with higher accuracy, DGPS is widely used. The basic idea is to determine the difference of coordinates between two points (say A and B), one of which (say A) is known precisely. In this way, one GPS receiver is on each point and both receivers simultaneously receive signals from the same set of satellites over a period of time. The reason why accuracy can be improved by such an arrangement is that the satellite clock error and atmospheric effects are cancelled if one subtracts the range (called pseudo-range) from point A to a satellite from the range from point B to the same satellite.

3.6.3 The Principles of Traditional Surveying Techniques

Traditional surveying techniques determine the position (coordinates) of a point through the measurement of distances and angles. The traditional instruments are theodolites and computerized total stations.

Figure 3.28 shows two typical examples of a traditional surveying problem. In Figure 3.28(a), the 3-D coordinates of point A are known. Through measurement of the vertical angle, distance (between A and P), and a bearing, the position of point P can be determined.

In Figure 3.28(b), the 3-D coordinates of points A and B are known. Through the measurement of horizontal angles and distances (between A and P and between B and P), the position of point P can be determined.

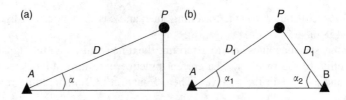

Figure 3.28 Two typical problems in traditional surveying. (a) From a known point A to determine the position of P. (b) From two known points A and B to determine the position of P.

Table 3.6 A Comparison of Various DTM Acquisition Methods

Acquisition Method	Accuracy of Data	Speed	Cost	Application Domain
Traditional surveying	High (cm–m)	Very slow	Very high	Small areas
GPS survey	Relatively high (cm–m)	Slow	Relatively high	Small areas
Photogrammetry	Medium to high (cm–m)	Fast	Relatively low	Medium to large areas
Space photogrammetry	Low to medium (m)	Very fast	Low	Large areas
InSAR	Low (m)	Very fast	Low	Large areas
Radargrammetry	Very low (10 m)	Very fast	Low	Large areas
LIDAR	High (cm)	Fast	High	Medium to large areas
Map digitization	Relatively low (m)	Slow	High	Any area size
Map scanning	Relatively low (m)	Fast	Low	Any area size

3.7 A COMPARISON BETWEEN DTM DATA FROM DIFFERENT SOURCES

It should be pointed out that these data acquisition methods all have advantages and disadvantages. Therefore, when choosing a method, various aspects such as the purpose, accuracy requirements, conditions of the equipment, and availability of source materials should be considered. To assist in decision making, a comparison between these methods in various aspects, such as efficiency, cost, and accuracy, would be useful (see Table 3.6).

In terms of the accuracy of measurement, a millimeter-level can be reached by ground survey and centimeter-level by photogrammetry and meter-level by digitization from maps. The accuracy of photogrammetric data depends on the images used. In the case of space photogrammetry using satellite images, the accuracy could be very low, depending on the resolution. For example, if SPOT images with 10-m resolution is used, then the accuracy is 5 to 10 m. InSAR is a good technique for deformation (i.e., relative change) measurement and with it an accuracy of 1 cm can be reached. However, for DTM data acquisition (i.e., absolute heights on terrain surface), the accuracy is only about 5 m. The accuracy of radargrammetric data is even lower. Both with ground survey and in photogrammetric (as well as InSAR and radargrammetric) techniques, terrain feature points can be obtained if desired. Therefore, high fidelity to the original surface can be preserved by the digital data.

In terms of efficiency, ground surveying is more labor intensive and therefore is only suitable for modeling a small area, when high accuracy is required. On the other hand, most of the processes in photogrammetric technique have already been automated nowadays. Therefore, data acquisition is more efficient. Indeed, photogrammetric technique is suitable for medium- and large-size areas. For cartographic digitization, the raster scanning process can easily be automated but human interference is still needed during the raster and vector conversions.

In terms of availability, in developed and most developing countries, some contours maps are available. Such maps are the major source for digital terrain modeling. In many countries, the national DTMs have been generated from existing contour maps.

CHAPTER 4

Digital Terrain Surface Modeling

In the previous chapter, techniques for the acquisition of DTM source data were discussed. Also, surface modeling could be applied for the reconstruction of terrain surface. This is the topic of this chapter.

4.1 BASIC CONCEPTS OF SURFACE MODELING

4.1.1 Interpolation and Surface Modeling

A digital terrain model is a mathematical (or digital) model of the terrain surface. It employs one or more mathematical functions to represent the surface according to some specific methods based on the set of measured data points. These mathematical functions are usually referred to as *interpolation* functions. The process by which the representation of the terrain surface is achieved is referred to as *surface reconstruction* or *surface modeling* and the actual reconstructed surface is often referred to as the DTM surface. Therefore, terrain surface reconstruction can also be considered as DTM surface construction or DTM surface generation. After this reconstruction, height information for any point on the model can be extracted from the DTM surface.

The concept of *interpolation* in DTM is a little different from that of *surface reconstruction*. The former includes the whole process of estimating the elevation values of new points, which may in turn be used for surface reconstruction, while the latter emphasizes the process of actually reconstructing the surface, which may not involve interpolation. To clarify this matter further, *surface reconstruction* only covers those topics concerned with "how the surface is reconstructed and what kind of surface will be constructed." For example, should it be a continuous curved surface or should it consist of a linked series of planar facets?

In contrast, interpolation has a much wider scope. It may include surface reconstruction and the extraction of height information from the reconstructed surface; it may also include the formation of contours either from randomly located points or from a measured set of elevation values obtained in a regular grid pattern. In both of these latter cases, the measured values are honored in the resulting DTM surface and the interpolation process takes place only after surface reconstruction, either to extract height information for specific points or to construct contoured plots. Interpolation methods will be discussed in Chapter 6.

4.1.2 Surface Modeling and DTM Networks

It will be discussed later that regular-grid networks and triangular irregular networks (TINs) have been widely used for surface modeling. Here, some clarifications need to be made before the detailed discussions.

A network is a data structure implemented in a special pattern for surface modeling. A network is concerned mostly with the inter-relationship of the data points in the positional (planimetric) sense but not necessarily in the third dimension. This is the main difference between network and the DTM surface that is constructed from the network and comprises a series of sub-surfaces that may or may not have continuity in the first derivative. The topological relation for a regular grid is built-in (i.e., it is implicit) due to the special characteristics of the regular grid itself so that this difference is not appreciated or shown clearly. In contrast, in the case of triangle-based modeling, the distinction is very clear — the topological relationship needs to be sorted out to form a triangular network; then, the third dimension can be added to the network to form a continuous surface comprising a series of contiguous triangular facets.

4.1.3 Surface Modeling Function: General Polynomial

To model an area on terrain surface, a mathematical function needs to be used. There are many possibilities as discussed in Chapter 1. The function can be expressed in frequency or in space domain. In space domain, the general mathematical expression

Table 4.1 Polynomial Function Used for Surface Reconstruction

Individual Terms	Order	Descriptive Terms	No. of Terms
$Z = a_0$	Zero	Planar	1
$+a_1 X + a_2 Y$	First	Linear	2
$+a_3 X^2 + a_4 Y^2 + a_5 XY$	Second	Quadratic	3
$+a_6 X^3 + a_7 Y^3 + a_8 X^2 Y + a_9 XY^2$	Third	Cubic	4
$+a_{10} X^4 + a_{11} Y^4 + a_{12} X^3 Y + a_{13} X^2 Y^2 + a_{14} XY^3$	Fourth	Quartic	5
$+a_{15} X^5 + a_{16} Y^5 + a_{17} X^4 Y + a_{18} X^3 Y^2 + a_{19} X^2 Y^3 + a_{20} XY^4$	Fifth	Quintic	6

DIGITAL TERRAIN SURFACE MODELING

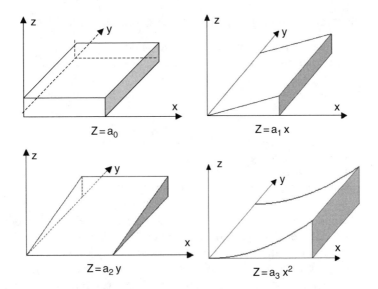

Figure 4.1 Surface shapes of the first 4 terms of general polynomial function.

is as follows:

$$Z = f(X, Y) \qquad (4.1)$$

The most widely used function for the realization of this expression is the general polynomial function as shown in Table 4.1 (Petrie and Kennie 1990).

A graphic representation of first 4 terms is shown in Figure 4.1. It is clear that each individual term of the general polynomial function has its own characteristics in terms of shape. A surface with unique characteristics can be constructed by using certain specific terms.

For the generation of the actual surface in a specific modeling program, it is not necessary (and in practice it is impossible) to use all of the terms inherent in this function. In practice, only a few terms are used, the selection of these being decided upon by the system designer and implementor. Only in a few cases is it possible for the user to select which terms in the function might be most appropriate for modeling the specific piece of terrain in question.

4.2 APPROACHES FOR DIGITAL TERRAIN SURFACE MODELING

After introducing these general concepts, alternative approaches for terrain surface modeling will be discussed.

4.2.1 Surface Modeling Approaches: A Classification

Surface modeling approaches may be classified based on various criteria, such as the basic geometric unit used for modeling, the type of source data used for modeling, and so on.

For the basic geometric unit used in modeling, the following approaches can be identified:

1. point-based modeling
2. triangle-based modeling
3. grid-based modeling
4. a hybrid approach combining any two of the above three items.

In actual applications, the triangle-based and grid-based modeling are more widely used and are considered as the two basic approaches. Since point-based modeling is not practical (and is therefore not widely used) and hybrid modeling is usually converted into the triangle-based approach, grid-based surface modeling is usually used to handle data covering rolling terrain over a large area. But it has less relevance (or application) for broken terrain with steep slopes, numerous break lines, sharp terrain discontinuities, etc.

According to the type of source data used, modeling can be divided into two types:

1. direct construction from measured data
2. indirect construction from derived data.

DTM surface can be constructed directly from (original) source data, for example, by using a square grid, by using regular triangles, or through triangulation in the case of randomly located data. In the case of DTM surface construction indirectly from derived data, an interpolation is applied to the source data to form a regular grid and then the surface is reconstructed from the grid data. Such an interpolation process is often referred to as *random-to-grid interpolation.*

4.2.2 Point-Based Surface Modeling

If the *zero order* term in the polynomial is used for DTM surface realization, then the result is a horizontal (or level) planar, as shown in Figure 4.2. At every point, a horizontal (or level) planar surface can be constructed. If the planar surface constructed from an individual data point is used to represent the small area around the data point (also referred to as the region of influence of this point in the context of geographical analysis), then the whole DTM surface can be formed by a series of such contiguous discontinuous surface. The resulting overall surface will be discontinuous (see Figure 4.2a).

For each individual horizontal planar sub-surface, the mathematical expression is simply as follows:

$$Z_i = H_i \tag{4.2}$$

DIGITAL TERRAIN SURFACE MODELING

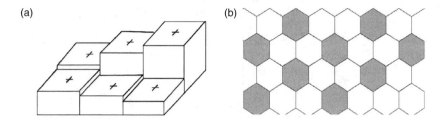

Figure 4.2 Discontinuous DTM surfaces resulting from point-based modeling: (a) sampled data with a square grid and (b) sampled data with a hexagon pattern.

where Z_i is the height on the level plane surface for an area around point I and H_i is the height of point I.

This approach is very simple. The only difficulty is to define the boundaries between the adjacent areas. The commonly used approach for boundary definition is to employ a Voronoi diagram of the data points, which will be discussed later in Section 4.3.2. Since this approach forms a series of sub-surfaces based on the height of individual points, the modeling based on this approach can be regarded as *point-based surface modeling*.

Theoretically, this approach is suitable for any data pattern, regular or irregular, since it only concerns individual points. However, as far as the process of determining the boundaries of the region of influence by each point is concerned, the computation will be much simpler if regular patterns such as a square grid, equilateral triangles, hexagons, etc. are used (Figure 4.2b). Although it would seem quite feasible to implement this approach in surface modeling, it is not really practical due to the resulting discontinuities in its surface. However, in certain applications (e.g., the calculation of total volumes of water, coal, etc.), this remains a valuable technique.

4.2.3 Triangle-Based Surface Modeling

If more terms are used, then a more complex surface can be constructed. Inspection of the first *three terms* (the two first-order together with the zero-order terms) shows that they form a planar surface. To determine the three coefficients of this particular polynomial, three data points are the minimum requirement. These three points can form a spatial triangle; then, a tilted planar surface can be defined and constructed.

If the surface determined by each triangle is used to represent only the area covered by the triangle, then the whole DTM surface can be formed by a linked series of contiguous triangles. The modeling based on this approach is usually referred to as *triangle-based surface modeling*. Figure 4.3(b) is an example of surfaces resulting from triangle-based modeling.

The triangle may be regarded as the most basic unit in all geometrical patterns, since a regular grid of square or rectangular cells or any polygon with any shape can be decomposed into a series of triangles. Therefore, triangle-based surface modeling is the approach that is feasible with any data pattern irrespective of whether it has resulted from selective sampling, composite sampling, regular grid sampling, profiling, or

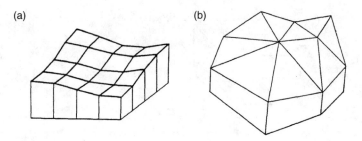

Figure 4.3 Continuous surfaces resulting from (a) grid- and (b) triangle-based surface modeling.

contouring. Since triangles have a great flexibility in terms of their shape and size, this approach can easily incorporate break lines, form lines, and other data. Therefore, the triangle-based approach has received increasing attention in terrain modeling practice and is regarded as the main approach to terrain surface modeling.

In fact, higher-order polynomials (usually second- or third-order) can also be used for triangle-based modeling to create curved facets. In this case, a linked series of triangles (e.g., a string of triangles centered at one point) is the basic unit for surface fitting.

4.2.4 Grid-Based Surface Modeling

If the first three terms, together with the term $a_3 XY$ of the general polynomial, are used for surface construction, then four data points are the minimum requirement to form a surface. The resulting surface is referred to as a *bilinear surface*. Theoretically, quadrilaterals of any shape such as parallelograms, rectangles, squares, or irregular polygons can be used. However, for practical reasons such as the resulting data structure and the final surface presentation, a regular square grid is the most suitable pattern. As in the case of triangle-based surface modeling, the result will consist of a series of contiguous bilinear surfaces (Figure 4.3a).

High-order polynomials can also be used for DTM surface generation (as shown in Figure 4.7). However, unpredictable oscillations in the resulting DTM surface can be created if too many terms of the polynomial are used, usually over a large area. In practice, in order to reduce the risk of this, a restricted number of terms — usually only the second- and third-order terms — are used. The minimum number of grid points necessary to construct the DTM surface will be governed by the number of terms used, but in any case, the number will be greater than four. In this case, different patterns and geometric figures (see Figure 4.1) other than the basic triangle or square grid cell can be considered for use in surface reconstruction. Nevertheless, because of the difficulties likely to be encountered in data structuring and handling, DTM source data that are evenly distributed, as in the case of regular grid and equilateral triangle patterns, are still important.

Grid data have many advantages in terms of data handling. Therefore, elevation grid data from regular grid sampling and progressive sampling, especially the square grid data, are particularly suitable. For this reason, some DTM software packages accept only gridded data. If this is the case, a preliminary data preprocessing operation (*random-to-grid interpolation*) is necessary to ensure that the input data are in grid form.

4.2.5 Hybrid Surface Modeling

The actual data structure implemented using a particular geometric pattern for surface modeling is usually referred to as a network. A DTM surface is usually constructed from one of the the two main types of network — grid or triangular. However, a hybrid approach is also widely used to construct DTM surfaces. For example, a grid network may be broken down into a triangular network to form a contiguous surface of linear facets. Going in the opposite direction, a grid network may also be formed by interpolation within an irregular triangular network.

In some software packages, hybrid surface modeling must have a basic grid of squares or triangles obtained by systematic grid sampling. If break lines and form lines are available for inclusion, the regular grid is broken into triangles and a local irregular triangular network is implemented. Figure 4.4 shows an example of hybrid surface modeling.

It might also be possible to combine point-based with grid-based or triangle-based modeling to form a hybrid approach. That is, the boundaries of the region of influence of a point can be determined using either a grid or a triangular network where the data are located in a regular pattern or based on a triangular network if the data are irregularly located.

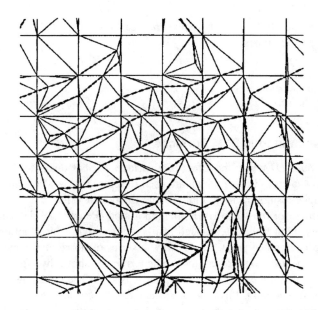

Figure 4.4 An example of surface modeling by hybrid surface (from HIFI Brochure).

4.3 THE CONTINUITY OF DTM SURFACES

After any of these modeling approaches is applied, a surface can be constructed. This section discusses the characteristics of the resultant DTM surface. Emphasis is given to continuity.

4.3.1 The Characteristics of DTM Surfaces: A Classification

The surfaces reconstructed from sampled points to represent terrain of the area can be categorized based on different criteria. Size of the area and continuity of the DTM surfaces are the two most widely used.

According to size of area (or coverage,) DTM surfaces can be classified as local, regional, and global.

1. A *local surface* refers to a DTM surface covering only a small area, based on the premise that the area to be reconstructed is complicated so that it must be processed piece by piece or that only a local area is of interest.
2. A *global surface* is a DTM surface covering the whole area, based on the understanding that this area contains very simple or regular terrain features so that it can be described by a single mathematical function. Alternatively, it may be used when only very general information about the terrain surface is needed for the purpose of reconnaissance.
3. A *regional surface* is a DTM surface with area size between local and global surfaces. That is, the whole area to be reconstructed is divided into larger pieces than local surfaces. This is a result of a compromise between the criteria given for using a global surface and those used to justify the use of a local surface.

According to the continuity between local surfaces, DTM surfaces can be classified into three types:

1. discontinuous surface
2. continuous surface
3. smooth surface.

4.3.2 Discontinuous DTM Surfaces

A *discontinuous DTM surface* refers to a surface that has discontinuity among the local surfaces, a collection of which are used to represent the whole area. A discontinuous surface results from the thought that the height value of a sampled point is a representative for the values of its neighborhood (Peucker 1972). Therefore, the height of any point to be interpolated can be approximated by adopting the height of the closest reference point. In this way, a series of horizontal planes (i.e., local surfaces) can be used to represent the whole terrain, as shown by Figure 4.2.

This type of surface is the result of point-based surface modeling. As discussed in point-based modeling, this type of surface can be constructed from any type of data set, irrespective of whether it is regular or irregular. From regular data, the determination of boundaries between the sub-surfaces is much easier. However,

DIGITAL TERRAIN SURFACE MODELING

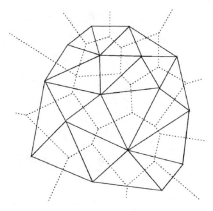

Figure 4.5 Voronoi diagram of a point set and its dual Delaunay triangulation.

whenever the data are irregularly distributed, the boundaries of the region of influence of each point need to be determined algorithmically. Normally, this is done by constructing the Thiessen polygons, which have been widely used in geographical analysis since this method was proposed by the climatologist A.H. Thiessen (Thiessen 1911; see also Brassel and Reif 1979). Actually, the Thiessen polygon is a region enclosed by an embedded series of perpendicular bisectors, each located midway between the point under consideration and each of its neighbors. The Thiessen polygons of all points in an area form a Thiessen diagram, also termed a Voronoi diagram, Wigner–Seitz cells, or Dirichlet tessellation. The actual term used seems to vary between different scientific disciplines, although the basic idea is common to them all. In recent years, the term Voronoi diagram seems to prevail in geographical information sciences and will therefore be used in this book from now on. The Thiessen polygon is also termed a Voronoi region. Figure 4.5 is example of the Voronoi diagram of a point set.

It can be seen from Figure 4.5 that the dual of the Voronoi diagram is a triangulation. This dual relationship was first recognized by Delaunay (1934). Therefore, such a triangulation is usually named after Delaunay. More detailed discussion on this topic will be conducted in Chapter 5, which is devoted to triangulation algorithms.

4.3.3 Continuous DTM Surfaces

A continuous DTM surface is a surface that has a series of local surfaces linked together to cover the terrain being modeled. This is based on the idea that each data point represents a sample of a single-value continuous surface. The boundary between two adjacent sub-surfaces may not be smooth, that is, not continuous in the first and higher derivatives.

The first derivative of a continuous surface can be either continuous or discontinuous. However, continuous surfaces here refer to only those that are discontinuous in the first derivative and those surfaces with continuous first derivative are referred to as smooth surface. Figure 4.3 shows two types of continuous DTM surfaces and Figure 4.6 illustrates the discontinuity problem in the first derivative.

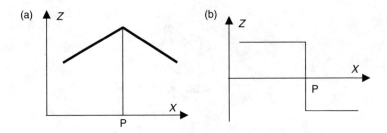

Figure 4.6 Discontinuity in the first derivative of a continuous surface: (a) a profile of a continuous surface and (b) the first derivative of the profile in (a).

The lack of continuity in the first derivative is, for some users, rather undesirable either in terms of modeling itself or in terms of the final graphic output. However, it is also worth noting that the lack of continuity in the first derivative resulting in a distinct boundary between adjacent patches, grid cells, or triangles is a feature that may not be disturbing in some, if not most, cases. Indeed, it may be deliberately sought after or introduced into the modeling process. This is particularly the case with data located along linear features such as rivers, break lines, faults, etc. acquired via selective or composite sampling, where this is indeed desirable so that interpolated contours change direction abruptly along such lines.

Furthermore, it can be found in the literature (e.g., Peucker 1972) that, in many cases, a continuous surface comprising a series of contiguous linear facets is the least misleading one although it may not look convincing or attractive visually.

4.3.4 Smooth DTM Surfaces

A *smooth DTM surface* is a surface that exhibits continuity in first- and higher-order derivatives. Usually, they are implemented regionally or globally. The generation of such a DTM surface is based on the following assumptions:

1. The resource data always contain a certain level of random error (or noise) in measurement so that the DTM surface does not need to pass through all the sampled data points.
2. The surface to be constructed should be smoother than (or at least as smooth as) the variation indicated by the source data.

For these conditions to be achieved, normally, a certain level of data redundancy is used and a least-squares method is implemented using a multi-termed polynomial to model the surface. Figure 4.7(a) shows examples of smooth surfaces.

For a single global surface based on a large data set, the whole of the surface is modeled by a single high-order polynomial. A huge amount of data may be involved, with an equation formed from each data point. There will be a substantial computational burden or overhead on the modeling operation. Also, the final resulting surface often exhibits unexpected and unpredictable oscillations among data points. These are highly undesirable in terms of both the surface modeling process itself

DIGITAL TERRAIN SURFACE MODELING

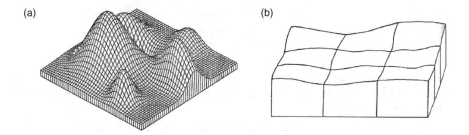

Figure 4.7 Examples of smooth surfaces: (a) a global (smooth) surface and (b) a smooth surface comprising a series of regional surfaces.

and the fidelity of the final results in terms of the actual representation of the terrain surface delivered to the user in the form of contour plots or perspective views.

The result of these considerations is that data sets are often divided into a series of continuous patches. The patches may be regular in terms of shape and size, as in the case of square grid cells or equilateral triangles, or they may be irregular both in shape and size, as in the case of the randomly distributed points normally encountered in a triangulation procedure. Within each data patch, a lower-order polynomial can be used to model the surface, again using the least squares method if there are redundant data. While the use of the polynomial ensures a smooth surface within each patch, a break in continuity will almost certainly occur along the boundaries between patches. The result of this is that continuity in the first and higher derivatives between adjacent patches will have to be built into the modeling system so that a smooth surface can be achieved without breaks or discontinuities along boundaries. In other words, a so-called seamless join must result between patches (Figure 4.7b). Needless to say, the successful implementation of such a requirement carries a heavy computation overhead.

4.4 TRIANGULAR NETWORK FORMATION FOR SURFACE MODELING

Triangular network is the most basic and can be applied to both regularly and irregularly located data. That is, a regular grid network can be formed by interpolation from a triangular network and either a continuous or a smooth surface can also be constructed from the same network. The formation of a triangular network will be discussed in this section and a discussion about the formation of a grid network will be discussed in Section 4.5.

4.4.1 Triangular Regular Network Formation from Regularly Distributed Data

The process of forming a triangular network is usually referred to as triangulation. Triangulation can be applied either to regularly distributed data (such as grid data) to form a *triangular regular network* (TRN) or to irregularly distributed data to form a TIN, which comprises a series of contiguous triangles of irregular size and shape.

If source data are acquired in a regular pattern, then this is the simplest network to form. For square grids, simple sub-division using one or two diagonals produces a series of regular triangles. Figure 4.8 shows three possible triangular patterns derived from a grid pattern. If the pattern is based on regular triangles (Figure 2.9), then the network is already triangular.

Of course, such an approach to form triangular networks from a square grid is sometimes very arbitrary. Figure 4.9 shows this. Figure 4.9(a) shows a bilinear surface constructed from a square grid. Figure 4.9(b) shows that a grid cell can be split into

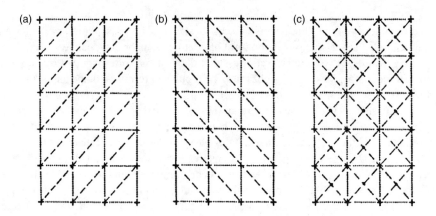

Figure 4.8 Triangular regular network from a regular grid.

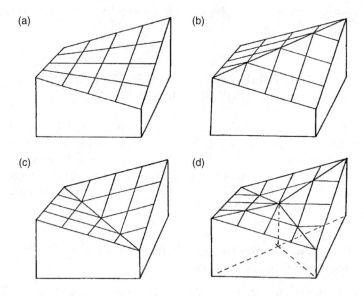

Figure 4.9 Possible types of linear facets constructed from a square grid.

DIGITAL TERRAIN SURFACE MODELING

two triangles by a single diagonal whose plan is shown in Figure 4.8(a). Similarly, Figure 4.9(c) shows the two triangles corresponding to those divided by the alternative single diagonal shown in Figure 4.8(b). Finally, those in Figure 4.9(d) correspond to the arrangement shown in Figure 4.8(c) with four center-point triangles formed by using both diagonals. It is apparent that the height values of points interpolated from these different surfaces shown in Figure 4.9(a) to Figure (d) will all be quite different, although the same height values have been used at the grid nodes in each of these four examples. This is a problem needing attention. However, even commercial packages carry out such a triangulation process blindly.

4.4.2 Triangular Irregular Network Formation from Regularly Distributed Data

In the triangular network formation process discussed above, there is no information loss on the surface. However, data redundancy can be a problem, as in the case of regular grid sampling. If so, some less important (or unimportant) points on the surface can be deleted from the data set. Alternatively, the VIPs are retained to form a TIN.

The key to the selection of VIPs is to assign a significance value to each point so that points with high significance values are selected. Chen and Guevara (1987) used the sum of the second differential values at a point in all four directions to represent the degree of significance. Suppose the height (H) of a point along a profile is a function of its position (x), as shown in Figure 4.10. The horizontal distance between X_{i-1}, X_i, and X_{i+1} is equal because of the regular grid sampling. Let the mathematical function of this profile be

$$H = f(x) \tag{4.3}$$

Then, its second differential value at point X_i is

$$\frac{d^2 H}{dX^2} = f''(X_i) = 2\left(f(X_{i-1}) - \frac{f(X_{i-1}) + f(X_{i+1})}{2}\right) \tag{4.4}$$

Actually, the distance AC in Figure 4.10 is the second differential value at point X_i. Chen and Guevara (1987) also consider four directions, that is, up–down, left–right,

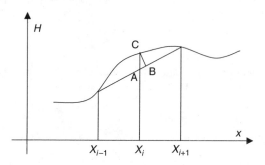

Figure 4.10 A terrain profile and its second differential value.

upper left–lower right, and lower left–upper right. For each point, the second differential values for all four directions are added to represent the degree of significance of this point. In their procedure, the number of points to be selected is specified first and then those points with the greatest significance are selected.

However, as one can imagine, the selection of points should relate to the required accuracy of the resulting DTM, instead of to a predefined number of points. Indeed, Li (1990) and Li et al. (1998) related the degree of significance to the DTM accuracy loss after VIP selection. In this case, a threshold for the significance values, instead of the number of points to be retained, is considered. Now the question arising is "what is the appropriate threshold for the significance value for a given allowable accuracy loss?"

To find such a threshold for point selection, a close examination of the distance AC in Figure 4.10 must first be undertaken. It can be found that AC is the error at $x = X_i$ if X_i is removed and the profile is constructed by linear interpolation between the elevation values at X_{i-1} and X_{i+1}. This represents the DTM error and a loss of accuracy will result from the selection of VIPs or removal of those points that have been regarded as insignificant.

The problem arising is "how much will the loss of accuracy be in terms of standard deviation or root mean square error (RMSE) if all the data points with a degree of significance smaller than a specific value are removed." In other words, the relationship between the accuracy loss and the specified critical value needs to be investigated. If the error distribution is known, then the relationship can easily be set out. However, the distribution is not exactly known although it is nearly normal (see Chapter 8). Therefore, such a relationship needs to be found out through experimental investigation. Indeed, Li (1990) and Li et al. (1998) found the DTM accuracy loss (σ_{loss}) after VIP selection and the threshold ($Sig_{Threshold}$) used for VIP selection:

$$\sigma_{loss} = \frac{Sig_{Threshold}}{3} \qquad (4.5)$$

This is the result obtained from two large testing areas with approximately 2000 check points.

Suppose that the required accuracy of the final DTM (after the VIP selection) in terms of variance is σ^2_{after} and the accuracy of initial DTM (before the VIP selection) is σ^2_{before}, then

$$\sigma^2_{after} = \sigma^2_{before} + \sigma^2_{loss} \qquad (4.6)$$

By combining Equations (4.5) and (4.6), the relationship between the threshold for VIP selection and the final DTM accuracy is as follows:

$$Sig_{Threshold} = 3\sigma_{loss} = 3\sqrt{\sigma^2_{after} - \sigma^2_{before}} \qquad (4.7)$$

Actually, the line of thought is similar to that used by Makarovic (1977, 1984), who called this a progressive rejection process and used a Laplacian operator for this

DIGITAL TERRAIN SURFACE MODELING

process as follows:

$$L = \begin{pmatrix} 0 & 1 & 0 \\ 1 & -4 & 1 \\ 0 & 1 & 0 \end{pmatrix} \quad (4.8)$$

After VIP selection, the resultant data will become irregular in distribution, then a TIN generation procedure is applied. The algorithms for TIN generation will be discussed in Chapter 5 and the general principle will be outlined in Section 4.4.3.

4.4.3 Triangular Irregular Network Formation from Irregularly Distributed Data

From irregularly distributed data, either a regular grid or an irregular triangular network can be formed. The formation of regular grid network will be discussed in Section 4.5.

The formation of a TIN from irregularly distributed data is not as easy as in the case of TRN from grids although there are a lot of algorithms available. Generally, there are three basic requirements for TIN formation:

1. For a given set of data points, the resulting TIN should be unique if the same algorithm is used, although one may start from different places, for example, the geometric center, upper-left corner, lower-left corner or other points.
2. The geometric shapes of resultant triangles are optimum, that is, each triangle is nearly equilateral, if there are no specific conditions.
3. Each triangle is formed with nearest neighbor points, that is, the sum of the three edges of the triangle is minimum.

In all the possible alternatives, Delaunay triangulation is the one most widely used because it satisfies all three requirements. A Delaunay triangulation is a set of linked but nonoverlapping triangles. The circumscribing circle (circumcircle) of each triangle would not include any other points, as this is one of the conditions used for construction of Delaunay triangulation. The Delaunay triangulation is a dual diagram of the Voronoi diagram (Figure 4.5) and thus can also be derived from the Voronoi diagram. Therefore, an alternative approach for the construction of Delaunay triangulation is first of all to construct a Vironoi diagram and then to derive the triangulation.

Delaunay triangulation is constructed by connecting three neighboring points, the corresponding Voronoi regions of which has a common vertex, and the vertex is the center of the Delaunay triangulation's circumscribing circle. Figure 4.5 shows that Delaunay triangulation obeys the Euler's theorem of planar graphs as follows:

$$N_{\text{regions}} + N_{\text{vertices}} - N_{\text{edges}} = 2 \quad (4.9)$$

Delaunay triangulation can be formed in either dynamic or static mode. Static triangulation means that the triangulation network that has already been constructed will not be altered by adding new points in the formation process. In contrast, in dynamic triangulation, the network already constructed will be changed if a new point is

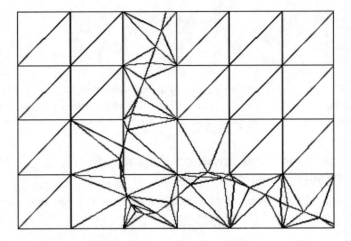

Figure 4.11 Delaunay triangulations of composite data.

added, so as to meet the Delaunay circumcircle principle. There are many algorithms for Delaunay triangulation in both modes and they will be discussed in Chapter 5.

4.4.4 Triangular Irregular Network Formation from Specially Distributed Data

Specially distributed data refer to two types of data, composite and contour data. Composite data result from composite sampling, that is, a regular grid plus features points and lines. Contour data can either be digitized from existing contour maps or measured on a stereo model by means of contouring.

To form a triangulation network from composite data, the grids are first split into regular triangles and Delaunay triangulation is then built-in grids containing features points. Figure 4.11 shows an example.

To form a triangular network from contour data by a Delaunay triangulation algorithm, special care needs to be taken, or else flat triangles (i.e., the heights at the three vertices of a triangle being the same) may result. Figure 4.12 shows flat triangles formed from two contours. This is because the three vertices are selected from the same contour line. This can be avoided by the constraint that "No more than two points can be selected from an individual contour line." An alternative is to produce the skeleton of the contour and to make use of the points along the skeleton together to form the triangulation. This method will be discussed in more detail in Chapter 5.

4.5 GRID NETWORK FORMATION FOR SURFACE MODELING

As discussed previously, grid-based modeling is another main approach for DTM surface modeling. If regular grid sampling method is used, then the resultant data have already had the grid structure and no special process is needed when all the points

DIGITAL TERRAIN SURFACE MODELING

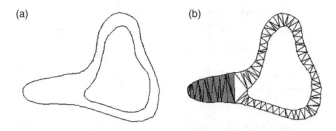

Figure 4.12 Flat triangles formed when a triangulation is constructed from contour data. (a) A map with two contour lines. (b) Flat triangles formed in the shaded area.

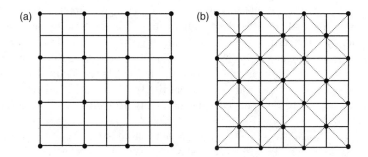

Figure 4.13 Simple resampling for the generation of coarse grids from fine grid (Li 1992b): (a) new grid intervals equal to two times the old and (b) new grid intervals equal to $\sqrt{2}$ times the old.

are used for modeling. Otherwise, resampling needs to be done to obtain a new grid. If the data are irregularly distributed, a random-to-grid interpolation is required.

4.5.1 Coarser Grid Network Formation from Finer Grid Data: Resampling

One simple resampling method without interpolation is to select certain points at specific locations without any interpolation (Li 1992b). Figure 4.13 shows new grids from simple resampling. In Figure 4.13(a), points in alternate rows and columns are selected for a new grid with intervals equal to two times the intervals of the original grid. Similarly, one could obtain new grids with intervals of three times, four times, ..., N times the intervals of the original. Figure 4.13(b) shows another possible selection, that is, to select points along the diagonal. In this case, a new grid with intervals equal to $\sqrt{2}$ times the intervals of the original grid is generated. Similarly, new grids with $2\sqrt{2}, 3\sqrt{2}, \ldots$, and $N\sqrt{2}$ times the intervals of the original grid can be generated.

If the desired intervals of the new grid network are always N or $N\sqrt{2}$ times the intervals of the original grid, then the matter is quite simple. In practice, however, this

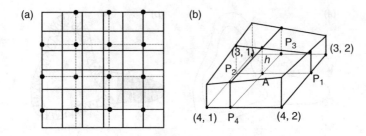

Figure 4.14 Formation of finer grid to coarser grid by resampling: (a) from 3-m grid to 5-m grids and (b) bilinear interpolation.

is not always the case. For instance, one may need a grid with intervals of 5 m from an original grid that had intervals of 3 m (Figure 4.14a). In this case, some points can be automatically selected from the original grid, whereas others need to be interpolated. Suppose the coordinates of the starting grid node are (0, 0), the positions at (0, 5), (5, 0), and (5, 5) will automatically become the nodes of the new grid. The coordinates of these points in the new grid network are (0, 3), (3, 0), and (3, 3). All other new grid points need to be interpolated. The commonly used interpolation methods are:

1. *Nearest point*: For example, point A is within the grid formed by four nodes with coordinates (3, 1), (3, 2), (4, 1), and (4, 2). As point A is the nearest node (3, 2), the elevation at (3, 2) will be assigned to point A.
2. *Bilinear interpolation*: Bilinear interpolation, as the name implies, is the linear interpolation in both X and Y (or row and column) directions. Figure 4.14(b) shows bilinear interpolation. If one wants to obtain the elevation (at position P) of the new grid point (i.e., point A), one first linearly interpolates the elevation for P_1 using nodes (3, 2) and (4, 2) and then linearly interpolates the elevation for P_2 using nodes (3, 1) and (4, 1). This is the linear interpolation in the column. Next, the elevation of point A can be obtained by linear interpolation using points P_1 and P_2. This is the linear interpolation in the row. Also, one can first interpolate the elevations for P_1 and P_2 in the row and then interpolate the elevation for point A using P_1 and P_2.
3. *Bicubic interpolation*: In both directions, a cubic function is used for interpolation, instead of a linear function. Normally, the bicubic function is applied to a patch, for example, consisting of a 3×3, or 4×4 grid.

A detailed discussion on interpolation and interpolation methods will be given in Chapter 6.

4.5.2 Grid Network Formation from Randomly Distributed Data

From randomly distributed data, grid networks can be formed in two ways, that is, direct random-to-grid interpolation and indirect interpolation from triangles through a triangulation process. Figure 4.15 illustrates these two solutions.

In indirect interpolation via triangulation, two types of surfaces can be constructed from the neighboring triangles. The first is simply the linear facet formed by the single

DIGITAL TERRAIN SURFACE MODELING

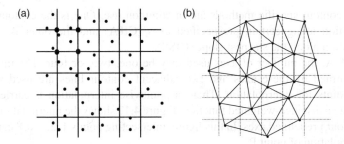

Figure 4.15 Grid network formation from randomly distributed data: (a) direct random-to-grid interpolation and (b) indirect interpolation via triangulation.

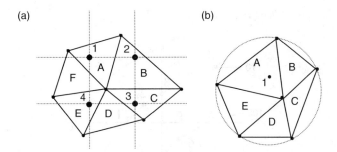

Figure 4.16 From random data to grid data via triangulation: (a) linear interpolation in triangular facets and (b) a curved surface formed from neighbor triangles.

triangle within which the interpolation point is located. Figure 4.16(a) shows where grid nodes 1, 2, 3, and 4 can be obtained by linear interpolation from triangular facets A, B, C, and E, respectively. The second is a curved surface constructed from a string of triangles neighboring the interpolation point. Figure 4.16(b) shows that grid node 1 may be interpolated by the curved surface formed from triangles A, B, C, D, and E.

4.5.3 Grid Network Formation from Contour Data

As discussed previously, contour data are one of the major sources for digital terrain modeling. To form a grid network from such a data set, three solutions are possible, that is,

1. to treat the contour points as randomly distributed points, and then to apply a random-to-grid interpolation
2. to form a triangulation from contour data and then to apply interpolation as discussed in the previous sub-section
3. to design a contour-specific interpolation method.

Two contour-specific methods are in common use. One is the contour-specific interpolation along certain prespecified axes (CIPA) and the other is the cubic interpolation along the steepest slope (CISS).

In CIPA, the number of axes used may be one, two, or four. The intersecting points formed by these axes and two adjacent contour lines are used as neighboring points for interpolation. Then, a pointwise interpolation is carried out by employing a distance-weighted function. Figure 4.17 shows the interpolation of point P using four predefined axes. In this figure, intersecting points 1, 2, ..., 8 can be used for interpolation of point P.

Alternatively, two points, each on one of two adjacent contours, along the steepest slope passing through the point to be interpolated could be used for linear interpolation. For example, points 5 and 1 in Figure 4.17 may be used for such an

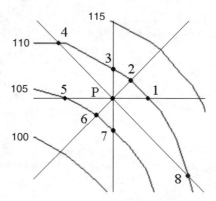

Figure 4.17 Contour-specific interpolation using predefined axes.

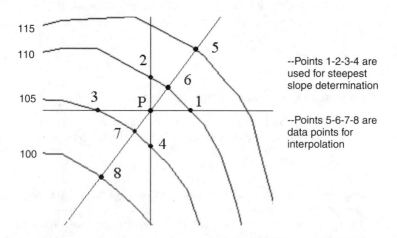

--Points 1-2-3-4 are used for steepest slope determination

--Points 5-6-7-8 are data points for interpolation

Figure 4.18 Selection of steepest slope direction and data points for interpolation by CISS.

interpolation. As discussed by Leberl and Olson (1982), all eight intersecting points are used to determine the direction of the steepest slope.

In fact, along the steepest slope, nonlinear interpolation is also possible. Clarke et al. (1982) used a cubic polynomial function. In this case, four known points are required for any interpolation point. That is, four adjacent contour lines are used. On each of them, an intersection results. Figure 4.18 shows this.

CHAPTER 5

Generation of Triangular Irregular Networks

In Chapter 4, digital terrain surface modeling was discussed. It was pointed out that grid- and triangle-based modeling approaches are more widely used than point-based approaches. For grid-based modeling, a grid network needs to be formed through a random-to-grid interpolation, if the original data are not in grid form. The discussion of interpolation methods will be conducted in Chapter 6. On the other hand, for triangle-based modeling, a triangular network needs to be formed through a triangulation procedure if the data are not in grid or triangular form. The formation of *triangular irregular network* (TIN) from irregularly distributed data is discussed in this chapter.

5.1 TRIANGULAR IRREGULAR NETWORK FORMATION: PRINCIPLES

There are a number of ways for the construction of a triangulation network from a given set of randomly (irregularly) distributed data. They are based on different principles. In this section, these principles will be presented.

5.1.1 Approaches for Triangular Irregular Network Formation

To form a TIN, there are two choices for making use of the data points. The first is to consider all the data to form an overall network. This is a *batch* (or static) approach for the Delaunay triangulation of a set of data points. The alternative is to allow the addition or removal of points during the triangulation process. This is a dynamic process and thus is called dynamic triangulation, as modifications to the structure can be made without reconstructing the whole network each time. It should be noted that "dynamic" does not mean that the points are considered to be moving — that is another property, usually known as kinetic (Guibas et al. 1991).

Spatial data can be in either vector or raster format. Therefore, the triangulation can be in either vector or raster mode. It is possible to convert vector data to raster and then triangulate in raster mode. Alternatively, it is possible to convert raster data into vector and then triangulate in vector mode.

As will be discussed later in this section, there are many possible criteria for the construction of triangles, thus leading to many alternative methods. The most widely used method, as was discussed in Chapter 4, is the Delaunay triangulation, which has a dual relationship with the Voronoi diagram. It implies that the Delaunay triangulation network can be formed either directly by algorithm or indirectly through the Voronoi diagram. However, the triangulation in raster mode is usually achieved via the Voronoi diagram because in raster space the construction of Voronoi diagrams is much easier than that of Delaunay triangulation. Therefore, the approaches for triangulation can be summarized as in Figure 5.1.

5.1.2 Principles of Triangular Irregular Network Formation

From a set of randomly distributed data, there are alternative ways to form triangular networks. Figure 5.2 illustrates the three alternative triangular networks generated from the same set of data. The question that arises is "which one is the best?" There must be some basic principles to guide the construction of triangular networks. This section discusses these principles.

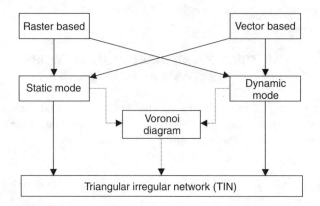

Figure 5.1 Approaches for triangular irregular network formation.

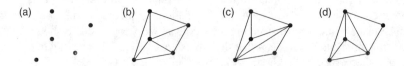

Figure 5.2 Triangular networks with different shapes constructed from the same data set: (a) a set of data; (b) result 1; (c) result 2; and (d) result 3.

As mentioned in Chapter 4, one of the basic characteristics of Delaunay triangulation is that no other data points are contained by the circumcircle of a Delaunay triangle. This is one of the basic principles for the generation of a Delaunay triangulation from a set of two-dimensional data points, referred to as empty circumcircle principle. Figure 5.3 illustrates this. In Figure 5.3(a), point D is within the circle circumscribing △ABC if point C is selected to form a triangle with points A and B. It means that point D instead of point C should be used to form a triangle with points A and B. Figure 5.3(b) shows this case, where point C does not fall into the circle circumscribing △ABD.

Local equiangularity is another principle suggested by Lawson in 1972 (see Tsai 1993) for Delaunay triangulation. It says that the triangular network is optimum if for every convex quadrilateral formed by two adjacent triangles, the swapping of diagonals will not cause a decrease in the minimum of the six interior angles concerned and at the same time will not cause an increase in the maximum angle. In this way, the minimum angle is maximized and the maximum angle is minimized for all the triangles. This is also called the *MAX–MIN* angle principle. The procedure for swapping diagonals is called a local optimization procedure (LOP) (Tsai 1993). Figure 5.4 illustrates this principle. In Figure 5.4(a), two triangles, △ABC and △ADC are used to form a convex quadrilateral. The minimum interior angle is ∠CAD and the maximum interior angle is ∠ADC. After swapping the diagonal, as shown in Figure 5.4(b), the minimum interior angle then becomes ∠CBD, which is larger than ∠CAD and the maximum interior angle is ∠ADB, which is smaller than ∠ADC. This means that the shape in Figure 5.4(b) is the optimal configuration.

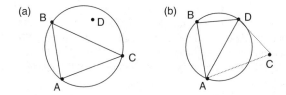

Figure 5.3 The *empty circumcircle* principle for Delaunay triangulation. (a) Circumcircle containing point D. (b) Point D is used to form the triangle.

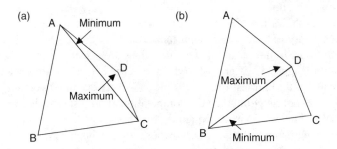

Figure 5.4 Illustration of the LOP process for local equiangularity: (a) before swapping the diagonal; and (b) after swapping the diagonal.

An intuitive principle is the minimum sum-distance, which refers to the sum of the distances from a new point to the two end points of a triangle baseline (Yeoli 1977). The corresponding algorithms are based on the criterion that the new point selected to construct a new triangle is the one that has the sum of its distances to the end points of the baseline as the smallest value.

Another simple principle is the minimum circumscribing circle radius (Elfick 1979). The corresponding algorithms are based on the criterion that the new point that is selected should form a triangle in which its circumscribing circle radius is the smallest value. The minimum distance from the center of the circumscribing circle to the base line has also been suggested (McLain 1976).

5.2 VECTOR-BASED STATIC DELAUNAY TRIANGULATION

As discussed in the previous section, various principles can be used for the implementation of TINs, leading to different types of algorithms. No attempt is made here to introduce all these algorithms. Instead, only the methods for the construction of Delaunay triangulations will be presented here.

5.2.1 Selection of a Starting Point for Delaunay Triangulation

An advantage of Delaunay triangulation is that the resulting triangulation network is independent of the starting point. Therefore, the selection of a starting point is only for the convenience of algorithm implementation. Some choices for the starting point are:

1. the geometric center of the data points (Elfick 1979)
2. the shortest of all possible lines between any two data points (Yeoli 1977)
3. a line segment on the imaginary boundary (McCullagh and Ross 1980)
4. a line segment on the boundary convex hull (Tsai 1993; Gosper 1998).

After choosing the starting point, another point, which is normally the nearest neighbor is selected to form the initial base. Then, a third point is searched to form the first triangle. Other triangles can then be formed by using the three sides of the initial triangle as these bases. The search for points to form triangles will be discussed in the next sub-section.

Figure 5.5 shows the triangulation process starting from the geometric center. It is not necessary to have a data point located exactly at the geometric center. The data point closest to this X and Y average values are selected as the starting point. Point 1 in this Figure 5.5 is closest to the geometric center and thus selected as the starting point. If the shortest of all possible lines between any two data points was selected as the starting point, then the computation of all distances between two data points would be heavy. As a consequence, this choice is not very popular.

Many triangulations start from anywhere on the boundary of the area to be modeled; for many applications, the area to be modeled is explicitly defined. However, in many other applications, the boundary is not explicitly defined, then the boundary needs to be sorted out first.

GENERATION OF TRIANGULAR IRREGULAR NETWORKS

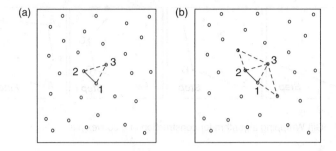

Figure 5.5 Delaunay triangulation starting from the geometric center: (a) generation of the first triangle and (b) generation of the second and third triangles.

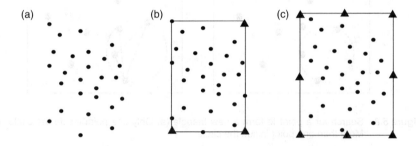

Figure 5.6 Delaunay triangulation starting from anywhere on the imaginary boundary box (the points indicated by triangles are the imaginary points). (a) A set of random points. (b) Minimum bounding rectangle. (c) Imaginary boundary box.

A set of imaginary points can be used to define the boundary box of the area to be triangulated. Figure 5.6 illustrates the triangulation from such an imaginary boundary box. Figure 5.6(a) shows a set of random points. The area can then be defined by the minimum bounding rectangle as shown in Figure 5.6(b) or by an imaginary boundary box containing all the data points (Figure 5.6c). Usually, on the imaginary boundary box, a few imaginary points (e.g., at the four corners and on the four sides) are added for convenience of point searching. From anywhere on the rectangle or on the box, the triangulation process can start, for example, from the upper/left corner.

Often, the convex hull of the data points is used to define the area of interest which is the smallest convex polygon containing all data points. A convex polygon means that a line segment connecting any two points must be completely within it (Tsai 1993). A number of algorithms are available for constructing the convex hull of a set of points on a 2-D plane such as Graham's scan, Jarvis' march (gift wrapping), and Quick hull. A detailed discussion of these algorithms can be found elsewhere (O'Rourke 1993; Gosper 1998). The *Gift Wrapping* algorithm is simple and popularly used. Figure 5.7 illustrates the working principle of this algorithm. The first step is to find the point with the minimum Y coordinate as a starting point; the second step is

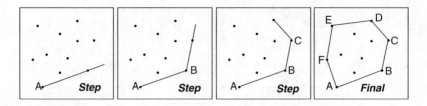

Figure 5.7 Gift Wrapping algorithm for construction of a convex hull.

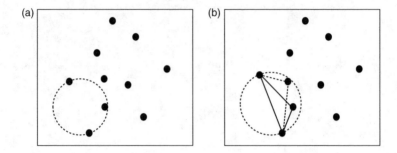

Figure 5.8 Search for a point to form a new triangle (a) Only one point inside the circle; (b) More than one point inside the circle.

to find B where all points lie to the left of line $\overline{AB}$ by scanning through all the points. That is, B makes the largest right-hand turn from A. In a similar way, points A, B, C, D, E and F could be found to form a convex hull.

5.2.2 Searching for a Point to Form a New Triangle

After the starting point (also called rotation point) is determined, the nearest point is selected as the known point to form an initial base. Then, a new point is selected as the vertex of a new Delaunay triangle, which is located to the right side (i.e., clockwise from the known point to the rotation point) of the base. A simple search method is to draw a circle from the middle point of the base with the base as the diameter. If there is only one point inside this circle, then the point will be picked up to form a new triangle. If there is more than one point inside the circle, then the point that has the largest angle subtended from the base from all possible choices around the starting point will be selected. If this does not succeed in finding a point, then the search circle is expanded using the base line as a chord and with progressively larger circles until the appropriate neighbors are found. Figure 5.8 shows this searching process. In this way, the most likely neighbors are first picked up and are then tested to find one with the largest angle.

GENERATION OF TRIANGULAR IRREGULAR NETWORKS

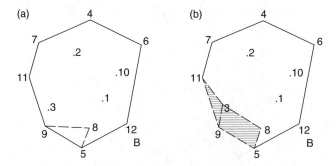

Figure 5.9 Triangulation by enclosing from convex hull boundaries. (a) First triangle starting from the boundary. (b) Subsequent triangles formed near the boundary.

To increase the efficiency of point searching, data points could be grouped into blocks beforehand according to their (x or y) coordinates.

5.2.3 The Process of Delaunay Triangulation

Once the starting point is determined and the search principle is specified, the triangulation can be carried out smoothly. An example of triangulation from the geometric center of the data points has already been given in Figure 5.5. Therefore, the examples given in this section are of triangulations starting from the boundaries.

The first example starts from the convex hull of the data area and moves along the boundary. It is illustrated in Figure 5.9(a). The procedure is as follows:

1. The starting point has the point with minimum Y coordinate, that is, point 5 in Figure 5.9(a).
2. The second vertex of the first triangle is the one nearest the starting point along the convex hull clockwise, that is, point 9.
3. The third vertex, point 8, is selected by using line $\overline{5,9}$ as the base line and the first triangle is formed by following the search procedures discussed in Section 5.2.2.
4. The triangulation proceeds along the convex boundary. That is, triangle edge $\overline{8,9}$ is used as the second base line for the formation of the second triangle.
5. The advancing front of the triangulation will be moved forward clockwise and gradually toward the center, until all data points are triangulated.

The second example is to start from an imaginary boundary. This is illustrated in Figure 5.10. The lower-left corner is selected as the starting point. This time the movement is counterclockwise. A shell of triangles is formed when the front hits the imaginary boundary. Instead of continuing along the boundary, a new shell will start from the imaginary boundary near the first base. In this way, triangulation for the whole data set will be completed shell by shell.

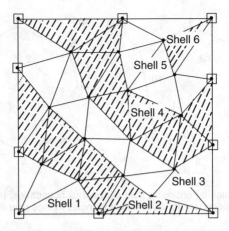

Figure 5.10 Delaunay triangulation starting from an imaginary boundary.

5.3 VECTOR-BASED DYNAMIC DELAUNAY TRIANGULATION

In the previous section, Delaunay triangulation in static mode was discussed. The search for points is usually an inefficient process if the amount of data is large. Therefore, triangulation is more often done dynamically by gradually adding new points into the network. This kind of dynamic operation is also referred to as incremental triangulation. There are many incremental triangulation algorithms available in the literature (e.g., Zhu and Chen 1998), but in this section, only the simplest and most robust one is described as a representative, that is the Bowyer–Watson algorithm (Bowyer 1981; Watson 1981), which is usually called the *simple incremental algorithm*.

5.3.1 The Principle of Bowyer–Watson Algorithm for Dynamic Triangulation

The Bowyer–Watson algorithm (Bowyer 1981; Watson 1981) is regarded as the most practical triangulation algorithm. The basic idea is to start with coarse triangles and then to add points sequentially into this coarse triangulation network. Figure 5.11 illustrates the refinement of triangles after adding more points.

The initial triangulation is usually very simple (e.g., two triangles of the bounding rectangle) enclosing all points in the area of concern. The insertion process of this algorithm is illustrated in Figure 5.12. When a point p is inserted into a triangle (i.e., $\triangle ABC$ in Figure 5.12a) this triangle is split into three, with the new point forming a vertex of each of the three new triangles (Figure 5.12b). Then, each of the three edges of the old triangle (i.e., $\triangle ABC$) is checked to see whether there is a need for swapping the edge with the alternative diagonal by applying the *empty circumcircle principle*. In this example, edge $\overrightarrow{AB}$ must be swapped with the alternative $\overline{Dp}$ and

GENERATION OF TRIANGULAR IRREGULAR NETWORKS

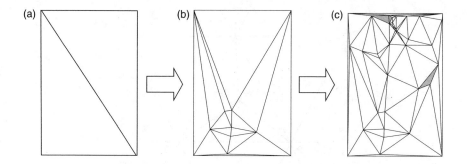

Figure 5.11 Dynamic Delaunay triangulation by the insertion of points into the initial coarse triangles.

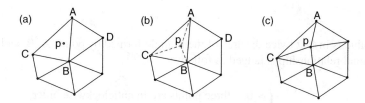

Figure 5.12 Delaunay triangulation by the Bowyer–Watson algorithm: (a) initial triangulation; (b) splitting the enclosing triangle; and (c) the "swap" operation.

the new triangulation after the insertion of point p is shown in Figure 5.12(c). More discussion on a numerical criterion for edge swapping is given in Section 5.3.3.

5.3.2 Walk-Through Algorithm for Locating the Triangle Containing a Point

For a large volume of data, locating the triangle where a point is to be inserted is done by a walking-through algorithm (Gold et al. 1977) to improve efficiency. In this algorithm, two problems need to be solved. The first is to set a numerical criterion, which tells whether a point is within the triangle. The second is to give a pointer to the next triangle to be examined if the current triangle does not contain the point.

The directional relationship between a point P and a directed line segment $\overrightarrow{AB}$ can be determined by the following formula:

$$D(A, B, P) = \begin{vmatrix} x_A & y_A & 1 \\ x_B & y_B & 1 \\ x_P & y_P & 1 \end{vmatrix} \quad (5.1)$$

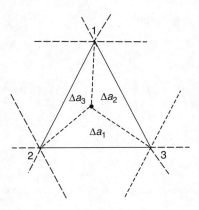

Figure 5.13 Local "area coordinates" to determine whether a point is inside a triangle.

This value is in fact twice the area of the triangle formed by points A, B, and P. The directional relationship is judged as follows:

$$D(A, B, P) \begin{cases} > 0, & \text{three points are in anticlockwise order,} \\ & \text{i.e., P on the leftside of line } \overrightarrow{AB} \\ = 0, & \text{three points are in a line} \\ < 0, & \text{three points are in clockwise order,} \\ & \text{i.e., P on the right side of line } \overrightarrow{AB} \end{cases} \quad (5.2)$$

By using Equations (5.1) and (5.2), the question of whether point P is within a triangle can then be answered. Figure 5.13 illustrates the principle. In this figure, points 1, 2, and 3 are the three vertices. For a point P to be checked, three equations can be established as follows:

$$\Delta a_1 = \begin{vmatrix} x_p & y_p & 1 \\ x_2 & y_2 & 1 \\ x_3 & y_3 & 1 \end{vmatrix}, \quad \Delta a_2 = \begin{vmatrix} x_1 & y_1 & 1 \\ x_p & y_p & 1 \\ x_3 & y_3 & 1 \end{vmatrix}, \quad \Delta a_3 = \begin{vmatrix} x_1 & y_1 & 1 \\ x_2 & y_2 & 1 \\ x_p & y_p & 1 \end{vmatrix} \quad (5.3)$$

These three area values are called the local area coordinates with respect to the vertices 1, 2, and 3. If the local area coordinates of P are all positive, point P is inside $\triangle 123$. If the point P is outside the triangle, then one or more area coordinates of P will be negative.

To find a triangle that contains the new point, the algorithm starts from any arbitrary triangle. If all the area coordinates of the point are positive, then the point falls within this triangle. If not, one crosses the edge that has a negative coordinate and repeats the same procedure for the new triangle. The walk continues until a triangle is found where all three coordinates of P are positive, which means that p is inside the triangle. Figure 5.14 illustrates the walk operation through the triangulation.

5.3.3 Numerical Criterion for Edge Swapping

After the triangle containing P has been found, the point is inserted into the triangle, splitting the old one into three, as in Figure 5.11(a) and Figure 5.11(b). The three exterior triangle edges now need to be tested, to see if they conform to the Delaunay (empty circumcircle) condition. This is computed as follows:

$$H(A, B, C, D) = \begin{vmatrix} x_A & y_A & x_A^2 + y_A^2 & 1 \\ x_B & y_B & x_B^2 + y_B^2 & 1 \\ x_C & y_C & x_C^2 + y_C^2 & 1 \\ x_D & y_D & x_D^2 + y_D^2 & 1 \end{vmatrix} \quad (5.4)$$

A, B, and C are the vertices of the triangle given in anticlockwise order, and D is the fourth point being tested. Then, the following condition holds true:

$$H(A, B, C, D) = \begin{cases} > 0, & \text{point D is inside the circumcircle of } \triangle ABC \\ < 0, & \text{point D is outside the triangle} \\ = 0, & \text{co-circular} \end{cases} \quad (5.5)$$

If the fourth vertex (obtained from the next triangle outward from the split triangle) is inside the split triangle's circumcircle, then their common edge must be exchanged with the other diagonal, making the inserted point a "neighbor" to the fourth point. The new exterior edges must be stacked for later testing, and this process continues until all exterior edges satisfy the Delaunay condition. Figure 5.15 illustrates the edge switching. Figure 5.15(a) shows the case of vertex D being inside the circumcircle of $\triangle ABC$, and thus being a neighbor of B, and Figure 5.15(b) shows vertex D outside — with the result that the diagonal $\overline{BD}$ has been switched and D is no longer a neighbor of B. If $H = 0$ the four points are co-circular and either configuration is acceptable.

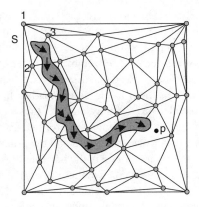

Figure 5.14 The walk operation in a dynamic Delaunay triangulation.

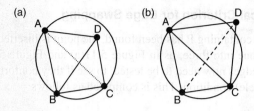

Figure 5.15 Empty circumcircle test.

To avoid "infinite loops" a zero value is considered to mean that D is "outside," and thus will not be switched again.

5.3.4 Removal of a Point from the Delaunay Triangulation

It is very useful to allow insertion of points in an existing triangulated network, especially in engineering design. It is equally important to allow the removal of points from the network for the consideration of alternative designs. The problem of removing points from the Delaunay triangulation can be considered as the inverse of the incremental insertion algorithm (Heller 1990). The potential triangle with the smallest circumcircle is removed by swapping the edge (the inverse of the insertion algorithm described previously) to reduce the set of neighbors by one, and the process is repeated until only three triangles are left. Again, as the inverse of the insertion algorithm, P is removed and the three triangles are merged. A triangle considered for removal is often called a "ear" and consists of adjacent triples of point P. Figure 5.16 illustrates an example of point deletion. In this figure, △DCB might have the smallest circumcircle (Figure 5.16a) and thus is removed first (Figure 5.16b). The triple with smallest circumcircle in the remaining set of neighbors is AED, and this is removed second, leaving only three triangles (Figure 5.16c). Point P is removed in the end and the final result is shown in Figure 5.16(d).

An alternative (Devillers 1999) to the method described earlier is to remove the ears in order of the power of P (Aurenhammer 1987). The power of P with respect to the ear with vertices v_1, v_2, and v_3 is simply $H(v_1, v_2, v_3)$ divided by the triangle area. Since it is necessary to select the ears for removal in order of their power, a priority queue data structure is required. This data structure saves ears in the order in which they are supplied, and returns them in the order smallest first. The ear that has the smallest power is guaranteed to be Delaunay, and may thus be removed. With a similar treatment shown in Figure 5.16, new triangles are formed by swapping the diagonals; powers for the two new triangles are computed and the priority queue updated. This process is repeated until three ears remain in the list. Finally, these last three triangles are collapsed and point P is removed. Devillers (1999) shows an example where Heller's algorithm does not give correct results. Mostafavi et al. (2003) give a similar algorithm where an ear is removed if its circumcircle contains none of the neighboring points of P. While less efficient in extreme cases, it does not require the calculation of

GENERATION OF TRIANGULAR IRREGULAR NETWORKS

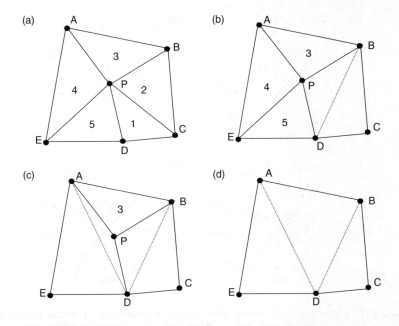

Figure 5.16 Deletion of a point from a triangular network: (a) neighbors of P (to be deleted); (b) triple with smallest circumcircle deleted; (c) only three triangles left; and (d) point P deleted from the triangulation.

the power of P (which requires a division), which may be an advantage where robust arithmetic is required for degenerate configurations.

5.4 CONSTRAINED DELAUNAY TRIANGULATION

As will be discussed later in Chapter 8, the accuracy of the resultant DTM surface will be much improved if the terrain F-S points and lines are all measured during the data acquisition. This section will discuss the special treatment of these lines during the triangulation process.

5.4.1 Constraints for Delaunay Triangulation: The Issue and Solutions

As discussed in Chapter 2, the terrain feature lines are special. Ridge lines are the connected lines of local maxima (points) and the ravine lines are the local minima. These lines are so special that they should not be broken by any triangle edges. Figure 5.17 shows the consequence of a ravine line being broken by a triangle edge.

In other words, special attention should be paid to these lines in the triangulation process. There are two possible solutions, the first and the simplest is to make the points on these lines very dense so that the lines will hopefully not be broken by

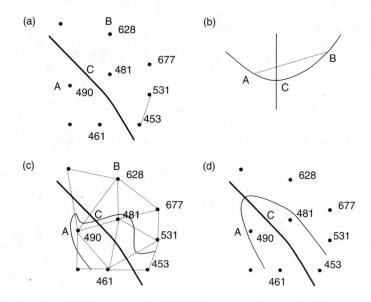

Figure 5.17 Consequence of triangulation without consideration of terrain feature lines: (a) a set of data with a ravine line; (b) a possible profile across ACB; (c) triangulation without considering the ravine; and (d) contouring after considering the ravine.

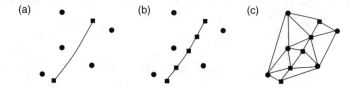

Figure 5.18 Densification of points on feature lines for triangulation: (a) data with a feature line; (b) point densification; and (c) triangulation result.

any triangle edges even though they are treated as ordinary points. In this method, the critical stage is the point densification process, which in effect transforms the connected feature lines into ordinary data points. Experience shows that the interval between the points along the feature lines has to be equal to or shorter than half of the average point intervals computed from the whole data set. This method can meet the demand very well in normal circumstances with simplicity, stability, and reliability although the data volume is increased and the original data set is modified by the densification. Figure 5.18 illustrates this treatment. A mathematically rigorous solution for point density along curves is given by Amenta et al. (1998). The other method is to treat each feature line as a constraint, ensuring that no triangle edge is

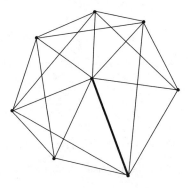

Figure 5.19 Inter-visibility of nine points and two constrained line segments.

allowed to cross the algorithm. This is a rigorous solution and will be discussed in the following sections.

5.4.2 Delaunay Triangulation with Constraints

Delaunay triangulation with consideration of a given constraint is constrained Delaunay triangulation, or CDT in short. A CDT is not truly a Delaunay triangulation as some of its triangles might not follow the Delaunay principle. For a given set of data points and a set of lines as constraints, CDT is the triangulation of vertices with the following properties:

1. the given constraint lines are included in the triangulation
2. the resulting triangulation is as close as possible to the Delaunay triangulation.

By "constrained" we mean that the predefined lines are not to be crossed by any triangle edges. To accommodate this, the empty circle principle is modified to apply to only those points that can be seen from at least one edge of the triangle where the predefined lines are treated as opaque. As a result, the constrained Delaunay principle becomes: *only when the circumcircle of the triangle does not contain any other points, and its three vertexes are visible to each other, is this triangle a CDT*. Here, visibility plays a central role. Figure 5.19 illustrates the inter-visibility of data points after the insertion of two predefined lines (i.e., constraints). The Lawson LOP is applied only if the constrained Delaunay principle holds.

A two-step method is commonly used for the construction of CDT as follows:

1. to construct the standard Delaunay triangulation with all the data including the data points on the predefined line segments (called constraint line segments)
2. to embed the constraint line segments, and adjust all triangles in the local areas where they exist through the diagonal swapping process.

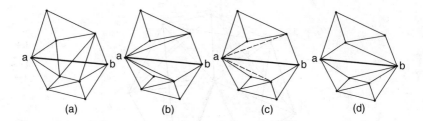

Figure 5.20 Constrained Delaunay triangulation. (a) Line segment $\overline{ab}$ as a constraint. (b) All the vertices of the influence polygon connected to a. (c) Constrained Lawson LOP exchange to optimize the triangulation. (d) Triangulation with constraints.

Figure 5.20 shows the process. In Figure 5.20(a), the standard Delaunay triangulation is completed and constraint line $\overline{ab}$ is inserted. In Figure 5.20(b), those triangles whose edges intersect the constraint line segment are identified. If two such triangles have a common edge, this edge is deleted. In this way, the so-called influence polygon of the constraint line segment is formed. All vertices of the influence polygon are connected to the starting point of the constraint line segment. In Figure 5.20(c), the Lawson LOP is applied to optimize the local areas but the constraint line segment is still an edge of some triangles. The final result is shown in Figure 5.20(d).

5.5 TRIANGULATION FROM CONTOUR DATA WITH SKELETONIZATION

Contour lines are a special type of feature lines. Three approaches can be used to form triangulation from contour data as follows:

1. treat contour lines as random points and apply Delaunay triangulation to form a triangulation network
2. treat all contour lines as constraint lines
3. a compromise between the above two approaches.

In the first approach, some undesirable effects may be created because this algorithm treats each data point separately, for example, three vertices of a triangle taken from the same contour line (leading to the so-called *flat triangles*) and some triangle edges crossing the contour lines (Figure 5.21). Therefore, this approach is seldom used. On the other hand, if all contour lines are treated as constraints, then the computation involved is heavier. A compromise is to add more points to avoid these two problems associated with the first approach, i.e. to derive the skeleton lines of the contour map and then to use these points for triangulation (Thibault and Gold 2000). This section will describe the extraction of skeletons from contour maps for the formation of more desirable triangulation networks.

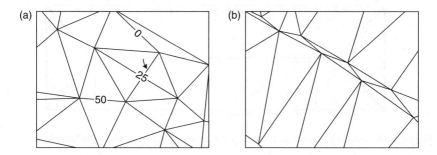

Figure 5.21 Unreasonable examples of generating triangles from the contour line: (a) an edge of a triangle crossing the 25-m contour line and (b) the three vertices of a triangle taken from the same contour line.

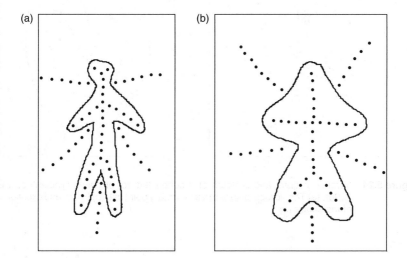

Figure 5.22 Endo- and exoskeletons of biological shapes (Reprinted from Blum 1967, with permission from MIT Press).

5.5.1 Extraction of Skeleton Lines from Contour Map

The skeleton or medial axis transform (MAT) of irregular "biological" shapes (see Figure 5.22) was first studied by Blum in 1967. Each point on the MAT of a continuous shape is the center of a disk touching the boundary at two or more locations — thus, the shape may be reconstructed from the union of all the MAT disks. (It should be noted that in the case of the discretely sampled skeleton the disks must touch at least three sample points.) The principle of skeletonization is illustrated in Figure 5.23, where the skeleton of a rectangle is extracted.

The skeleton of a connected set of points can be extracted by means of Voronoi diagrams (Aumann et al. 1991; Amenta et al. 1998; Thibault and Gold 2000) if the curve is sampled with sufficient density (better than 0.42 times the distance between the curve

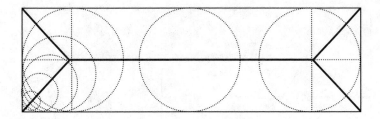

Figure 5.23 Skeleton formed by a locus of the center of a disk touching the boundary.

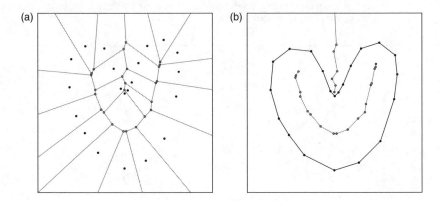

Figure 5.24 Voronoi diagram and skeleton of a connected set of points (Thibault and Gold 2000). (a) The Voronoi diagram of a boundary point set. (b) Endo- and exoskeletons of the shape (a).

and the skeleton) (Gold 1999; Gold and Snoeyink 2001). Figure 5.24 shows the process. Figure 5.24(a) shows the Voronoi diagram of a curved line that is approximated by a set of points. The medial axis (skeleton) of the curve is approximated by a subset of the Voronoi edges. The "crust" is the fully connected set of points along the curve, formed from edges of the Delaunay triangulation. The crust and skeletons are shown in Figure 5.24(b). Figure 5.25 shows a contour map with the extracted skeleton lines.

After the skeleton lines are extracted, they will be used for the triangulation and then for surface reconstruction. However, unlike the contour lines, the heights of the points on skeleton lines are not known and therefore the next step is to estimate the height for each point on the skeleton lines.

5.5.2 Height Estimation for Skeleton Points

As can be seen from Figure 5.25, some parts of the skeleton lie simply in the middle of the two adjacent contour lines, therefore the heights of the points on these parts

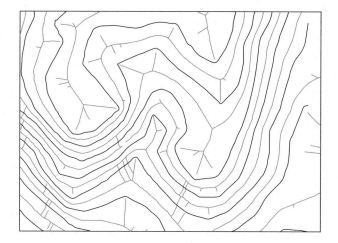

Figure 5.25 Contour map with extracted skeleton lines (Thibault and Gold 2000).

is simply the average of the heights of these two contours. The difficult part is to estimate the heights for these small branches.

Figure 5.26 shows the estimation process. Figure 5.26(a) is an enlarged diagram of a valley, indicated by a small branch of the skeleton line. The estimation method is illustrated in Figure 5.26(b). The height of the contour line is marked as 20 m and the height of a major part of the skeleton line is 15 m. The height of a point on this branch of skeleton must be from 15 to 20 m. As each of these points is a Voronoi vertex, its circumcircle (called skeleton circle here) touches both sides of the contour re-entrant. The minor valleys become narrower toward their heads; the circle radius of any skeleton point on the re-entrant, compared with the skeleton circle at the junction of the branch with the main medial skeleton, gives the elevation of skeleton point as a proportion of the elevations of the medial contour and the one forming the re-entrant. The mathematical formula is as follows:

$$Z_i = Z_c - \frac{R_i}{R_r} \times \frac{Z_c - Z_b}{2} \tag{5.6}$$

where Z_c is the elevation of the contour with the re-entrant; Z_b is the elevation of the other contour; Z_i is the elevation of the skeleton point to be estimated; R_r is the radius of the reference circle; and R_i is the radius of the skeleton point.

Figures 5.27 shows two examples of values estimated in this way. For summits or pits the ratio of the circumcircles of a terminal node on the summit skeleton and its associated skeleton node on the other side of the contour line is used — see Thibault and Gold (2000) for details. These enriching points may now be inserted into a new triangulation along with the original data.

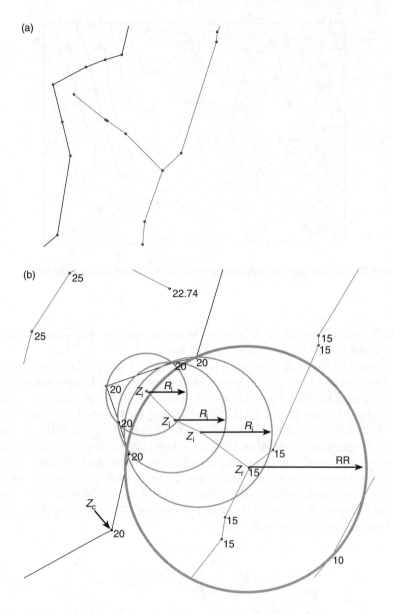

Figure 5.26 Height of each point on this branch estimated based on circle ratio. (a) A branch of skeleton for the valley; (b) Skeleton point height estimation using circle ratios.

5.5.3 Triangulation from Contour Data with Skeletons

The points on the skeleton lines are also added to the contour data for triangulation to avoid the problems mentioned at the beginning of Section 5.5. Figure 5.28 shows the differences between simple triangulation from contour data (Figure 5.28a) and that from the enriched contour data (Figure 5.28b).

GENERATION OF TRIANGULAR IRREGULAR NETWORKS 107

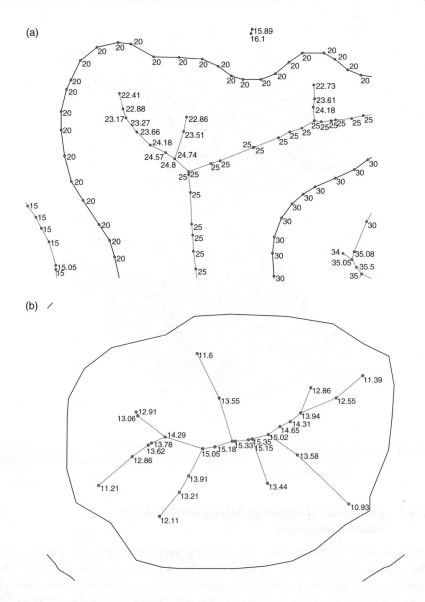

Figure 5.27 Two examples of height estimation for skeleton lines: (a) for the case with contour re-entrant and (b) for the case of a summit.

5.6 DELAUNAY TRIANGULATIONS VIA VORONOI DIAGRAMS

It was discussed in Section 5.1 that Delaunay triangulations can also be constructed indirectly from Voronoi diagrams because they have a dual relationship. This will be addressed although it is more popular to derive Voronoi diagrams from Delaunay triangulations.

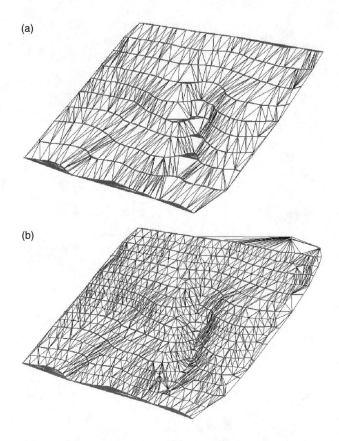

Figure 5.28 A comparison between triangulation from original contour data and that from enriched contour data: (a) triangulation from original data and (b) triangulation from enriched data.

5.6.1 Derivation of Delaunay Triangulations from Voronoi Diagrams

Triangulation from Voronoi diagrams is simple. Figure 5.29 shows the process. After the Voronoi diagram (Figure 5.29b) is constructed from a set of data (Figure 5.29a), any two points that share a common Voronoi boundary are joined to form a triangle edge (Figure 5.29c). These triangles are then searched for, recorded sequentially, and finally form a Delaunay triangulation. The important issue in indirect triangulation via the Voronoi diagram is the construction of the Voronoi diagram, which will be discussed later.

5.6.2 Vector-Based Algorithms for the Generation of Voronoi Diagram

The development of efficient and robust methods for the computation of Voronoi diagrams has been considered challenging and has attracted much attention from

GENERATION OF TRIANGULAR IRREGULAR NETWORKS

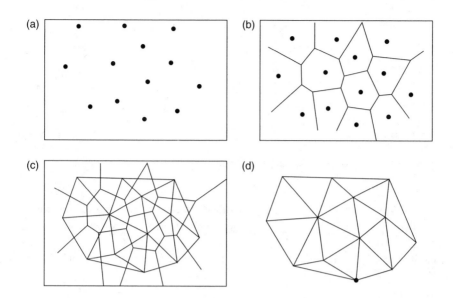

Figure 5.29 Derivation of Delaunay triangulation from Voronoi diagram: (a) a set of date points; (b) Voronoi diagram of the set; (c) dual relationship; and (d) triangulation of the set.

researchers. As a result, many algorithms for computation of Voronoi diagrams are available (Green and Sibson 1977; Brassel and Reif 1979; Bowyer 1981; Lee and Drysdale 1981; Miles and Maillardet 1982; Ohya et al. 1984a,b; Klein 1988; Masser 1988; Sugihara 1992). Efforts have also been made on the development of dynamic and kinetic Voronoi diagrams (Zaninetti 1990; Gold and Condal 1995). A comprehensive survey of such algorithms has been covered by Aurenhammer (1991) and Okabe et al. (2000). No attempt has been made to provide a similar coverage here in this section because in practice indirect triangulation via Voronoi diagram is not a popular approach, although it is the best method for manual triangulation of data: first sketch the Voronoi diagram, then the Delaunay triangulation.

From the viewpoint of computational geometry, a Voronoi diagram is essentially "a partition of the plane into N polygonal regions, each of which is associated with a given point. The region associated with a point is the set of points closer to that point than to any other given point" (Lee and Drysdale 1981). Suppose there are N distinct points $P_1, P_2, \ldots, P_n$ in the plane. Each point will have a Thiessen polygon. All these Thiessen polygons (or Voronoi region) together form a pattern of packed convex polygons covering the whole plane (no gap or overlap). This is the Voronoi diagram of the point set (Figure 5.29).

Usually, a Voronoi region can be used to define the spatial proximity for each spatial point object, while such a region is required to meet the nearest-neighbor rule formulated as follows (Okabe et al. 2000):

$$V(p_i) = \{X \mid \| X - X_i \| \leq \| X - X_j \|, \text{ for } j \neq i, j \in I_n\} \quad (5.7)$$

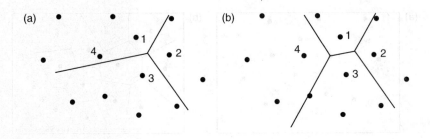

Figure 5.30 Incremental method for the computation of Voronoi diagram.

where $V(p_i)$ denotes the Voronoi region of a point object p_i; P is a set of points and $P = \{p_1, p_2, \ldots, p_n\}$; p is an arbitrary point in 2-D space with coordinates (x, y) or a location vector X; $\|X - X_i\| = d(p, p_i) = \sqrt{(x - x_i)^2 + (y - y_i)^2}$ is a distance function; and $I_n = \{1, 2, \ldots, n\}$, $2 \leq n < \infty$. The Voronoi diagram of the data set is

$$V = \{V(p_1), V(p_2), \ldots, V(p_n)\} \tag{5.8}$$

A simple method for constructing Voronoi diagram is the so-called incremental method (Forture 1975; Green and Sibson 1977; Bowyer 1981; Lee and Drysdale 1981; Ohya et al. 1984a,b). The basic idea of the incremental method is to expand the Voronoi diagram incrementally, that is, to add one point at a time. Figure 5.30 illustrates the principle. First of all, the Voronoi diagram of the first three points, that is, 1, 2, and 3, is constructed and point 4 is to be considered. It is found that point 4 is located within the Voronoi region of point 1. Then, a perpendicular bisector between points 1 and 4 is drawn. This perpendicular bisector intersects the common boundary of Voronoi regions $V(p_1)$ and $V(p_3)$. Then, a perpendicular bisector between points 4 and 3 is drawn. As a result, part of $V(p_1)$ and part of $V(p_3)$ together form $V(p_4)$. The process continues until the Voronoi diagram of N points is computed through Voronoi diagram of $N - 1$ points by adding the last point.

The incremental method can also be implemented dynamically by inserting points into the existing Voronoi diagram. This would be useful both in terms of Voronoi diagram generation and in terms of query of Voronoi. Figure 5.31 illustrates the insertion of a point into an existing Voronoi diagram. First, point 7 is inserted into a Voronoi region $V(p_3)$, then the perpendicular bisector of line $\overline{3,7}$ is drawn. This line intersects the common boundary between $V(p_3)$ and $V(p_1)$ and that between $V(p_3)$ and $V(p_5)$. Then, the perpendicular bisectors of line $\overline{1,7}$ and line $\overline{5,7}$ are drawn. These two lines intersect the common boundary between $V(p_1)$ and $V(p_2)$ and that between $V(p_5)$ and $V(p_6)$. The perpendicular bisectors of lines $\overline{6,7}$ and $\overline{2,7}$ are drawn and the last two intersections with the two common boundaries with $V(p_4)$ are found. By joining these intersections a new polygon is formed, which is the Voronoi region of point 7, that is, $V(p_7)$ (i.e., the shaded area of Figure 5.31b). Indeed, this new point has "stolen" some of the area of each of these neighbors.

GENERATION OF TRIANGULAR IRREGULAR NETWORKS

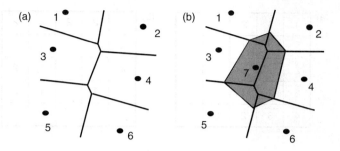

Figure 5.31 Insertion of a point into an existing Voronoi diagram.

5.6.3 Raster-Based Algorithms for the Generation of Voronoi Diagram

Although the potential of computing Voronoi diagrams via distance transformation has been recognized by researchers (Borgefors 1986; Tang 1989; Okabe et al. 2000), a detailed discussion of methods was not available until the paper by Li et al. (1999). Recently, methods for the generation of Voronoi diagrams in raster mode from quadtree structures (Zhao et al. 2002) and on spherical features (Chen et al. 2003) have also been presented.

As the Voronoi boundary of point P is formed by perpendicular bisectors between P and its close neighbors, distance is a key concept in the generation of Voronoi diagrams. In vector mode, "distance" means the Euclidean distance. The distance between two points $P_1(X_1, Y_1)$ and $P_2(X_2, Y_2)$ is defined as follows:

$$D(P_1, P_2) = f(X_1, X_2, Y_1, Y_2) = \sqrt{(X_1 - X_2)^2 + (Y_1 - Y_2)^2} \qquad (5.9)$$

However, in raster mode, the coordinates are defined by the integer numbers of rows and columns of raster pixels. Suppose there are two points $P_1(i, j)$ and $P_2(m, n)$, then the Euclidean distance between them is defined as follows:

$$D(P_1, P_2) = f(i, j, m, n) = \sqrt{(i - m)^2 + (j - n)^2} \qquad (5.10)$$

The unit is number of pixels. For example, if the two points are at (2, 2) and (3, 3), then the result is $\sqrt{2}$ (=1.414) pixels. This result in decimal form is inconvenient to use in raster mode. Distance in an integer is more desirable and thus normally used. The problem arising is "how to find an integer number for every possible distance between two points that is the best approximation of the Euclidean distance." In the example given above, either 1 or 2 would be the best choice as the raster distance to approximate the Euclidean distance of $\sqrt{2}$. However, other integers (e.g., 3 in the case of the Chamfer 2-3 function) could also be used, depending on the definition given. Figure 5.32 shows two definitions of raster distance, that is, city block and chessboard.

With the definition of raster distance given, the distance from each pixel to a given set of points can be obtained by distance transformation. Figure 5.33(a) shows the distance transformation of a set of points (A, B, C, D, E, F, and G). In this figure, the

(a)

4	3	2	3	4
3	2	1	2	3
2	1	0	1	2
3	2	1	2	3
4	3	2	3	4

(b)

2	2	2	2	2
2	1	1	1	2
2	1	0	1	2
2	1	1	1	2
2	2	2	2	2

Figure 5.32 Definitions of raster distance: (a) city block distance and (b) chessboard distance.

(a)

1	1	2	3	3	2	1	1	1	2	3	4	5	6	7
A	1	2	3	3	2	1	**B**	1	2	3	4	5	6	7
1	1	2	3	3	2	1	1	1	2	3	4	5	6	7
2	2	2	2	2	2	2	2	2	2	3	4	5	6	7
2	2	2	1	1	1	2	2	2	2	3	4	5	6	7
1	1	2	1	**G**	1	1	1	1	2	3	4	5	6	7
F	1	2	1	1	1	1	**C**	1	2	3	4	5	6	7
1	1	2	2	2	2	1	1	1	2	3	4	5	6	7
2	2	2	2	2	2	2	2	2	2	3	4	5	6	7
2	1	1	1	2	2	1	1	1	2	3	4	5	6	7
2	1	**E**	1	2	2	1	**D**	1	2	3	4	5	6	7
2	1	1	1	2	2	1	1	1	2	3	4	5	6	7
2	2	2	2	2	2	2	2	2	2	3	4	5	6	7
3	3	3	3	3	3	3	3	3	3	3	4	5	6	7
4	4	4	4	4	4	4	4	4	4	4	4	5	6	7

(b)

A	A	A	A	B	B	B	B	B	B	B	B	B	B	B
A	A	A	A	B	B	B	**B**	B	B	B	B	B	B	B
A	A	A	G	G	G	B	B	B	B	B	B	B	B	B
A	A	**G**	G	G	G	B	B	B	B	B	B	B	B	B
F	F	G	G	G	G	C	C	C	C	C	C	C	C	C
F	F	G	G	**G**	G	C	C	C	C	C	C	C	C	C
F	F	G	G	G	G	C	**C**	C	C	C	C	C	C	C
F	F	F	G	G	C	C	C	C	C	C	C	C	C	C
F	E	E	E	E	D	D	D	D	D	D	D	D	D	D
E	E	E	E	E	D	D	D	D	D	D	D	D	D	D
E	E	**E**	E	E	D	D	**D**	D	D	D	D	D	D	D
E	E	E	E	E	D	D	D	D	D	D	D	D	D	D
E	E	E	E	E	E	D	D	D	D	D	D	D	D	D
E	E	E	E	E	E	D	D	D	D	D	D	D	D	D
E	E	E	E	E	E	E	D	D	D	D	D	D	D	D

Figure 5.33 Distance transformation for Voronoi diagram generation. (a) Distance transformation (chessboard). (b) Voronoi regions formed from distance map.

number in each pixel indicates the minimum distance from this pixel to its neighboring points. Naturally, the large number of the pixel is mid way between two points (e.g., 3 between A and B). There could be 20 different distance values for a single pixel if there are 20 points in the set. However, in any case, only the distance with smallest value is taken for that pixel. The result is a distance contour map. These influence regions are in fact the Voronoi regions of the set of points, as shown in Figure 5.33(b).

In fact, distance transformation can also be achieved by operators developed in mathematical morphology (Serra 1982). There are two basic operators developed in mathematical morphology, that is, *dilation* and erosion. These can be compared to "+," "−," "×," and "÷" in ordinary algebra and they are defined as follows (see Serra 1982; Haralick et al. 1987):

Dilation:

$$A \oplus B = \{a + b : a \in A,\ b \in B\} = \cup_{b \in B} A_b \tag{5.11}$$

Erosion:

$$A \ominus B = \{a : a + b \in A,\ b \in B\} = \cap_{b \in B} A_b \tag{5.12}$$

GENERATION OF TRIANGULAR IRREGULAR NETWORKS

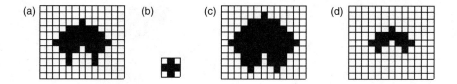

Figure 5.34 Two basic morphological operators: dilation and erosion. (a) Original image A. (b) Structuring element B. (c) A dilated by B. (d) A eroded by B.

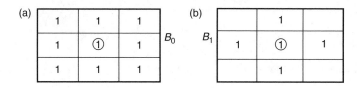

Figure 5.35 Two structuring elements mimicking chessboard and city block distances.

where A is the original image with features and B is the structuring element. Examples of dilation and erosion are given in Figure 5.34.

With the basic concepts and operators in mathematical morphology introduced, the next step is to employ an appropriate structuring element so that distance contours can be obtained using the dilation operator repeatedly. By a close examination of possible structuring elements, it is found that structuring elements B_1 and B_0, as shown in Figure 5.35 are the right choices for chess board and city block distance, respectively. These two types of distance contours can be expressed mathematically as follows:

Chessboard distance:
$$D_n = D_{n-1} \oplus B_0 \qquad (5.13)$$

City block distance:
$$K_n = K_{n-1} \oplus B_1 \qquad (5.14)$$

By using these two structuring elements for dilation operations, distance maps similar to Figure 5.32 can be obtained. Then, the Voronoi diagrams can easily be derived. In fact, it is easier to use a raster-based method for the generation of Voronoi diagrams with line and area features as constraints. Figure 5.36 is a Voronoi diagram of a data set with points, lines, and area features, generated in raster mode.

It is understandable that the approximation of raster distance to Euclidean distance would be poor if point intervals are large. To overcome this, dynamic distance transformation using morphological operators is also possible to ensure that the distance error is within a pixel (Li et al. 1999).

It should be noted here that there are other alternatives for Delaunay triangulation in raster mode. Instead of using the dilation operator to generate a distance

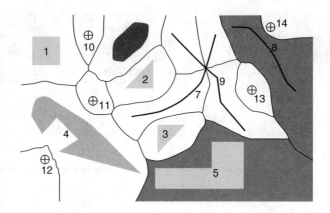

Figure 5.36 Voronoi diagram of point, line, and area features.

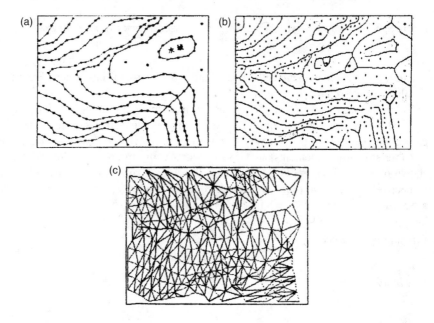

Figure 5.37 Delaunay triangulation by morphological skeletonization (Chen 1991) (a) Original contour lines; (b) Skeletinization; (c) Triangulation network.

contour map for the extraction of Voronoi regions for the data points, a skeletonization algorithm could be used to derive the skeleton of the complement (i.e., a raster area) of data points. The skeleton lines are then the boundaries of the Voronoi regions of the data points (Chen 1991). This approach is particularly effective with contour data that consist of line features (Su et al. 1998). Figure 5.37 shows the triangulation constructed from contour line data by this skeletonization approach.

CHAPTER 6

Interpolation Techniques for Terrain Surface Modeling

6.1 INTERPOLATION TECHNIQUES: AN OVERVIEW

Interpolation is an approximation problem in mathematics and an estimation problem in statistics. Interpolation in digital terrain modeling is used to determine the height value of a point by using the known heights of neighboring points. There are two implicit assumptions behind interpolation techniques: (a) the terrain surface is continuous and smooth and (b) there is a high correlation between the neighboring data points. Interpolation is one of the core techniques in digital terrain modeling because it is involved in the various stages of the modeling process such as quality control, surface reconstruction, accuracy assessment, terrain analysis, and applications.

Interpolation techniques can be classified according to different criteria and they can be used for different purposes. Table 6.1 attempts to provide a simple classification.

By the size of the area for interpolation, two approaches are identified (Petrie 1990a), that is, *area based* and *point based*. In the area-based approach the surface is constructed by using all the reference (known) points within this area and the height of any point within this area can be determined by using this constructed surface. Area-based interpolation could be either global or local. *Global interpolation* involves the construction of a single complex 3-D surface from the complete data set of measured points, from which the height values of all other points can be obtained. This is an extreme approach. The usefulness of this type of technique depends on the complexity of the terrain surface and the actual size of the area. A more adaptive solution is to divide a large area into a series of patches of identical shape and size. This is called *local* or *patchwise interpolation*. A surface is constructed for each patch by using all the reference points and heights of all points (to be interpolated) within this patch are obtained from this patch surface. The size of the patch is determined by the complexity of the area and there could be

Table 6.1 A Classification of Interpolation Techniques

Criteria	Interpolation Techniques
Size of area for interpolation	Point based, area based (patchwise or global)
Exactness of the surface	Exact fitting, best fitting
Smoothness of the surface	Linear, nonlinear
Continuity of the surface	Step, continuous
Preciseness of the function	Precise, approximate
Certainty of the problem	Functional, stochastic
Domain of interest	Spatial, spectral (i.e., frequency)
Complexity of the phenomenon	Analytical, numerical iteration

a certain degree of overlap between neighboring patches to ensure smooth connection between patches. At the other extreme, a surface could be constructed for the interpolation of each point, leading to *pointwise interpolation*. This requires heavy computation.

In the construction of a surface for interpolation, (whether global, patchwise, or pointwise), the surface may or may not pass through all the reference points. If it does pass through all the points, it is an exact reconstruction and is called *exact fitting*. However, the surface may not pass through all the reference points, due to errors in the reference points. In this case, there might be a deviation from each of these reference points. If such deviations are limited to a minimum, then the surface is a best fit. This type of interpolation is called *best fitting*.

A surface could be in the form of *steps*, as shown in Figure 4.2. This is a discontinuous surface. In many applications, continuity is a requirement, at least visually, thus a *continuous surface* can be constructed. The continuous surface may or may not be smooth. For example, as discussed in Chapter 4, a surface comprising a set of contiguous linear facets is not continuous in the first derivative and thus is not smooth. Both triangular facets and bilinear surfaces are linear surfaces. Usually, a smooth surface is constructed by using some kind of higher-order polynomials.

It is also possible to use an approximate function for interpolation if the original function is too complex or the approximate function is good enough but much simpler. For example, when x is a very small angle, the value of $\sin(x)$ can be nicely approximated by x itself. In other words, $y = x$ is a very good approximation of the function $y = \sin(x)$ under this condition. In fact, it is also possible (but not in terrain modeling) that the nature of a phenomenon is too complex and one is not able to establish an analytical function for the problem, so numerical approximation by iteration is used.

A problem could be deterministic or probabilistic. For the former, a deterministic function can be used and for the latter a stochastic model must be used. Sometimes the small variations on the terrain are so complex that the variation is then treated as a stochastic process.

Interpolation can take place in either the space or the spectral domain. Most interpolation techniques for terrain modeling are for spatial interpolation. However, it is also possible to transform the data into a frequency domain and perform interpolation there.

INTERPOLATION TECHNIQUES FOR TERRAIN SURFACE MODELING

It should be emphasized that interpolation techniques are well discussed within the mathematical community and a large body of literature is available. Therefore, only those widely used in digital terrain modeling are described in this chapter. In other words, some interpolation functions, such as *Kriging*, are omitted here due to their complexity. Furthermore, in photogrammetric community, it was found that sampling is the single vital step in digital terrain modeling because information lost at the sampling stage can never be reconstructed by whatever interpolation function. This is why no complicated interpolation is used in digital terrain modeling.

6.2 AREA-BASED EXACT FITTING OF LINEAR SURFACES

This section discusses interpolation of a linear surface that fits exactly to the reference points. In other words, each reference point is honored.

6.2.1 Simple Linear Interpolation

As discussed in Chapter 4, a plane can be determined by three points on it and a triangular facet is a typical example of such a surface. The mathematical function of a plane is as follows:

$$z = a_0 + a_1 x + a_2 y \qquad (6.1)$$

where a_0, a_1, and a_2 are the three coefficients and (x, y, z) is the set of coordinates of a surface point. To compute these three coefficients, three reference points with known coordinates, for example, $P_1(x_1, y_1, z_1)$, $P_2(x_2, y_2, z_2)$, and $P_3(x_3, y_3, z_3)$, are required to establish three equations as follows:

$$\begin{bmatrix} a_0 \\ a_1 \\ a_2 \end{bmatrix} = \begin{bmatrix} 1 & x_1 & y_1 \\ 1 & x_2 & y_2 \\ 1 & x_3 & y_3 \end{bmatrix}^{-1} \begin{bmatrix} z_1 \\ z_2 \\ z_3 \end{bmatrix} \qquad (6.2)$$

Once the coefficients a_0, a_1, and a_2 are computed, then the height z_i of any point i with a given set of coordinates (x_i, y_i) can be obtained by substituting (x_i, y_i) into Equation (6.1).

6.2.2 Bilinear Interpolation

Bilinear interpolation from a square grid has been mentioned in Section 4.5 and illustrated in Figure 4.14. Bilinear interpolation can be performed for any four points (not along a line). The mathematical function is as follows:

$$z = a_0 + a_1 x + a_2 y + a_3 xy \qquad (6.3)$$

where a_0, a_1, a_2, a_3 is the set of coefficients. They are to be determined by four equations that are formed by making use of the coordinates of four reference points, say, $P_1(x_1, y_1, z_1)$, $P_2(x_2, y_2, z_2)$, $P_3(x_3, y_3, z_3)$, and $P_4(x_4, y_4, z_4)$. The mathematical formula is as follows:

$$\begin{bmatrix} a_0 \\ a_1 \\ a_2 \\ a_3 \end{bmatrix} = \begin{bmatrix} 1 & x_1 & y_1 & x_1 y_1 \\ 1 & x_2 & y_2 & x_2 y_2 \\ 1 & x_3 & y_3 & x_3 y_3 \\ 1 & x_4 & y_4 & x_4 y_4 \end{bmatrix}^{-1} \begin{bmatrix} z_1 \\ z_2 \\ z_3 \\ z_4 \end{bmatrix} \quad (6.4)$$

Once the coefficients a_0, a_1, a_2, and a_3 are computed, then the height z_i of any point i with a given set of coordinates (x_i, y_i) can be obtained by substituting (x_i, y_i) into Equation (6.3).

If data (reference) points are distributed in the form of square grids, then the following formula can be used:

$$z_p = z_1 \left(1 - \frac{\Delta x}{d}\right)\left(1 - \frac{\Delta y}{d}\right) + z_2 \left(1 - \frac{\Delta y}{d}\right)\left(\frac{\Delta x}{d}\right) + z_3 \left(\frac{\Delta x}{d}\right)\left(\frac{\Delta y}{d}\right)$$
$$+ z_4 \left(1 - \frac{\Delta x}{d}\right)\left(\frac{\Delta y}{d}\right) \quad (6.5)$$

In the formula, points 1, 2, 3, and 4 are the four nodes of the square grid, and d is the length of the grid interval (Figure 6.1a).

In fact, interpolation on a triangular facet can also be done in a similar way to grid-based bilinear interpolation. As shown in Figure 6.1(b), the height of point $p(x_p, y_p, z_p)$ can be interpolated from points 1 and 2 as follows:

$$z_p = z_1 + (z_2 - z_1) \times (x_p - x_1)/(x_2 - x_1) \quad (6.6)$$

and

$$z_1 = z_A + (z_B - z_A) \times (x_1 - x_A)/(x_B - x_A) \quad (6.7a)$$

$$z_2 = z_A + (z_C - z_A) \times (x_2 - x_A)/(x_C - x_A) \quad (6.7b)$$

where $y_p = y_1 = y_2$, and points 1 and 2 lie on lines $\overline{AB}$ and $\overline{AC}$, respectively. Alternatively, the local area coordinates of Figure 5.13 may be used for linear interpolation using a weighted average, that is,

$$z_p = \frac{z_1 \times \Delta a_1 + z_2 \times \Delta a_2 + z_3 \times \Delta a_3}{\Delta a_1 + \Delta a_2 + \Delta a_3} \quad (6.8)$$

This guarantees continuity between adjacent triangles. Indeed, if the distribution of the reference points is not good (e.g., nearly along a straight line), then Equation (6.2) is not stable and the use of Equation (6.6) is recommended in such a case.

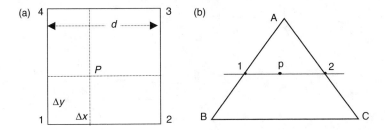

Figure 6.1 Bilinear interpolation: (a) for square grids and (b) for triangles.

6.3 AREA-BASED EXACT FITTING OF CURVED SURFACE

Bilinear interpolation is widely used in DTM interpolation because it is simple, intuitive, and reliable. But the resulting surface is not smooth. To make the surface smooth, a polynomial surface might be fitted to a set of contiguous linear surfaces. Alternatively, exact fitting of curved surfaces is also possible, such as a bicubic spline function.

6.3.1 Bicubic Spline Interpolation

To overcome the shortcomings of bilinear functions, a bicubic spline function can be used to construct a smooth DTM surface over a patch consisting of four grid nodes, for example, points A, B, C, and D in Figure 6.2. The mathematical function of a bicubic function is as follows:

$$\begin{aligned}
z = f(x,y) &= \sum_{j=0}^{3}\sum_{i=0}^{3} a_{i,j} x^i y^j \\
&= a_{00} + a_{10}x + a_{20}x^2 + a_{30}x^3 \\
&\quad + a_{01}y + a_{11}xy + a_{21}x^2y + a_{31}x^3y \\
&\quad + a_{02}y^2 + a_{12}xy^2 + a_{22}x^2y^2 + a_{32}x^3y^2 \\
&\quad + a_{03}y^3 + a_{13}xy^3 + a_{23}x^2y^3 + a_{33}x^3y^3
\end{aligned} \quad (6.9)$$

where $a_{00}, a_{01}, a_{10}, \ldots, a_{33}$ are the 16 coefficients to be determined.

Sixteen equations are needed to solve the 16 coefficients. With the coordinates of the four grid nodes known, four equations can be established. Therefore, another 12 equations are needed and will come from the conditions for the connections between patches, that is,

1. the slopes at each node (i.e., the joint between four adjacent patches) should be continuous in x, y directions
2. the torque of the joint of adjacent patches is also continuous.

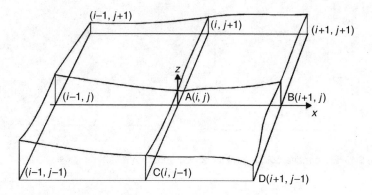

Figure 6.2 Bicubic spline interpolation.

Let R and S be the slopes in x and y directions, respectively, and T be the torque, then

$$R = \frac{\partial z}{\partial x}, \quad S = \frac{\partial z}{\partial y}, \quad T = \frac{\partial^2 z}{\partial x \partial y} \qquad (6.10)$$

As the reference points are located at square-grid nodes, the height differences can be used to compute these values as follows:

$$\begin{aligned} R_{i,j} &= \frac{\partial z}{\partial x} = \frac{z_{i+1,j} - z_{i-1,j}}{2} \\ S_{i,j} &= \frac{\partial z}{\partial y} = \frac{z_{i,j+1} - z_{i,j-1}}{2} \\ T_{i,j} &= \frac{\partial^2 z}{\partial x \partial y} = \frac{(z_{i-1,j-1} + z_{i+1,j+1}) - (z_{i+1,j-1} - z_{i-1,j+1})}{4} \end{aligned} \qquad (6.11)$$

There will be 12 such equations for a patch with four nodes as there are three equations for each node.

After these coefficients are solved, then for a point P with coordinates (x_p, y_p), the height can be computed by Equation (6.9).

The connection requirements between patches are adopted from elastic mechanics and the results of such interpolation may be not as desirable as expected because terrain patches are not elastic crusts in the narrow sense.

6.3.2 Multi-Surface Interpolation (Hardy Method)

Multi-surface interpolation is also known as the Hardy method (Hardy 1971). The basic idea is that any (regular or irregular) continuous curved surface can be approximated by the sum of a series of simple surfaces (i.e., single-value mathematical surfaces), with a desired accuracy. This might be regarded as an analogy of the Fourier series. The actual process establishes a curved surface for each reference point using a basic function (called kernel function) and the height of any point between reference points will take a weighted average from these curved surfaces. In this way, the final surface will be continuous and pass through all reference points. The mathematical expression of multi-surface overlapping is:

$$z = f(x, y) = \sum_{i=1}^{n} k_i Q(x, y, x_i, y_i)$$
$$= k_1 Q(x, y, x_1, y_1) + k_2 Q(x, y, x_2, y_2) + \cdots + k_n Q(x, y, x_n, y_n) \quad (6.12)$$

where $Q(x, y, x_i, y_i)$ is the simple (single-value) mathematical surface, called the kernel function in multi-surfaces; n is the number of simple mathematical surfaces (or the number of surface layers) the value of which is equal to the number of reference points within the patch; and k_i ($i = 1, 2, 3, \ldots, n$) is the coefficient, that represents the contribution of the ith kernel function to the final surface. To make computation simple, the kernel functions are usually simple functions of the same type and formed by rotating around an axis (which just passes through the reference point). Examples of such simple functions are:

1. *Conic function*:

$$Q_1(x, y, x_i, y_i) = C + [(x - x_i)^2 + (y - y_i)^2]^{1/2} \quad (6.13)$$

where $[(x - x_i)^2 + (y - y_i)^2]^{1/2}$ is the horizontal distance between the interpolation point (x, y) and the reference point (x_i, y_i).

2. *Hyperbolical function*:

$$Q_2(x, y, x_i, y_i) = [(x - x_i)^2 + (y - y_i)^2 + \sigma]^{1/2} \quad (6.14)$$

where σ is a nonzero parameter. Equation (6.14) represents a curved surface that is formed through the rotation of a hyperbola curve around a vertical axis. When $\sigma = 0$, this curved surface degenerates to become a conic surface.

3. *Cubic function*:

$$Q_3(x, y, x_i, y_i) = C + [(x - x_i)^2 + (y - y_i)^2]^{3/2} \quad (6.15)$$

4. *Geometric function*:

$$Q_4 = 1 - \frac{D_i^2}{a^2} \quad (6.16)$$

5. *Exponential function*:

$$Q_5 = C_0 \times e^{-a^2 D_i^2} \quad (6.17)$$

where C_0 and a are the two parameters.

The following kernel functions are well known and widely used (Li 1988):

1. *Arthur function*:

$$Q(d) = e^{-25d^2/a^2} \quad (6.18)$$

where d is the distance between two points and a is the longest distance among various data points.

2. *Lu function*:
$$Q(d) = 1 + d^3 \qquad (6.19)$$

3. *Wild function*:
$$Q_2(x, y, x_i, y_i) = \left(1 + \frac{(x - x_i)^2 + (y - y_i)^2}{(d_{ki})^2_{\min}}\right)^{1/2} \qquad (6.20)$$

where $(d_{ki})^2_{\min}$ represents the distance between data point i and its closest data point k. When $n = m$, Q matrix is an asymmetric matrix, because each data point has its own reference $(d_{ki})_{\min}$.

The Wild function is the result of modifying Equation (6.14). This is because the surface obtained by using Equation (6.14) will become smoother as σ becomes larger. Figure 6.3 shows this trend, where a set of values for σ (i.e., 0, 0.6, and 10) were

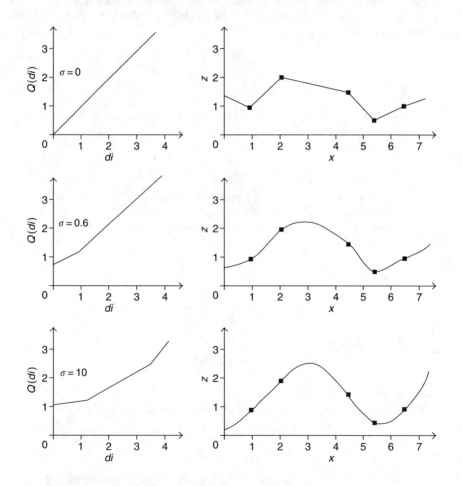

Figure 6.3 Various σ values and interpolation curves for Equation (6.14).

INTERPOLATION TECHNIQUES FOR TERRAIN SURFACE MODELING

used. The advantage of the Wild function is that the terrain feature points and lines can be used during the interpolation process even with very small $(d_{ki})_{\min}$ as long as the set of points are dense enough. As a result, a steep slope could be produced on the top of the curved surface. In this way, abrupt changes in the terrain surface can be accommodated.

If the number of reference points m is larger than the number of kernel surfaces, then a least-square solution is taken, which will be discussed later in this chapter.

One advantage of multi-surface interpolation is that different types of kernel surfaces could be designed to accommodate different features and terrain surfaces. This is useful when the density of sampled reference points is low but the accuracy of such points is relatively high. On the other hand, the process is rather complicated and inefficient, and thus this method is not widely used.

6.4 AREA-BASED BEST FITTING OF SURFACES

It is understandable that if the terrain surface is complicated, it is impossible to use any mathematical function to completely describe it. Instead, one uses an interpolation function to approximate the terrain surface. The accuracy of such approximation will be discussed in Chapter 8. It must be noted here that a surface passing through all the reference points is not necessarily a better approximation than other surfaces. If the area is big and there are many reference points available, one needs to use very high-order polynomials to achieve exact fitting of the surface. Indeed, it is dangerous to do so because unstable oscillation can be caused by such a high-order polynomial function. Figure 6.4 illustrates such an oscillation. Therefore, best fitting (instead of exact fitting) of curved surfaces is a method widely in use and will be described in this section. The theory behind best fitting is that small variations on the terrain surface are so complex that they can be treated as a stochastic process.

6.4.1 Least-Squares Fitting of a Local Surface

There are many possibilities for best fitting depending on the definition of "best." A simple definition could be that the sum of the absolute values of the errors is at a minimum. Another popular definition is the sum of the square errors being minimum, leading to the least squares, which is widely used in error theory. The mathematical

Figure 6.4 Oscillation of high-order polynomial surface.

expressions are as follows:

$$\sum_{i=1}^{n} |e_i| = \min \qquad (6.21)$$

$$\sum_{i=1}^{n} e_i^2 = \min \qquad (6.22)$$

where e_i is the deviation of the ith reference point from the fitting surface and n is the total number of reference points.

For a set of reference points and a fitting function, there is an infinite number of fitting. Figure 6.5 shows such a case by using linear surface as the fitting function. This figure shows that there can be a deviation at each reference point from the fitted surface and the deviation is also called residual in error theory. The best-fitting result is the one with the smallest sum of square residuals.

The surface fitted to the data could be linear (Figure 6.6a) or a smooth curved surface (Figure 6.6b). There are different types of curved surfaces as discussed in the previous section. For the same set of data, if the surface fitted is different, then the residual at each reference could be different. In Figure 6.5, there are three surfaces (one linear and two curved) fitted to the same set of data and three sets of residuals can be obtained, that is,

$$\text{Sum}_{\text{Linear}} = \sum_{i=1}^{n} \Delta z_{i,\text{L}}^2$$

$$\text{Sum}_{\text{Curved}-1} = \sum_{i=1}^{n} \Delta z_{i,\text{c}-1}^2$$

$$\text{Sum}_{\text{Curved}-2} = \sum_{i=1}^{n} \Delta z_{i,\text{c}-2}^2$$

where $\Delta z_{i,\text{L}}$ is the residual at the ith point in the case of linear surface (i.e., the vertical distance from the point to the linear line); $\text{Sum}_{\text{Linear}} = \sum_{i=1}^{n} \Delta z_{i,\text{L}}^2$ is the sum of the squares of the residuals; and $\Delta z_{i,\text{c}-1}$ is the residual at the ith point for the first type of curved surface (i.e., the vertical distance from the point to the curved surface). The least-squares condition says that the surface that produces the least sum of square errors is the best. In this example, among $\text{Sum}_{\text{Linear}}$, $\text{Sum}_{\text{Curved}-1}$, and $\text{Sum}_{\text{Curved}-2}$, if $\text{Sum}_{\text{Curved}-2}$ is the smallest, then curve 2 is regarded as the best fit.

The above discussion is about which types of surface to be considered. The commonly used functions for curved surface fitting are the second-order and third-order polynomials and bicubic functions. No matter which function it is, the principles and procedures of the least-square solution are identical. Therefore, the simpler

INTERPOLATION TECHNIQUES FOR TERRAIN SURFACE MODELING 125

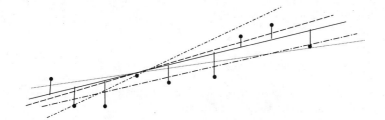

Figure 6.5 Residuals at reference points on the same type of surfaces but with different fitting.

(a)　　　　　　　　　　　　　　　(b)

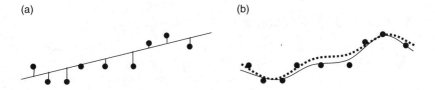

Figure 6.6 Residuals at reference points on different surfaces.

second-order polynomial is used for illustration:

$$z = f(x, y) = a_0 + a_1 x + a_2 y + a_3 xy + a_4 x^2 + a_5 y^2 \quad (6.23)$$

where $a_0, a_1, a_2, \ldots, a_5$ are the six coefficients. They need to be determined by making use of reference points. If there are n (>6) reference points, then there are n equations as follows:

$$\begin{bmatrix} z_1 \\ z_2 \\ \vdots \\ z_n \end{bmatrix} = \begin{bmatrix} 1 & x_1 & y_1 & x_1 y_1 & x_1^2 & y_1^2 \\ 1 & x_2 & y_2 & x_2 y_2 & x_2^2 & y_2^2 \\ \vdots & \vdots & \vdots & \vdots & \vdots & \vdots \\ 1 & x_n & y_n & x_n y_n & x_n^2 & y_n^2 \end{bmatrix} \begin{bmatrix} a_0 \\ a_1 \\ \vdots \\ a_5 \end{bmatrix} \quad (6.24)$$

The error functions can be written as follows:

$$\begin{bmatrix} v_1 \\ v_2 \\ \vdots \\ v_n \end{bmatrix} = \begin{bmatrix} 1 & x_1 & y_1 & x_1 y_1 & x_1^2 & y_1^2 \\ 1 & x_2 & y_2 & x_2 y_2 & x_2^2 & y_2^2 \\ \vdots & \vdots & \vdots & \vdots & \vdots & \vdots \\ 1 & x_n & y_n & x_n y_n & x_n^2 & y_n^2 \end{bmatrix} \begin{bmatrix} a_0 \\ a_1 \\ \vdots \\ a_5 \end{bmatrix} - \begin{bmatrix} z_1 \\ z_2 \\ \vdots \\ z_n \end{bmatrix} \quad (6.25)$$

and simplified as

$$\underset{n \times 1}{V} = \underset{n \times 6}{X} \underset{6 \times 1}{A} - \underset{n \times 1}{Z} \quad (6.26)$$

According to the least-squares solution,

$$\underset{6\times 1}{A} = \left(\underset{6\times n}{X^T} \underset{n\times 6}{X} \right)^{-1} \left(\underset{6\times n}{X^T} \underset{n\times 6}{X} \underset{n\times 1}{Z} \right) \tag{6.27}$$

After the coefficients are computed, the height z_p of any point P at location (x, y) can be obtained by substituting x, y into Equation (6.23).

It must be noted here that this is a simple regression method. There are other more sophisticated treatments using the least-squares concept such as the *least-squares collocation* developed by Moritz (1980).

6.4.2 Least-Squares Fitting of Finite Elements

Finite element is a method widely used in mechanics. It divides a large piece of material into small units (cells) for treatment. In the case of digital terrain modeling, a surface over a large area can be divided into small-area units such as grid or triangular cells. Then, a simple mathematical function is used to approximate the surface over each small cell. In other words, the large surface consists of a finite number of small-area units (Ebner et al. 1980).

In fact, bilinear and bicubic interpolations do employ the concept of finite element analysis, especially in the case of exact fitting of bicubic spline. However, in the case discussed here, the grid nodes are unknown, and thus have to be interpolated. Figure 6.7 shows such a case. In this figure, the height of point P could be determined by making use of the heights at the four grid nodes, that is, $z_{i,j}$, $z_{i+1,j}$, $z_{i+1,j+1}$, and $z_{i,j+1}$. That is,

$$z(x, y) = z_{i,j}\left(1 - \frac{\Delta x}{d}\right)\left(1 - \frac{\Delta y}{d}\right) + z_{i+1,j}\left(1 - \frac{\Delta y}{d}\right)\left(\frac{\Delta x}{d}\right)$$
$$+ z_{i+1,j+1}\left(\frac{\Delta x}{d}\right)\left(\frac{\Delta y}{d}\right) + z_{i,j+1}\left(1 - \frac{\Delta x}{d}\right)\left(\frac{\Delta y}{d}\right) \tag{6.28}$$

Indeed, this equation is identical to Equation (6.5). Let the increments $\delta x = \Delta x/d$ and $\delta y = \Delta y/d$, then

$$z(x, y) = z_{i,j}(1 - \delta x)(1 - \delta y) + z_{i+1,j}(1 - \delta y)\delta x$$
$$+ z_{i+1,j+1}\delta x \delta y + z_{i,j+1}(1 - \delta x)\delta y \tag{6.29}$$

As P is a known point, and an observation equation can be obtained, the error equation is:

$$v_p = z_{i,j}(1 - \delta x)(1 - \delta y) + z_{i+1,j}(1 - \delta y)\delta x$$
$$+ z_{i+1,j+1}\delta x \delta y + z_{i,j+1}(1 - \delta x)\delta y - z_p \tag{6.30}$$

In order to ensure the smoothness of the constructed surface, the second derivatives (more precisely, the second differences) in both x and y directions can be used to

INTERPOLATION TECHNIQUES FOR TERRAIN SURFACE MODELING

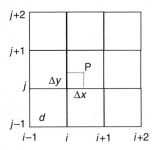

Figure 6.7 Finite element interpolation of grid nodes.

construct two virtual error equations as follows:

$$\begin{aligned} v_x(i,j) &= z_{i-1,j} - 2z_{i,j} + z_{i+1,j} = 0 \\ v_y(i,j) &= z_{i,j-1} - 2z_{i,j} + z_{i,j+1} = 0 \end{aligned} \quad (6.31)$$

Weights can also be introduced for errors. The simplest treatment of virtual observation values is to assume that they are not correlated and all have a weight 1. If the weight for known points is w_p, then the condition is

$$\sum_{k=1}^{S} v_k^2 w_p + \sum_{i=2}^{n-1} \sum_{j=1}^{m} v_x^2(i,j) + \sum_{i=1}^{n} \sum_{j=2}^{m-1} v_y^2(i,j) = \min \quad (6.32)$$

where S is the total number of reference points and m and n are the numbers of rows and columns, respectively, of the DTM grid that is to be interpolated.

6.5 POINT-BASED MOVING AVERAGING

In the previous section, some area-based methods were introduced. In this and the coming sections, some point-based interpolation methods will be introduced. This section describes the moving averaging.

6.5.1 The Principle of Point-Based Moving Averaging

One of the point-based interpolation methods is moving averaging, which is normally seen as a smoothing method. Figure 6.8 shows the three moving average lines of the Hang Seng Index of Hong Kong Stocks, one for 20 days and the other for 50 days, which show a smoothing effect on the Hang Seng Index.

A similar technique is in common use in digital terrain modeling. It is used to interpolate a point by making use of a number of reference points nearby. The mathematical

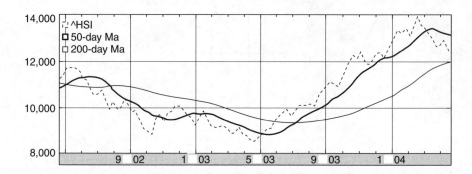

Figure 6.8 Moving averages of Hang Seng Index over last 3 years.

expression is as follows:

$$z = \frac{\sum_{i=1}^{n} z_i}{n} \quad (6.33)$$

where n is the total number of reference points used for the averaging operation and z_i ($i \in 1, n$) is the height of the ith reference point. For example, if five neighbor points are selected as reference points for interpolation of point P, and their heights are 4, 5, 6, 3.5, and 4.5, then the height value of P is $(4 + 5 + 6 + 3.5 + 4.5)/5 = 4.6$. This is a simple averaging. It means that no matter how close a reference is to the interpolation point, the weight is still the same as that of the others. This equal weighting seems unfair to those reference points that are closer to the interpolation point. This leads to a weighted moving averaging as follows:

$$z = \frac{\sum_{i=1}^{n} w_i z_i}{\sum_{i=1}^{n} w_i} \quad (6.34)$$

where w_i is the weight for the ith point, which may or may not be different from that of the others.

Two questions arising now are (a) which points should be used as reference points and (b) how to assign a weight to each reference point. These two problems will be discussed in the following two subsections.

6.5.2 Searching for Neighbor Points

Neighbor points should be close to the point to be interpolated. Using distance as a criterion, a circle or a rectangle can be drawn around the interpolation point and all points within this area are selected. Figure 6.9(b) shows this. If too many points are closely located, then the number may also be considered. For example, only the six closest points are selected, as shown in Figure 6.9(a).

In such a selection, there could be a danger that most of the points are located in a single direction. Figure 6.10(a) shows this case. An alternative is to consider the point distribution. That is, the points are partitioned into four (Figure 6.10b)

INTERPOLATION TECHNIQUES FOR TERRAIN SURFACE MODELING 129

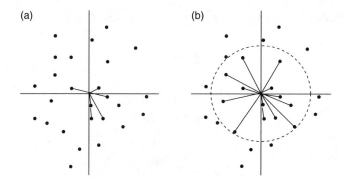

Figure 6.9 Selection of the neighbor points for interpolation: (a) based on number of points and (b) based on a search range.

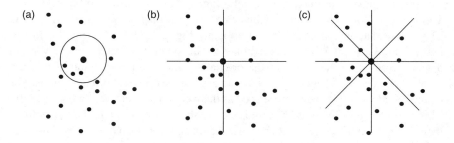

Figure 6.10 Selection of some points from each sector as reference points: (a) no sectors; (b) four sectors; and (c) eight sectors.

or eight (Figure 6.10c) sectors and a number of points from each sector are then selected.

Another alternative, similar to the idea of partitioning, is to generate a Voronoi diagram of the data points, and then those points whose Voronoi regions have common boundaries with the Voronoi region of interpolation points are selected for interpolation. Of course, one may argue that since the Voronoi diagram of the point set has already been constructed, why should we not perform interpolation on the triangular network.

6.5.3 Determination of Weighting Functions

The fundamental idea of assigning a weight to a reference point is to consider the influence of the reference point. A distance function is widely used. That is, the smaller the distance, the larger the weight.

It is obvious that the closer together the observed points are located, the greater their similarity; the farther apart they are, the smaller the similarity. Therefore, distance affects the degree of influence of different points on the elevation of the to-be-interpolated point. So, in moving averaging, one usually adopts a weighting function related to distance. The commonly used weighting functions are as below:

$$w_i = \frac{1}{d_i^2} \tag{6.35}$$

$$w_i = \left(\frac{R - d_i}{d_i}\right)^2 \tag{6.36}$$

$$w_i = e^{-d_i^2/K^2} \tag{6.37}$$

where w_i is the weight for the reference point i; R is the radius of the circle; d_i is the distance of the reference point to the interpolation point; and K is a constant.

An alternative is to make the weight a function of area instead of distance. The Voronoi region of each reference point can be used to construct a weighting function. This determination can be explained by using Figure 5.31 (in Chapter 5). In this figure, suppose point 7 is to be interpolated. This point is inserted into the Voronoi diagram formed by the points of the orignal set and a Voronoi region of the new point is obtained by "stealing" a piece from each of the original Voronoi regions. The amount of area being stolen can be used as the weight for interpolation (Gold 1989). The area corresponding to each point is clearly shown in this figure. For any point in the reference list, if no area of its Voronoi region is stolen, then it exerts no influence on the interpolation point. The intersection of the Voronoi region of the interpolation point and the old Voronoi region of a reference point, say p_i, is the area stolen and is used as the weight for p_i. Mathematically,

$$w_i = V(p_i)_{\text{old}} \cap V(p_{\text{new}}) \tag{6.38}$$

Here, the notation in Section 5.6.2 is followed. $V(p_i)_{\text{old}}$ is the Voronoi region of point p_i before the new point (i.e., the point to be interpolated) is added; $V(p_{\text{new}})$ is the Voronoi region of the new point; and w_i is the weight for point p_i. Sibson (1980, 1981) first described this weighted-average interpolant, which is smooth everywhere except at data points, and has a set of weights that map directly to the set of Voronoi neighbors. Dakowicz and Gold (2002) describe how to use this technique to generate surfaces with meaningful slopes.

6.6 POINT-BASED MOVING SURFACES

In moving averaging, the average value of a number of neighbor points is assigned to the interpolation. In fact, by moving averaging, a moving surface has also been created. Therefore, most of the principles discussed in the previous section also apply in this section. Various types of surfaces can be created for point-based interpolation.

6.6.1 Principles of Moving Surfaces

For a given set of data, different types of surface can be created. If the height of the reference point nearest the interpolation point is assigned to the interpolation point, it is called nearest-neighbor interpolation, although no real interpolation occurs. In this case, a horizontal plane is created using a deterministic function. Figure 6.11(a) shows this type of surface. The mathematical function is:

$$z = z_i, \quad \text{if } d_i = \min(d_1, d_2, \ldots, d_i, \ldots, d_n) \tag{6.39}$$

The graphic illustration of moving averaging is shown in Figure 6.11(b). In fact, it is also a horizontal plane but a stochastic model. The fitting condition is:

$$\sum_{i=1}^{k} \Delta z_i = 0 \tag{6.40}$$

In fact, any function can be used as the model for a moving surface. If the first three terms of the polynomial function are used, then a tilted linear surface is created (Figure 6.11c). But the curved surface (Figure 6.11d) is in common use, such as the second-order polynomial surface

$$z = a_0 + a_1 x + a_2 y + a_3 x^2 + a_4 xy + a_5 y^2 \tag{6.41}$$

The condition for surface fitting is the least squares.

After a surface model is selected, for any interpolation point, say p, a surface is constructed from a set of neighboring reference points by least-squares conditions.

To make the computation more efficient, it is normal practice that a new coordinate system with the interpolation point $p(x_p, y_p)$ as the origin is used. That is,

$$\begin{aligned} x_{i,\text{new}} &= x_i - x_p \\ y_{i,\text{new}} &= y_i - y_p \end{aligned} \tag{6.42}$$

After such a treatment, from Equation (6.41), it can be noted that the height for the interpolation point is

$$z = a_0 \tag{6.43}$$

6.6.2 Selection of Points

The principles for the selection of reference points discussed in the previous section are also applicable in this case. The only thing special here is that the number of reference points must be larger than the number of coefficients involved in the mathematical model. But too many points will be computationally expensive. A compromise is to select about ten points. An adaptive circle radius can be used for point selection.

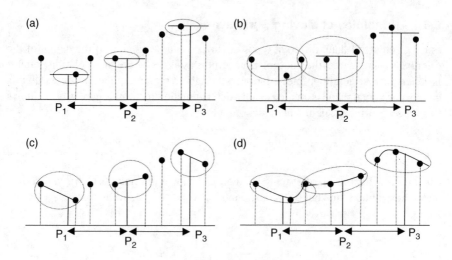

Figure 6.11 Different types of moving surfaces for interpolation: (a) nearest; (b) averaging; (c) linear surface; and (d) curved surface.

The idea is to start from the average density of the data points to determine the radius R of a circle within which there are approximately ten points:

$$\pi R^2 = 10(A/N) \tag{6.44}$$

where N is the total number of the points and A is the total area. This method takes into consideration the two elements for point selection, that is, the number of points and the range.

CHAPTER **7**

Quality Control in Terrain Data Acquisition

Like industrial production, there must be some procedures or methodology for quality management and control in digital terrain modeling.

7.1 QUALITY CONTROL: CONCEPTS AND STRATEGY

7.1.1 A Simple Strategy for Quality Control in Digital Terrain Modeling

The quality of DTM data is usually measured by the accuracy of position and height. However, updatedness (or currency) has also become an important issue. This importance can be illustrated by the generation of DTM from a pair of aerial photographs taken 10 years ago. Although the DTM is of great fidelity to the original terrain, the result may not necessarily be useful if there have been a lot of changes. In this context, it is assumed that the source materials used for digital terrain modeling are not out of date. Therefore, only accuracy is of concern in this chapter.

Quality control is complicated. To build a DTM of high quality, one has to take care of each of the processes in digital terrain modeling so as to eliminate, reduce, or minimize the magnitude of errors that could be introduced. A simple strategy is

1. to minimize errors introduced during data acquisition
2. to apply procedures to eliminate errors and reduce the effect of errors
3. to minimize errors introduced in the surface modeling process.

This chapter is only concerned with the first two. Error propagation in the modeling process will be discussed in Chapter 8.

7.1.2 Sources of Error in DTM Source (Raw) Data

Measured data will always contain errors, no matter which measurement methods are used. The errors in data come from

1. errors in the source materials
2. inaccuracy of the equipment for data acquisition
3. human errors introduced in the acquisition process
4. errors introduced in coordinate transformation and other data processing.

For DTM source data acquired by photogrammetry, errors in source materials include those in aerial photographs (e.g., those caused by lens distortion) and those at control points. Inaccuracy of equipment refers to the limited accuracy and precision of the photogrammetric instrument as well as the limited number of digits used by a computer; human errors include errors in measurement using float marks and typing mistakes; coordinate transformation errors include those introduced in relative and absolute orientation and image matching if automated method is used.

7.1.3 Types of Error in DTM Source Data

Generally speaking, three types of errors can be distinguished, namely,

1. random errors
2. systematic errors
3. gross errors (i.e., mistakes).

In classic error theory, the variability of serious measurements of a single quantity is due to observational errors. Such errors do not follow any deterministic rule, thus leading to the concept of random errors. Random errors are also referred to as random noise in image processing and as white noise in statistics. Random errors have a normal distribution. For such errors, a filtering process is usually applied to reduce their effects. This is the topic of Section 7.3 in this chapter.

Systematic errors usually occur due to distortions in source materials (e.g., systematic distortion of a map), lack of adequate adjustment of the instrumentation before use, or physical causes (e.g., photo distortion due to temperature changes). Alternatively, systematic errors may be the result of the human observer's limitations, for example, stereo acuity or carelessness such as failing to execute correct absolute orientation. Systematic errors may be constant or counteracting. They may appear as a function of space and time. Most practitioners in the area of terrain data acquisition are aware of systematic errors and strive to minimize them.

Gross errors are, in fact, mistakes. Compared with random and systematic errors, they occur with a small probability during measurement. Gross errors occur when, for example, the operator records a wrong reading on the correct point or observes the wrong point through misidentification, or if the measuring instrument is not working properly when an automatic recorder is used. Indeed, gross errors often occur in automatic image matching (due to mismatching of image points).

From a statistical point of view, gross errors are specific observations that cannot be considered as belonging to the same population (or sampling space) as the other observations. Therefore, gross errors should not be used together with the other observations from the population. Consequently, measurement should be planned and observational procedures designed in such a way as to allow for the detection of gross errors so that they can be rejected and removed from the set of observations. The detection and removal of gross error will be discussed in Section 7.4 to Section 7.8.

7.2 ON-LINE QUALITY CONTROL IN PHOTOGRAMMETRIC DATA ACQUISITION

On-line quality control is to examine the acquired data during the process of data acquisition and to correct errors immediately if any. Visual inspection is an approach often widely used in practice. Four methods will be introduced in this section. However, the last three can be used for either on-line quality control or off-line quality checking.

7.2.1 Superimposition of Contours Back to the Stereo Model

In practical applications, on-line quality control in photogrammetric data acquisition is achieved by superimposition of contour lines back to the stereo model to examine whether there is any inconsistency between contour lines and the relief on the stereo model. The contour lines are generated from the data just acquired from the stereo model. If no inconsistency is found, it means that no gross errors occurred. However, if there is a clear inconsistency somewhere, it means that there are gross errors and it will be necessary to edit the data and remeasure some data points. Ostman (1986) called such an accessory system a graphic editor for DTMs.

Another method is to superimpose the contours interpolated from the measured DTM data onto the orthoimages to inspect whether there is mutation of contours, or to compare them with topographic maps and terrain feature points and lines. When there is relatively great difference of the landforms or elevations of the points, they need to be measured again and to be edited till the DTM data meet the requirements. This method is limited to the inspection of the gross errors.

7.2.2 Zero Stereo Model from Orthoimages

An alternative is to compare the orthoimages made from both left and right images. If both orthoimages are made using the DTM obtained from the same stereo pair through image matching and there are no obstacles (i.e., buildings and trees) in this area, then the two orthoimages will form a zero stereo model (i.e., no height information in the model) if the DTM used for orthoimage generation is free of errors. Zero stereo model also indicates that no x-parallax can be observed anywhere on the model. If parallax does exist, it may result from:

1. errors in the orientation parameters, leading to inconsistency of the left and right orthoimages
2. something wrong with the image matching of the orthoimage pair
3. errors in the DTM data used for the orthoimage generation.

If the first two possibilities are excluded, then any parallax appearing on the orthoimage pairs is the direct reflection of the errors of the DTM data.

7.2.3 Trend Surface Analysis

Most terrains follow certain natural trends, such as continuous gradual spatial change. The shape of the trends may vary with the genesis of landforms. The continuous change of terrain surfaces may be described by smooth mathematical surfaces, referred to as trend surfaces.

A typical trend surface analysis will reveal the greatest deviations from the general trend in the area. As data points with gross errors appear to be abnormal, the deviations of data values from the general trend will be obvious. In other words, gross errors are detected if great deviations from the trend surface are found.

There are different types of trend surfaces. One of them is the least-square trend surface as follows:

$$Z(x,y) = \sum_{k=0}^{j} \sum_{i=0}^{k} a_{ki} x^{k-i} y^i, \quad j = 1, 3 \qquad (7.1)$$

The number of terms to be used for this polynomial function should be selected according to the size and complexity of the area of interest. For example, in a large and complex area, a higher-order polynomial should be used.

The critical issue is to set a threshold so that any point with a deviation larger than this threshold will be suspected of having gross errors. In practice, a value of three times the standard deviation is often regarded as a reliable threshold. However, due to the instability of higher-order polynomial surfaces over rough terrain with irregularly distributed data points, a large deviation does not necessarily mean gross errors at the point. This is the limitation of trend analysis.

7.2.4 Three-Dimensional Perspective View for Visual Inspection

A fourth method is to create a 3-D surface from the DTM data for interactive visual inspection. In this way, those points that look unreasonable can be regarded as gross errors and removed from the data set.

The visualization of a 3-D surface from DTM data is an important application of DTM and will be addressed in Chapter 12. To create a 3-D surface for visual inspection, a TIN model can be constructed directly from all of the original data points so as to ensure that all the analyses will be based on the original data. For efficiency, it is recommended that a wire net perspective display based on the TIN is created for a local area around the point to be inspected. Figure 7.1 shows the visual inspection of contour data. The spikes indicate that wrong height values have been assigned to some contour points. Such an inspection is intuitive but the results are likely to be reliable.

7.3 FILTERING OF THE RANDOM ERRORS OF THE ORIGINAL DATA

Because DTM products are obtained after a series of processes from the DTM source (raw) data, the quality of the source data will greatly affect the fidelity of the DTM

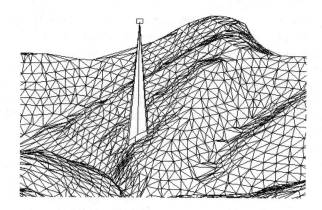

Figure 7.1 Wire net perspective view of the area around the suspected point.

surfaces constructed from such data and the products derived from the DTM. The quality of the source data can be judged by using their three attributes (i.e., density, distribution, and accuracy) as criteria. The quality of a set of data can be considered as being poor if the data are not well distributed, for example, very few scattered points in areas of rough and steep terrain but with a high density of points on relatively smooth and flat areas. However, these first two factors, that is, density and distribution, are related to sampling and the problems can be solved by employing an appropriate sampling strategy.

An important factor for the quality of DTM source data is its inherent accuracy. The lower the accuracy, the poorer the data quality. Accuracy is primarily related to measurement. After a set of data points have been measured, an accuracy figure can be obtained or estimated. The accuracy figure obtained for any measured data set is the overall result of different types of errors. The purpose of this section is to devise filtering techniques to eliminate or reduce the effects of some of these errors so as to improve the quality of the final DTM and thus its products.

7.3.1 The Effect of Random Noise on the Quality of DTM Data

Any spatial data set can be viewed as consisting of three components: (a) regional variations, (b) local variations, and (c) random noise. In digital terrain modeling, the first is of most interest, because it defines the basic shape of the terrain. Interest in the second varies with the scale of the desired DTM products. For example, at a large scale, it is extremely important. However, if a small-scale contour map covering a large region is the desired product, then this component may be regarded as random noise because less detailed information about the terrain variations is needed. By contrast, the third component is always a matter of concern since it may distort the real picture (appearance) of both regional and local variations on the terrain, but especially the latter. As a matter of fact, it is difficult to clearly define these three components. Generally speaking, the high-frequency component of the data can be regarded as random noise.

It is important to separate the main components of the data set that are of interest to the user from the remainder of the information present in the data set, which is regarded as random noise. The technique used for this purpose is referred to as filtering and the device or procedure used for filtering is referred to as a filter. A digital filter can be used to extract a particular component from a digital data set, thus ensuring that all other components are filtered out. If a digital filter can separate the large-scale (low-frequency) component from the remainder, this filter is called as a low-pass filter. By contrast, if a digital filter can separate the small-scale (high-frequency) component from the remainder, then this filter is referred to as a high-pass filter. However, here only the low-pass filter is of interest since it is the high-frequency component that needs to be filtered out.

Before discussing how to filter out random noise, it is necessary to know how random noise affects the quality of the DTM and its products.

Ebisch (1984) discussed the effect of round-off errors found in grid DTM data on the quality of the contours derived from the DTM data, and he also demonstrated the effect of random noise in the DTM data on the contours produced from it. Ebisch first produced smooth contours (Figure 7.2a) with 1.0-m intervals from a conical surface represented by a grid of 51 by 51 points. Then, he rounded-off the grid heights to the nearest 0.1 m to produce another contour map (Figure 7.2b) to show the effect of round-off error. After that, he added random noise with a maximum amplitude of ±0.165 m to the grid, producing a contour map with zigzag and meandering contour lines (Figure 7.2c). Figure 7.2(d) shows the contours produced from the DTM data with both round-off and added random errors. This figure shows the effects of random noise on the quality of DTM source data and the quality of the contours derived from these data.

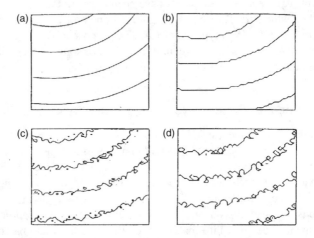

Figure 7.2 Effect of round-off errors and random noise on the contours produced from the data set (Reprinted with permission from Ebisch 1984). (a) Contours produced from the original data set (a smooth surface). (b) Contours produced from the data set after rounding off the decimal fraction of original DTM data. (c) Contours produced from the data set with a random noise of magnitude ±0.165 m added. (d) Contours produced from the data set with both random noise and round-off errors included.

QUALITY CONTROL IN TERRAIN DATA ACQUISITION

7.3.2 Low-Pass Filter for Noise Filtering

A low-pass filter is usually implemented as a convolution procedure, which is an integral expressing the amount of overlap of one function (X) as it is shifted over another function (f). Convolution can take place either as a 1-D or a 2-D operation. However, the principles are the same in both cases. Therefore, for simplicity, the 1-D convolution is presented here.

Suppose $X(t)$ and $f(t)$ are two functions, and the result of convolving of $X(t)$ with $f(t)$ is $Y(t)$. Then, the value of $Y(t)$ at position u is defined as:

$$Y(t) = \int_{-\infty}^{+\infty} X(t) f(u - t) dt \tag{7.2}$$

In DTM data filtering, $X(t)$ refers to the input terrain data containing random errors; $f(t)$ can be considered as a normalized weighting function; and $Y(t)$ comprises the low-frequency components of the terrain variations present in the input data and is the remaining part after filtering out random noise. Practically, it is not necessary to have the integration from negative to positive infinity for Equation (7.2). In most cases, an integral that operates over a restricted length will do.

Certain functions such as a rectangular function, a triangular function, or a Gaussian pulse can be used as the weighting function for this purpose. The Gaussian function is more widely used. The expression is:

$$f(t) = e^{(-t^2/2\sigma^2)} = \exp(-t^2/2\sigma^2) \tag{7.3}$$

The definition of convolution given in Equation (7.3) applies to continuous functions. However, in DTM practice, the source data are only available in a discrete form. Therefore, only the discrete convolution operation is of interest here. The principle of the operation is to use a symmetric function as a weighting function. It will be used here as the weighting function, since the Gaussian function is symmetric. Its principle as applied in 1-D is explained below. Suppose,

$$X(t) = (A_1, A_2, A_3, A_4, A_5, A_6, A_7)$$

$$f(t) = (W_1, W_2, W_3, W_4, W_5)$$

and

$$Y(t) = (B_1, B_2, B_3, B_4, B_5, B_6, B_7)$$

Then, the discrete convolution operation can be illustrated in Table 7.1. To explain how it works, the result for B_4 can be taken as an example,

$$B_4 = W_1 \times A_2 + W_2 \times A_3 + W_3 \times A_4 + W_4 \times A_5 + W_5 \times A_6$$

The size of the window and the weights selected for the various data points lying within the window have a great effect on the degree of smoothing achievable by the

Table 7.1 Discrete Convolution Operation

$X(t)$	0	0	A_1	A_2	A_3	A_4	A_5	A_6	A_7	0	0	
Operation	×	+ ×	+ ×	+ ×	+ ×	+ ×	+ ×	+ ×	+ ×	+ ×	+ ×	**Results**
	W_1	W_2	W_3	W_4	W_5							B_1
		W_1	W_2	W_3	W_4	W_5						B_2
			W_1	W_2	W_3	W_4	W_5					B_3
$f(t)$				W_1	W_2	W_3	W_4	W_5				$= B_4$
					W_1	W_2	W_3	W_4	W_5			B_5
						W_1	W_2	W_3	W_4	W_5		B_6
							W_1	W_2	W_3	W_4	W_5	B_7

Table 7.2 Sample Values of the Gaussian Function as Weights for Convolution

t	$0.0 \times \sigma$	$0.5 \times \sigma$	$1.0 \times \sigma$	$1.5 \times \sigma$	$2.0 \times \sigma$	$3.0 \times \sigma$
$f(t)$	1.0	0.8825	0.6065	0.3247	0.1353	0.0111

convolution operation. For example, if only one point is within the window, then no smoothing will take place. Also, the smaller the differences in the weights given to the points lying within the window, the larger the smoothing effect it will have. For example, if the same weight is given to each point within the window, then the result is simply the arithmetic average. Table 7.2 lists some of the values for the Gaussian pulse expressed by Equation (7.3). From these values, a variety of weighting matrices may be constructed. The weight matrix can also be computed directly from Equation (7.3) using predefined parameters.

7.3.3 Improvement of DTM Data Quality by Filtering

Li (1990) conducted a test on the improvement of DTM data quality by noise filtering. The source data was generated by using a completely digital photogrammetric system. The digital photos used in the system were formed from a pair of aerial photos taken at a scale of 1 : 18,000 using a scanning microdensitometer with a pixel size of 32 μm. The data were measured in a profiling mode with a 4-pixel interval between measured points; thus, the interval between any two data points is 128 μm on the photo. The data points acquired from image matching produced a data set only approximately in a grid form in this test area, with grid intervals of about 2.3 m. The data are very dense. In an area of 1 cm × 1 cm at photo scale, approximately 8588 (113 × 76) points were measured. This data set provides very detailed information about the terrain roughness. The check points used for this study were measured from the same photos in hardcopy form using an analytical instrument.

A filter based on the convolution operation described above was used for this test. Since the original data were not in an exact grid, a 1-D convolution was carried out on each of the two grid directions rather than a single 2-D operation. Therefore, for each point, the average of the two corresponding values was used as the final result. The window size comprises five points in each grid direction. The five weights for

QUALITY CONTROL IN TERRAIN DATA ACQUISITION

Table 7.3 Accuracy Improvement with Random Noise Filtering (Li 1990)

Parameters	Before Filtering	After Filtering
+Maximum residual (m)	+3.20	+2.67
−Maximum residual (m)	−3.29	−2.76
Mean (m)	0.12	−0.02
Standard deviation (m)	±1.11	±0.98
RMSE (m)	±1.12	±0.98
No. of check points	154	154

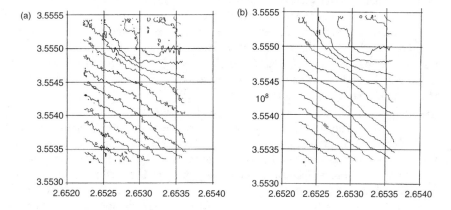

Figure 7.3 Improvement of data quality by using a low-pass filter (Li 1990). (a) Contours generated from the original data. (b) Contour generated from the smoothed data.

these five points were computed according to Equation (7.3) individually since the point intervals varied. These values before normalization were approximately:

$$f(t) = (0.135,\ 0.6065,\ 1.0,\ 0.6065,\ 0.1353) \tag{7.4}$$

In computing the value for each of these five weights corresponding to each of the five points lying within the window, the distance of the point to the central point of the window was used as the value of the variable t in Equation (7.2). Also, the average value of the intervals between each pair of data points (i.e., 2.3 m) was used as the variable σ. Table 7.3 shows the comparison between the accuracy of the experimental DTM data before and after filtering. It is clear that the improvement in RMSE was about 17%. Figure 7.3 shows the corresponding contours before and after filtering. It can be seen clearly that the small fluctuations in the contour shapes arising from noise in the data have to a large extent been removed after the filtering. Therefore, the presentation of the contours after filtering is also much better visually.

7.3.4 Discussion: When to Apply a Low-Pass Filtering

The data set used in this study was very dense. Realistically, such a dense data set can only be obtained from devices equipped with automated or semiautomated techniques, for example, using image-matching techniques based on automatic image correlation. In such a data set, loss in the fidelity of the representation of terrain topology is not likely to be a serious problem. By contrast, the effect of random errors involved in the measuring process and of any other random noise on the data quality is considerable at the local or detailed level.

From the study it is clear that the availability of too detailed information about the roughness of the terrain topography, coupled with the measuring errors likely to be encountered with image-matching techniques, can have a significant negative effect on DTM data quality and thus on the quality of derived DTM products. Therefore, with dense data, a filter such as that based on a convolution operation can be used to smooth the digital data set and improve the quality.

An important question arising is: "when should a filtering process be applied to digital data?" That is also to say, "under what circumstances is it necessary to apply a filtering process to the data?" This is a question very difficult to answer. The magnitude of random errors occurring during measurement needs to be taken into consideration. From the literature it can be found that 70 to 90% of photogrammetric operators are measuring with a precision (RMSE) within the range ± 10 to 20 μm (Schwarz 1982). This could be a good indicator. Alternatively, according to Kubik and Roy (1986), 0.05‰ of H (flying height) might be regarded as an appropriate value. Therefore, a rough answer to this question might be that if the accuracy loss arising from data selection and reconstruction (topographic generalization) is much larger than this value (0.05‰ of H), then a filtering process is not necessary. In contrast, if random noise does form an important part of the error budget, then a filtering process may be applied to improve data quality.

7.4 DETECTION OF GROSS ERRORS IN GRID DATA BASED ON SLOPE INFORMATION

Often the presence of gross errors will distort the image (i.e., the appearance) of the spatial variation present in DTM data sets much more seriously than that resulting from random noise. In some cases, totally undesirable results may be produced in the DTM and in the products derived from it, due to the existence of such errors. Therefore, methods are needed to detect this type of errors in DTM data set and to ensure their removal from the data set. In Section 7.2, some on-line methods were described and from this section onward, some off-line methods will be presented.

DTM source data may be either in a regular grid or irregularly distributed. Regular grid data have a certain special characteristic. That is, they can be stored in a concise and economic form in a height matrix. This can also help in designing an algorithm for gross error detection. However, an algorithm suitable for application to grid data is unlikely to suit irregularly distributed data. Therefore, different approaches need to be taken for the detection of gross errors in each of these two cases. In this section, algorithms for the detection of gross errors in a regular grid data set are developed while two algorithms for detecting gross errors in irregularly distributed data will be presented in Section 7.5 and Section 7.6.

QUALITY CONTROL IN TERRAIN DATA ACQUISITION

7.4.1 Gross Error Detection Using Slope Information: An Introduction

To develop algorithms for gross error detection, the first question is "what kind of information can be used for this purpose?" Slope is the fundamental attribute of a terrain point and, therefore, slope information can probably serve as the basis for the development of suitable algorithms. The second problem to be considered is the feasibility of obtaining the slope information from the data set. The computation of the slope of each grid point in different directions does not present a real problem. In this view, it appears promising to make use of slope information as the basis for developing algorithms for detection of gross errors. Hannah (1981) and Li (1990) both have used slope information for such a purpose.

Hannah (1981) developed an algorithm for the reduction of gross errors, based on the absolute slope values. The principle of Hannah's algorithm can be described briefly as follows. As a first step, the slopes between the point under investigation, say P, and its neighbors (eight if not located on the boundary) are computed. Once this has been done for the whole data set, three tests are carried out on the slopes.

1. The first test, called a *slope constraining test*, checks the (eight) slopes immediately surrounding P to see if they are reasonable, that is, whether they exceed the predefined threshold value or not.
2. The second, called the *local neighbor slope consistency test*, checks the four pairs of slopes crossing P to see if the absolute value of the difference in slope in each pair exceeds the given threshold value.
3. The third, called the *distant neighbor slope consistency test*, is similar to the second test. This test checks whether pairs of slopes approaching a point across each of the eight neighbors are consistent.

The results of these three tests are used as the basis to judge whether a point is accepted or rejected. It has been found (Li 1990) that this algorithm produces an over-smoothing result in areas of rough terrain in order to detect the gross errors or other unnatural features of relatively smooth terrain.

Li (1990) further pointed out that the most serious demerit of Hannah's algorithm is that all the criteria for acceptance or rejection of data are expressed in an absolute sense. Obviously, absolute slope values and slope differences will vary from place to place. For example, in an area with rough terrain, absolute slope differences will be larger than those found in smooth areas. Absolute values of slopes in steep areas will be larger than those found in flat areas. That is, it is not feasible for an absolute threshold value to be suitable overall for an area of interest except in a homogenous area. For this reason, Li (1990) tried to make relative thresholds for his algorithm development and his algorithm will be presented in the following sub-sections.

7.4.2 General Principle of Gross Error Detection Based on an Adaptive Threshold

The algorithm developed by Li (1990) is based on the concept of slope consistency. Instead of absolute values of slope and slope changes, relative values are considered.

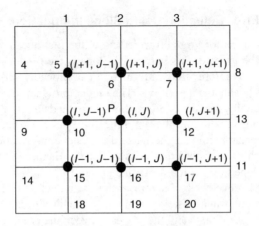

Figure 7.4 Point P in the original grid data and its neighbors.

Furthermore, a statistic is taken from these relative values and is then used as the threshold value to measure the validity of a data point instead of using a predefined value. Thus, this algorithm is adaptive to any data set.

As shown in Figure 7.4, data point P can be defined by its row and column number, (I, J) within the height matrix. Its eight immediate neighbors — points 5, 6, 7, 10, 12, 15, 16, and 17 — can also be defined by row and column as $(I+1, J-1)$, $(I+1, J)$, $(I+1, J+1)$, $(I, J-1)$, $(I, J+1)$, $(I-1, J-1)$, $(I-1, J)$, and $(I-1, J+1)$. From these eight points and point P itself, six slopes can be computed in both the row (i.e., I) and the column (i.e., J) directions. Taking the row direction as an example, six slopes — those between points 5 and 6, 6 and 7, 10 and P, P and 12, 15 and 16, as well as 16 and 17 — can be computed. From the set of six slope values, three slope changes can then be computed. For example, the slope changes at points 6, P, and 16 can be computed from those values mentioned above. These initial values are given in an absolute sense and will vary from place to place. Therefore, some relative values need to be computed from them.

If there is no gross error at point P, then for the same direction (e.g., the row direction), the difference in the slope change (DSC) at point P and that at its immediate neighbor (e.g., point 6 or 16) located in the row direction will be consistent, even though the absolute values of slope and slope change may vary from place to place. Therefore, these differences in slope change are the relative values that are being searched for and can be used as the basis of a method for detecting gross errors.

That is, for each point except those along the boundary, two DSC values can be computed from the three slope changes in each direction. The DSC values from all the data points will be used as the basic information for this algorithm. From these DSC values, a statistic will be computed and it will then be used to construct the required threshold value. Then, this threshold value will be used as the basis on which a judgment is made as to whether or not a point has a gross error in elevation. For example, if all four DSC values centred at P exceed the threshold value, then P will be suspected of containing gross errors.

7.4.3 Computation of an Adaptive Threshold

In this computation, first of all, the DSC values in both row and column directions are computed and then these DSC values are used to compute an adaptive threshold. The computation of the slope (Slope) in the J direction, for example, is as follows:

$$\text{Slope}_j(I+1, J-1) = \frac{Z(I+1, J) - Z(I+1, J-1)}{\text{Dist}(J-1, J)} \quad (7.5)$$

where $\text{Dist}(J-1, J)$ is the distance between the nodes at $(I+1, J)$ and $(I+1, J-1)$, that is, equal to the grid interval.

Similarly the values $\text{Slope}_j(I+1, J)$, $\text{Slope}_j(I, J-1)$, $\text{Slope}_j(I, J)$, $\text{Slope}_j(I-1, J-1)$, and $\text{Slope}_j(I-1, J)$ can be calculated. The computation of slopes in the other (I) direction is similar. After calculating the slopes, three slope changes (SlopeC) in each direction can be computed. For example, in the J direction, the computation is as follows:

$$\text{SlopeC}_j(I, J) = \text{Slope}_j(I, J) - \text{Slope}_j(I, J-1) \quad (7.6)$$

Also, $\text{SlopeC}_j(I+1, J)$ and $\text{SlopeC}_j(I-1, J)$ can be computed similarly. The computation of slope changes in the I direction is also similar. After this, two differences in slope change (DSlopeC) for the point (I, J) in each direction can be computed as follows:

J direction:

$$\begin{aligned}\text{DSlopeC}_j(I, J, 1) &= \text{SlopeC}_j(I, J) - \text{SlopeC}_j(I+1, J) \\ \text{DSlopeC}_j(I, J, 2) &= \text{SlopeC}_j(I, J) - \text{SlopeC}_j(I-1, J)\end{aligned} \quad (7.7)$$

I direction:

$$\begin{aligned}\text{DSlopeC}_j(I, J, 1) &= \text{SlopeC}_j(I, J) - \text{SlopeC}_j(I, J+1) \\ \text{DSlopeC}_j(I, J, 2) &= \text{SlopeC}_j(I, J) - \text{SlopeC}_j(I, J-1)\end{aligned} \quad (7.8)$$

All DSC values calculated from all the data points will be used for computation of a statistic, which will then be used as threshold for acceptance or rejection of the point. For such a statistic, the absolute mean, the range (biggest minus smallest), the mode, the RMSE, as well as the standard deviation and mathematical mean are all possible options. Li (1988) made a thorough analysis on the possible statistics and Li (1990) made some observations from experiments and then concluded that RMSE is as good as the combination of the mathematical mean and standard deviation. Thus, the threshold value is simply K times RMSE of the DSC values, that is,

$$\text{DSC}_T = K \times \text{RMSE}(\text{DSC}) \quad (7.9)$$

where K is a constant. It has been found that the DSC values are quite normally distributed and thus a value of 3 has been used for the constant K. There are three possible ways to compute RMSE values:

1. Compute the only RMSE value from all DSC values at all the data points in all directions.
2. Compute four RMSE values from the DSC values at all the data points, one for each of the four directions (above, left, below, and right) defining each data point.
3. Compute two RMSE values, one of which is related to the row (i.e., the I) direction and the other to the column (i.e., the J) direction. In this case, the two DSC values of each point in the same direction, say the J direction, are added together to get a new value and the RMSE value can be computed from these new values.

Theoretically, the last method is most reasonable because the absolute value of a sum of the two DSC values at the same point (e.g., P) in the same direction (e.g., in the J direction) will become smaller (approaching 0) if the slope change is consistent, and it will become larger if it is not consistent. Li (1990) tried different criteria and his results proved this point.

7.4.4 Detection of Gross Error and Correction of a Point

All the methodology described above is designed to judge whether or not a point has a gross error. A particular threshold value for an individual direction is used as the basis for judgment. If the threshold is exceeded, then this point is regarded as unnatural in the neighborhood and is suspected of having a gross error.

The procedures used for detecting a gross error in all the methods described above are similar. The only difference is to compare the DSC values with the overall RMSE or a particular RMSE value. Taking the second method as an example, if the absolute value of the DSC at a point along a single side is greater than the threshold value — K times the RMSE of this side — then the point is suspected of being unnatural compared to the values in the neighborhood. If all four sides around the point are suspect, then this particular point will be suspected of having a gross error. In most cases, if three sides of a point are suspected, then again it is regarded as having a gross error. For the last method, if a point is suspected in both the row and the column directions, then it is regarded as containing a gross error.

It is possible that some gross errors have not been detected in a single run if they are located close together, in that case, a point that has a gross error may still be considered natural if its neighbors also have gross errors of a similar magnitude. This means that a further detection of the remaining gross errors may be necessary. To improve the results in the next run, correction of those points found to have gross error must be done first. The principle of data correction used in this algorithm is as follows: In Figure 7.4, suppose that point P is the point containing a gross error, and points 1 to 20 are its neighbors. In the process of detecting gross errors, the slope and slope change values at all these points have been calculated (except those points near boundaries). From points 6, 16, 10, and 12, four estimates have been made. The estimate from Point 10 may be taken as an example. The average of the slope

QUALITY CONTROL IN TERRAIN DATA ACQUISITION

change values at points 5 and 15 (in the J direction) are taken as the estimated slope change at point 10 (in the same direction). The new slope at point 10 (to P) can then be computed as follows:

$$\text{Slope}(10, J) = \text{Slope}(9, J) + \frac{\text{SlopeC}(5, J) + \text{SlopeC}(15, J)}{2} \qquad (7.10)$$

where $\text{Slope}(10, J)$ and $\text{Slope}(9, J)$ denote the slopes at points 10 and 9 in the J direction, respectively, and $\text{SlopeC}(5, J)$ and $\text{SlopeC}(15, J)$ denote the slope changes at points 5 and 15 in the J direction, respectively.

These slope values are used to compute the height values of point P. Finally, the average of four such estimates is used as the height estimate for point P. If either points 9 or 10 is suspected of having a gross error, or if other neighbors in this side (points 4, 5, 6, 14, 15, and 16) are suspected of having gross errors, then any estimate from this side will be unreliable and should not be used. It is also possible that no reliable estimate can be made for point P in a single run. Therefore, some form of interactive processing is needed.

7.4.5 A Practical Example

Figure 7.5 shows how this algorithm works. Figure 7.5(a) shows the contours produced from a set of original DTM data. Clearly, some residual errors exist producing unnatural features in some of these contours. After applying gross error detection procedures, the corresponding contour plot, as shown in Figure 7.5(b), illustrates that these unnatural features have been removed.

7.5 DETECTION OF ISOLATED GROSS ERRORS IN IRREGULARLY DISTRIBUTED DATA

In an irregularly distributed data set, the information that is conveniently available to users is the set of X, Y, Z coordinates of the data points. Therefore, the height for every data point and its neighbors can be used to assess the validity of the data elevations. The algorithm to be described in this section (Li 1990) is based on this height information.

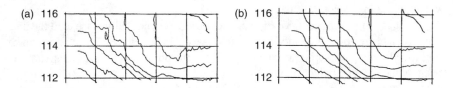

Figure 7.5 An example of contours produced from the data set before and after gross error removal: (a) contours generated from original data and (b) contours generated after removal of gross errors.

Gross errors may be scattered as isolated occurrences or they might occur in clusters. In the latter case, the situation is more complicated. In this section, an algorithm for detecting individual gross errors scattered in the data set will be discussed. Then, in Section 7.6, this algorithm will be modified to suit the detection of gross errors occurring in clusters.

7.5.1 Three Approaches for Developing Algorithms for Gross Error Detection

Depending on the size of the area, three approaches can be distinguished for the development of algorithms for gross error detection, namely global, regional, and local approaches.

Any method using a global approach must involve the construction of a global surface through all points in the data set using a high-order polynomial function and then checking the deviation of each data point from the constructed surface. If the deviation is greater than the threshold value, then this point is considered to have a gross error. The threshold value might be predefined or it may be computed from the deviations of the heights of the data points from the global surface. This was discussed in Section 7.2 as tread surface analysis. Global techniques, as Hannah (1981) pointed out, "have the drawback that they give identical treatment to all areas." However, terrain is rarely uniform in roughness, so the uniform application of a global technique to an area may result in too many points being regarded as having gross errors in rough areas, whereas in fact they do not, while failing to detect gross errors in relatively smooth areas. That is to say, the final result could be totally undesirable or misleading.

The methods employed in a regional approach could be similar to those used in the global approach, that is, constructing a regional surface by use of a polynomial function and then checking the deviations of the data points from the specific surface. The difference is the size of the area of the surface. The adequacy of this approach depends partly on this.

A major drawback in using a polynomial function to fit the terrain surface, regardless of the size of the area that such a surface covers, is that those points that have gross errors will also have been used to construct the DTM surface. In this case, all points near the particular point with a large gross error may have large deviations from the constructed DTM surface due to the large influence of the erroneous point on the constructed surface. Thus, they may all be identified as having gross errors when in fact this is not the case.

If a local approach is employed, then the use of a polynomial surface to fit the data points can be avoided. A method similar to that used in pointwise interpolation can be employed. This involves comparing the height of the point with a representative value such as the average height derived from the heights of its neighbors. As a result, if the difference is larger than a certain threshold value, then this point can be regarded as having a gross error.

The principle of the pointwise method is very simple and intuitive and the computation is also not complex. A simple algorithm that has been developed by

Li (1990) will be described in this section. Felicísimo (1994) has also developed a similar algorithm but it is omitted here.

7.5.2 General Principle Based on the Pointwise Algorithm

For a specific point P, a window of a certain size is first defined centered on P. Then, a representative value will be computed from all the points located within this window. This value is then regarded as an appropriate estimate for the height value of the point P. Or this value can be regarded as the *true value* of point P. By comparing the measured value of P with the representative value estimated from the neighbors, a difference in height can be obtained. If this difference is larger than the computed threshold value, then this point is suspected of having a gross error. The computation of the threshold will be discussed later.

In this method, the height of point P is not taken into consideration when computing the representative value for P. Therefore, the height of point P has no influence on the estimated value derived from the neighbors. This provides more reliable information about the relationship between P and its neighbors.

7.5.3 Range of Neighbors (Size of Window)

The range of the area within which neighboring points will be searched for is specified by a window centered on point P. This can be specified by defining either an area or the number of nearest points required. The former can be expressed as follows:

X range:

$$X_P - D_x < X_i < X_P + D_x \tag{7.11a}$$

Y range:

$$Y_P - D_y < Y_i < Y_P + D_y \tag{7.11b}$$

where X_P and Y_P are the coordinates of P — the point under inspection; X_i and Y_i are the X and Y coordinates of the ith point in the neighborhood; and D_x and D_y are the half-window sizes in the X and Y directions, respectively.

Also, a combination of both criteria can be used. The average window size can be computed according to the total number of points and the coordinate range of the area. This average value can be used as the initial window size. In an area with a higher density of points, the number of points lying within a window of this size will be larger than average. However, in a lower-density area, only a few points may be located inside such a window. Therefore, a minimum number of points may also need to be specified. If the number of points within a window is smaller than the specified value, then the window is enlarged a little until the specified number of points is reached.

7.5.4 Calculating the Threshold Value and Suspecting a Point

In this algorithm, the average height of the neighbors is used as the representative value. This value can be computed in either of two ways. One is to simply take the arithmetic mean and the other is to use a weight for every point according to its distance from the point, making the weight inversely proportional to the distance.

The weighted mean should be closer to the real value of P if there are no gross errors in the neighborhood. However, if a point with a large gross error is close to P, then the weighted mean will be greatly affected by this point, thus producing an unreliable value. Therefore, the simple arithmetic mean may be more desirable. In fact, practical tests confirm this. In addition, the calculation of the simple arithmetic mean takes much less computation time. It is, therefore, used in this algorithm.

In this algorithm, the height differences of all points are used to compute a statistical value, which will then serve as the basis for determining a threshold value. Suppose M_i is the arithmetic mean of the neighboring points centered at the ith point in the data set and the difference between the M_i and the height value of this (ith) point (H_i) is V_i, then

$$V_i = H_i - M_i \tag{7.12}$$

If the data set has N points, then the total number of V values is also N. The required statistical value can be computed from these values of V. In this study, the mathematical mean (μ) and standard deviation (σ) are computed from V values and are then used as the basis for calculating the threshold value:

$$V_{\text{Threshold}} = K \times \sigma \tag{7.13}$$

where K is a constant and in this algorithm $K = 3$.

After the threshold value has been set, every point in this data set can be checked. For any point i, if the absolute value of $(V_i - \mu)$ is larger than this threshold value, it is suspected of containing a gross error.

7.5.5 A Practical Example

The distribution of the data set and the contours generated from it are shown in Figure 7.6. Figure 7.6(a) shows the irregular distribution of the data points. Figure 7.6(b) (the corresponding contour plot) shows clearly that there are gross errors in the data set, especially in the upper-left corner. The size of this area is about 4.5 cm × 4.5 cm on the photo and about 800 m × 800 m on the ground. Within this area, the height value of 3496 points was measured by image matching.

In this example, the simple arithmetic mean was used as the representative value derived from the neighboring points while the window size was defined by the combination of specifying an area size and a certain number of points. The minimum number of points was initially defined as five. As a result, the algorithm did not work well. The number was gradually increased and it was found that a number between 15 and 20 gave the best results.

QUALITY CONTROL IN TERRAIN DATA ACQUISITION

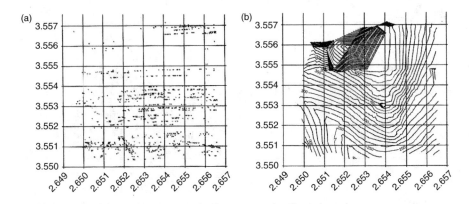

Figure 7.6 Distribution of acquired points with gross errors. (a) The distribution of data points. (b) Unnatural contour produced.

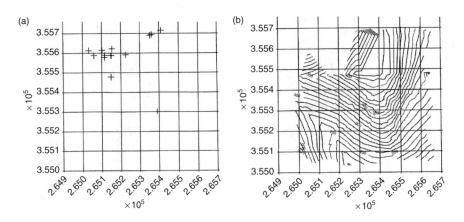

Figure 7.7 Detection and removal of gross errors from a data set: (a) gross error points detected and (b) the data set after removal of gross errors.

After applying this algorithm, those points that generated the unnatural contours were detected and their locations plotted in Figure 7.7(a). The data points after removal of those erroneous points were used to produce the contour map shown in Figure 7.7(b). It can be seen that this algorithm worked well.

7.6 DETECTION OF A CLUSTER OF GROSS ERRORS IN IRREGULARLY DISTRIBUTED DATA

7.6.1 Gross Errors in Cluster: The Issue

It must be pointed out that the algorithm described in the previous section works only in the case of isolated gross errors. When the gross errors are in cluster, the algorithm

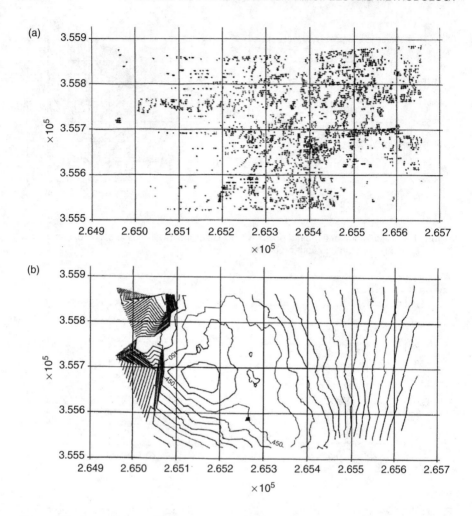

Figure 7.8 A set of data with gross errors in cluster. (a) Distribution of the original data set. (b) Contours from the original data set with gross errors.

will fail to work. Figure 7.8(a) shows the distribution of a set of data obtained from image matching. The corresponding contour plot is shown in Figure 7.8(b). The size of this area is about 4.0 cm × 2.2 cm on the photo and about 700 m × 400 m on the ground. In total, the elevation values of 4733 points were available within this area. Some gross errors are present in the data set, as can be seen clearly from Figure 7.8(b).

The algorithm with all the parameters and window size set in the previous example was applied and the result is shown in Figure 7.9. From the contour plot, it can be seen that there are still some gross errors left in the data set. A much larger window size (containing 60 points) was used, but the algorithm still failed to detect gross errors because the remaining gross errors occurred in clusters. In this section, an algorithm for detection of such errors in cluster developed by Li (1990) will be presented.

QUALITY CONTROL IN TERRAIN DATA ACQUISITION

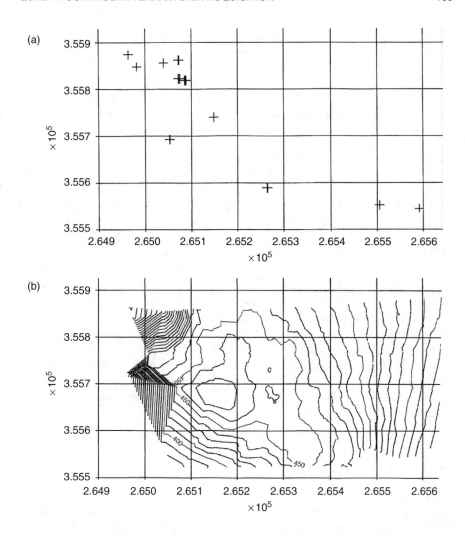

Figure 7.9 Result after removal of gross errors in isolation. (a) Gross error points detected. (b) Contours from the data set after removal of isolated gross errors.

7.6.2 The Algorithm for Detecting Gross Errors in Clusters

Theoretically, the use of an increased window size should solve the problem of detecting gross errors in clusters. However, an algorithm may still fail to work with a window size of up to 60 points, as discussed in the previous section. If the window size is increased more, it might work in some cases, but the results may not be satisfactory since the representative value derived from the neighboring points may then have deviated from what it should be, thus leading to an erroneous conclusion being made. There should be an alternative solution.

The idea behind the development of this algorithm is to find those points that have the most influence on the representative value, that is, the average value in this case.

Then, these points would not be used to compute the representative value. The method used for point data snooping in a window is similar to the idea used in the previous algorithm. The procedure used is as follows.

Take the first point out of the window and calculate a new value of the representative value from the remaining points; then, compute and record the difference between this average and the original one. This procedure is then applied to every point in the window. Suppose there are N points in the window, then N differences can be obtained as follows:

$$V_i = M_{P_i} - M_P \qquad (7.14)$$

where M_P is the average value computed from all the points in the window; M_{P_i} is the average value computed from all the remaining neighboring points in the window other than the ith point, which was taken out of the window; and V_i is the difference. The rest of the procedure is the same as was used for the previous algorithm. That is, the M values of V are used to compute a single statistical value, which is then used as the basis on which to construct a threshold value for snooping data points within the window. After that, every value of V can be checked. If any value of V, say V_j, exceeds this threshold value, then point J will be suspected of having gross errors and excluded from this window. In this way, all points that appear to make a great change in the representative value in a window will be excluded.

This point data detecting technique is then applied to each window. After this has been done, the rest of the procedure is exactly the same as the procedures described in the previous section, that is, computing a representative value, constructing a threshold value, and identifying suspect points.

7.6.3 A Practical Example

This algorithm was applied to the data set shown in Figure 7.8. The gross errors detected by this algorithm are plotted in Figure 7.10(a) and the contours produced from the data set after removal of gross errors are shown in Figure 7.10(b). It can be seen from Figure 7.10(b) that there is still a point with a small gross error located in the northwest of the test area since it produces an unnatural contour in that part of the plot. The reason why this point was not detected by this algorithm could be due to the fact that, in applying this algorithm, a larger window size needs to be used. For example, a minimum number of 35 points was specified. However, the use of a large window size resulted in a decrease in the sensitivity of this algorithm to gross error.

Inspection of Figures 7.10(a) and Figure 7.8(a) reveals that the majority of the gross errors detected by these two algorithms are identical. However, each may miss one or more points for the reasons discussed previously. Therefore, using both algorithms together may produce a more desirable result, because all points detected by both of them should be deleted from the data set. Figure 7.11 shows the gross errors detected by both algorithms. It can be found that a much more reasonable result was produced after removing them.

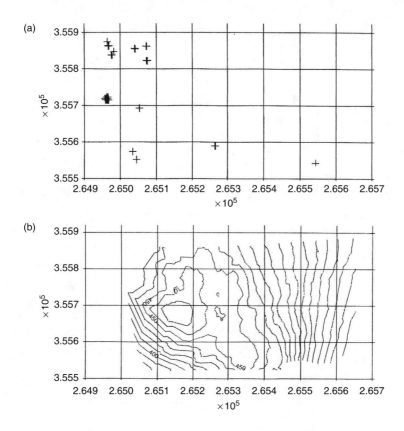

Figure 7.10 Results obtained by algorithm for detecting a cluster of gross errors: (a) gross error points detected and (b) contours from the data set after removal of gross errors.

7.7 DETECTION OF GROSS ERRORS BASED ON TOPOLOGIC RELATIONS OF CONTOURS

7.7.1 Gross Errors in Contour Data: An Example

As has been discussed previously, contour data are one of the main sources for digital terrain modeling. From analog map to digital data, one has to digitize the contour lines. During the course of digitization, elevation values of contours are normally entered by the operator. Quite often, the data points are recorded as follows:

1. N_1 (number of points in contour 1), H_1 (height of contour 1)
   - X_1 (X coordinates of point 1), Y_1 (Y coordinates of point 1)
   - X_2 (X coordinates of point 2), Y_2 (Y coordinates of point 2)
   - ....
   - X_{N_1} (X coordinates of point N_1), Y_{N_1} (Y coordinates of point N_1)
2. N_2 (number of points in contour 2), H_2 (height of contour 2)
   - X_1 (X coordinates of point 1), Y_1 (Y coordinates of point 1)
   - ....

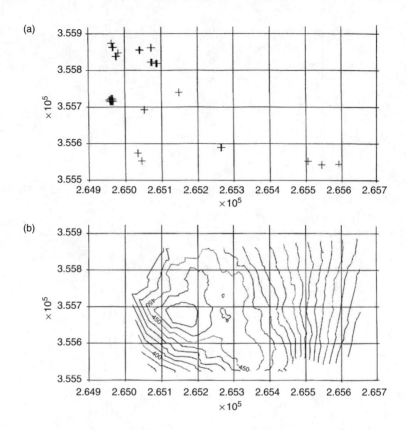

Figure 7.11 Results obtained from the complementary use of these two algorithms: (a) gross error points detected and (b) contours from the data set after removal of gross errors.

Such a manual operation is subject to mistakes, as shown in Figure 7.12. In this figure, the contour interval is 10 m. The elevation values are 50, 60, 70, 80, 90, 100, and 110 m. It is obvious that the elevation value of the third contour as indicated in this figure, that is, the 170 m, must be a mistake.

It should be pointed out that there might be cases where contour lines are broken (Figure 7.13). For example, when the slope is nearly vertical and the contours overlap, some contour lines will be broken. Other cases are the space required for indexing and areas with escarpment or faults. In such cases, the elevations of broken lines are more prone to error.

7.7.2 Topological Relations of Contours for Gross Error Detection

There are two possible approaches for the detection and removal of gross errors in a contour data set. One is to regard all the contour data as random points and then to apply the algorithms described in the previous sections. The other is to employ the topological relations between neighboring contour lines so as to detect and remove gross errors.

QUALITY CONTROL IN TERRAIN DATA ACQUISITION

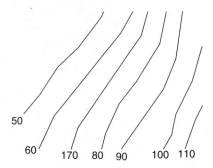

Figure 7.12 Contours digitized from a 1:10,000 scale map, with a gross error introduced.

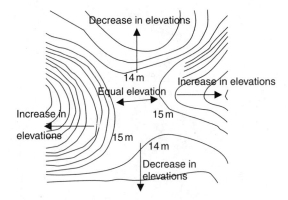

Figure 7.13 Relations between elevation values of adjacent contours.

Consider the fact that the elevation values of all points on the contour are wrong if the elevation of a contour line is erroneously given. If a contour is long, it is not efficient to employ the algorithms described in the previous sections. Therefore, considering the topological relations of contours is more reasonable and efficient.

There are three possible relations between elevation values of neighboring contours:

1. increase in elevation
2. decrease in elevation
3. equal elevation.

Figure 7.13 illustrates these three cases. According to these relations, it can be judged whether the elevation of a contour is wrong.

CHAPTER 8

Accuracy of Digital Terrain Models

The accuracy of DTMs is of concern to both DTM producers and users. For a DEM project, accuracy, efficiency, and economy are the three main factors to be considered (Li 1990). Accuracy is perhaps the single most important factor to be considered because, if the accuracy of a DEM does not meet the requirements, then the whole project needs to be repeated and thus the economy and efficiency will ultimately be affected. For this reason, this chapter is devoted to this topic.

8.1 DTM ACCURACY ASSESSMENT: AN OVERVIEW

8.1.1 Approaches for DTM Accuracy Assessment

A DTM surface is a 3-D representation of terrain surface. Unavoidably, some errors will be present in each of the three dimensions of the spatial (X, Y, Z) coordinates of the points occurring on DTM surfaces. Two of these (X and Y) are combined to give a planimetric (or horizontal) error while the third is in the vertical (Z) direction and is referred to as the elevation (or height) error.

The assessment of DTM accuracy can be carried out in two different modes, that is,

1. the planimetric accuracy and the height accuracy can be assessed separately
2. both can be assessed simultaneously.

For the former, accuracy results for the planimetry can be obtained separately from the accuracy of these results in a vertical direction. However, for the latter, an accuracy measure for both error components together is required.

There are four possible approaches for assessing the height accuracy of the DTM (Ley 1986), namely,

1. *Prediction by production (procedures)*: This is to assess the likely errors introduced at the various production stages together with an assessment of the vertical

accuracy of the source materials. The accuracy of the final DTM is the consequence or concatenation of the errors involved in all these stages.
2. *Prediction by area*: This is based on the fact that the vertical accuracy of contour lines on a topographic map is highly correlated with the mean slope of the area.
3. *Evaluation by cartometric testing*: This is about experimental evaluation. It is argued by many that the entire model rather than the node should be tested. For such a test, a set of checkpoints is required.
4. *Evaluation by diagnostic points*: A sample of heights is acquired from the source materials at the time of data acquisition and this set of data is used to check the quality of the model. This can be conducted at any intermediate stage as well as at the final stage.

There are three approaches for assessing the planimetric accuracy of DTM (Ley 1986), namely:

1. *No error*: It is argued that a DTM provides use of a set of heights with planimetric positions, which are inherently precise.
2. *Predictive*: Similar to the prediction by area used for vertical accuracy.
3. *Through height*: To fix the positions of node heights by comparing a series of points.

However, as he also mentioned, it is difficult to bring these into practice. This is perhaps the reason why the issue of planimetric accuracy is rarely addressed.

An alternative approach is to simultaneously assess the vertical and horizontal accuracies. In doing so, a measure capable of characterizing the accuracy in three dimensions is required. Ley (1986) suggested using a comparative measure of the mean slopes between the DTM surface and the original terrain surface. Others have also considered the use of other geomorphometric parameters as well as terrain feature points and lines. However, there is no consensus. Most people follow the practice of assessing the contour accuracy, that is, assessing the vertical accuracy only.

8.1.2 Distributions of DTM Errors

In the field of DTM data acquisition, it is usually assumed that errors in spatial data are normally distributed. However, it is not necessarily the case for DTM errors, as shown in Figure 8.1. These two sets of data were obtained from an experimental test conducted by Li (1990). Figure 8.1(a) is the result for the Sohnstetten area with a sample size of 1892 points and Figure 8.1(b) is the result for the Spitze area with a sample size of 2115 check points. Some information about these experimental tests is given in Section 8.2.

To understand the distributions better, the frequency of occurrence of large errors was also recorded. Table 8.1 lists the results (Li 1990). To show how the distributions deviate from the normal contribution, the theoretical values for the occurrence frequency of large errors are also listed. From this table, it is clear that curves of the distribution of DTM errors are flatter than the standard normal distribution $N(0,1)$.

ACCURACY OF DIGITAL TERRAIN MODELS

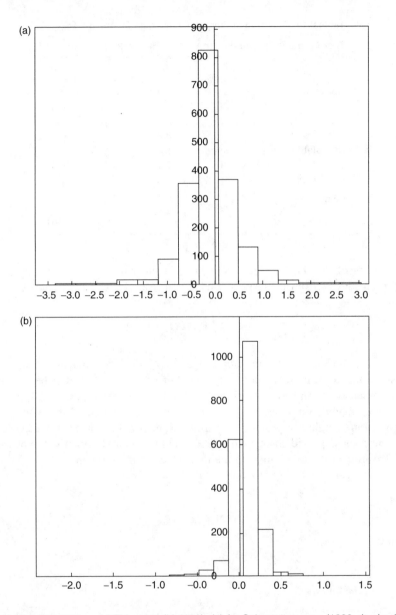

Figure 8.1 Distribution of DTM errors (Li 1990): (a) for Sohnstetten area (1892 checkpoints) and (b) for Spitze area (2115 checkpoints).

8.1.3 Measures for DTM Accuracy

Let $f(x, y)$ be the original terrain surface and $f'(x, y)$ be the constructed DTM surface, then the difference, $e(x, y)$, where

$$e(x, y) = f'(x, y) - f(x, y) \tag{8.1}$$

Table 8.1 Occurrence Frequency of Large Errors in DTM

Test Area	Grid Interval (m)	>2σ (%)	>3σ (%)	>4σ (%)
Uppland	$\sqrt{2} \times 20$	4.5	1.0	0.3
	40	5.1	1.1	0.3
	$\sqrt{2} \times 40$	5.2	1.3	0.3
	80	5.6	1.2	0.3
Sohnstetten	20	5.6	1.7	0.8
	$\sqrt{2} \times 20$	6.0	1.5	0.6
	40	6.6	1.5	0.3
	$\sqrt{2} \times 40$	6.1	1.5	0.3
Spitze	10	5.0	2.3	1.5
	$\sqrt{2} \times 10$	5.8	2.7	1.2
	20	5.4	2.7	1.4
$N(0,1)$		4.6	0.3	0.01

is the error of the DTM surface. Following a similar treatment by Tempfli (1980), the mean square error (mse) can be used as a measure for DTM accuracy, where

$$\text{mse} = \iint e^2(x, y)\, dx\, dy \quad (8.2)$$

$e(x, y)$ is a random variable in statistical terms (Li 1988) and magnitude and spread (dispersion) are the two characteristics of random variable. To measure the magnitude, some parameters can be used such as the extreme values ($e_{\max}$ and $e_{\min}$), the mode (the most likely value), the median (the frequency center), and the mathematical expectation (weighted average). To measure the dispersion, some parameters such as the range, the expected absolute deviation, and the standard deviation can be used.

To summarize, in addition to the mse which is in common use, the following parameters can also be used to measure DTM accuracy:

Range:
$$R = e_{\max} - e_{\min} \quad (8.3)$$

Mean:
$$\mu = \frac{\sum e}{N}$$

Standard deviation:
$$\sigma = \sqrt{\frac{\sum (e - \mu)^2}{N - 1}} \quad (8.4)$$

The use of range, R, may lead to a specification of DTM accuracy something like the American National Map Accuracy Standard. But some characteristics of this measure might be objectionable, that is,

1. The value R depends on only two values of the random variable and others are all ignored.
2. The probability of the values in $e(x, y)$ is ignored.

ACCURACY OF DIGITAL TERRAIN MODELS

Therefore, the combination of mean and standard deviation is preferred although the distribution of DTM errors is not necessarily normally, as shown in Figure 8.1.

This is because most of the probability distribution is massed with 4σ distance from μ, according Chebyshev's theorem (Burington and May 1970). Chebyshev's theorem states that the probability is at least as large as $1 - 1/k^2$ that an observation of a random variable (e) will be within the range from $\mu - k \times \sigma$ to $\mu + k \times \sigma$, or

$$P(|e - \mu| > k \times \sigma) < \frac{1}{k^2} \tag{8.5}$$

where k is any constant greater than or equal to 1. If the normal distribution is used to approximate the distribution of $e(x, y)$, the standard deviation computed from Equation (8.4) has the special meaning that is familiar to us.

8.1.4 Factors Affecting DTM Accuracy

The accuracy of the DTM is a function of a number of variables such as the roughness of the terrain surface, the interpolation function, interpolation methods, and the three attributes (accuracy, density, and distribution) of the source data (Li 1990, 1992a). Mathematically,

$$A_{DTM} = f(C_{DTM}, M_{Modeling}, R_{Terrain}, A_{Data}, D_{Data}, DN_{Data}, O) \tag{8.6}$$

where A_{DTM} is the accuracy of the DTM; C_{DTM} refers to the characteristics of the DTM surfaces; $M_{Modeling}$ is the method used for modeling DTM surfaces; $R_{Terrain}$ is the roughness of the terrain surface itself; A_{Data}, D_{Data}, and DN_{Data} are the three attributes (accuracy, distribution, and density) of the DTM source data; and O denotes other elements.

The roughness of the terrain surface determines the difficulty of DTM representation of terrain. If the terrain is simple, then only a few points need to be sampled and the surface to be used for reconstruction will be very simple. For example, if the terrain is flat, only three points are essential and a plane can be used for modeling this piece of terrain surface. On the other hand, if the surface is complex, then more points need to be measured and higher-order polynomials may have to be used for modeling this terrain. The descriptors for the complexity of terrain surfaces have already been introduced in Chapter 2. Among the various descriptors, slope is the most important one widely used in the practice of surveying and mapping and will be used later in the development of the DTM accuracy model.

A DTM surface can be constructed by two methods. One is to construct it directly from the measured data and the other is indirect. In the latter, the DTM surface is constructed from grid data that are interpolated via a random-to-grid interpolation. The accuracy of the DTM surface constructed indirectly will be lower than the accuracy of that constructed directly, due to accuracy loss in the random-to-grid interpolation process.

As discussed in Chapter 4, three types of DTM surfaces are possible, discontinuous, continuous, and smooth. It has been found that the continuous surface consisting of a series of contiguous linear facets is the least misleading (or the most trustable).

The three attributes of the source data (distribution, accuracy, and density) will also have a great influence on the accuracy of the final DTM. If there are a lot of points in the smooth or flat areas and few points in the rough areas, then the result will not be satisfactory. This is the combined effect of distribution and density, which was discussed in Chapter 2. The third attribute, the accuracy of the source data, will be discussed in detail in this section. Undoubtedly, errors in source data are propagated to the final DTM during the modeling process.

It has already been discussed in Chapter 3 that aerial photographs and existing topographical maps are the main data sources for digital terrain modeling. The accuracy of photogrammetric data is affected by the following factors:

1. the quality and scales of the photographs
2. the accuracy and physical conditions of the photogrammetric instruments used
3. the accuracy of measurement
4. the stereo geometry of aerial photographs.

Generally, the accuracy of photogrammetric data is 0.07 to $0.1H‰$ if acquired by using an analytical photogrammetric plotter or 0.1 to $0.2H‰$ if acquired by using an analog photogrammetric plotter. Here, H is the flying height, that is, the height of the aerial camera when the photographs were taken (usually with a wide-angle camera with a focal length of 152 mm and a frame of 23 cm × 23 cm). It refers to the accuracy of static measurement. However, if the measurement is dynamic (e.g., contouring and profiling), the accuracy is much lower. The speed of measurement is also an important factor. Various experimental tests (e.g., Sigle 1984) reveal that the accuracy of photogrammetrically measured data is about $0.3H‰$. Some experiments (e.g., Gong et al. 2000) also reveal that the accuracy of photogrammetric data acquired by a fully digital photogrammetric system is not as high as that from an analytical plotter.

The accuracy of contouring data obtained from digitization is affected by the following factors:

1. the accuracy and physical condition of the digitizer
2. the quality of the original map
3. the accuracy of measurement.

The accuracy of contours can be written as:

$$m_c = m_h + m_p \times \tan \alpha \tag{8.7}$$

where m_h refers to the accuracy of height measurement; m_p is the planimetric accuracy of the contour line; α is the slope angle of the terrain surface; and m_c is the overall height accuracy of the contours, including the effect of planimetric errors.

Usually, the accuracy specifications for contours all appear in the form of Equation (8.7). A summary of such specifications is given in Table 8.2. Accuracy loss during the digitization process is about 0.1 mm in point mode and 0.2 to 0.25 mm in stream mode. In any case, the overall accuracy of digitized contour data will be still within a 1/3 contouring interval.

ACCURACY OF DIGITAL TERRAIN MODELS

Table 8.2 Some Examples of Contour Accuracy Specifications

Country	Scale	Accuracy of Contours (m)
France	1:5000	$0.4 + 3.0 \times \tan\alpha$
Switzerland	1:10,000	$1.0 + 3.0 \times \tan\alpha$
Britain	1:10,560	$\sqrt{1.8^2 + (3.0 \times \tan\alpha)^2}$
Italy		$1.8 + 12.5 \times \tan\alpha$
France	1:25,000	$0.8 + 5.0 \times \tan\alpha$
Finland		$1.5 + 3.0 \times \tan\alpha$
America	1:50,000	$1.8 + 15 \times \tan\alpha$
Switzerland		$1.5 + 10 \times \tan\alpha$

Table 8.3 Comparison of the Accuracy of DTM Data Obtained by Different Techniques

Methods of Data Acquisition	Accuracy of Data
Ground measurement (including GPS)	1–10 cm
Digitized contour data	About 1/3 of contouring interval
Laser altimetry	0.5–2 m
Radargrammetry	10–100 m
Aerial photogrammetry	0.1–1 m
SAR interfereometry	5–20 m

For convenience of reference, the accuracy of DTM source data from various sources is summarized in Table 8.3.

8.2 DESIGN CONSIDERATIONS FOR EXPERIMENTAL TESTS ON DTM ACCURACY

8.2.1 Strategies for Experimental Tests

As stated previously, the accuracy of a DTM is the result of many individual factors, that is,

1. the three attributes (accuracy, density, and distribution) of the source data
2. the characteristics of the terrain surface
3. the method used for the construction of the DEM surface
4. the characteristics of the DEM surface constructed from the source data.

Accordingly, six strategies for an experimental testing of DEM accuracy are possible (Li 1992a), in each of which only one of the six factors is used as the independent variable and the other five as controlled variables:

1. *The accuracy of the source data* could be varied while all the other factors remain unchanged. This can be achieved by using different data acquisition techniques

such as GPS, photogrammetry, and other methods. It can also be achieved by using the same type of data acquisition techniques but with different accuracies.

2. *The density of the source data* could be varied while all other factors remain unchanged. This can be achieved by using different sampling intervals or data selection methods. Alternatively, resampling without involvement of interpolation, as discussed in Chapter 4, can be applied to a set of data with finer resolutions (i.e., smaller intervals) to coarser resolution (i.e., larger intervals).
3. *The distribution of source data* could be varied while all other factors remain unchanged. This can be achieved by using different sampling patterns or data selection methods. In digital terrain modeling practice, grid and contour data are the two types of basic data patterns that have been widely used. Another two types of data are also widely used, that is, with or without feature points (i.e., top of hills, bottom of valleys, points along ridge lines, points along ravine lines, points along the edge of terrace, saddle points, etc.).
4. *The type of terrain* could be varied while all other factors remain unchanged. This is achieved by using terrain surface with various types of relief.
5. *The type of DTM surface* could be varied while all other factors remain unchanged. This is achieved by using different types of discontinuous, continuous, and smooth surfaces for DTM surface reconstruction.
6. *Two types of modeling methods* are used to construct two types of surfaces, that is, direct modeling using triangulated networks and indirect modeling using a random-to-grid interpolation to form a grid network.

8.2.2 Requirements for Checkpoints in Experimental Tests*

In experimental tests on DTM accuracy, a set of checkpoints is used as the *ground truth*. Then, the points interpolated from the constructed DTM surface are checked against the corresponding checkpoints. After that, the difference between the two heights at each point is obtained. These differences are used to compute statistical values, as discussed in Section 8.1. It is clear that the final DTM accuracy figures are definitely affected by the characteristics of the set of checkpoints. In other words, the final estimates may be affected by the three attributes of the set of checkpoints, that is, accuracy, sample size (number of points), and distribution, because the three attributes can be used to characterize the set of checkpoints (Li 1991).

First, the required sample size (number) of the set of checkpoints will be considered. From statistical theory it can be found that this is related to the following two factors:

1. the degree of accuracy required for the accuracy figures (i.e., the mean μ and standard deviation σ) to be estimated
2. the variation associated with the random variable, that is, the height differences in the case of DTM accuracy tests.

The smaller the variation, the smaller the sample size needed to achieve a given degree of accuracy required for accuracy estimates. For an extreme example, if the σ of the height differences is 0, then one checkpoint is enough no matter how large

---

* Largely extracted from Li 1991, with permission from ASPRS

ACCURACY OF DIGITAL TERRAIN MODELS

the test area or the size of the data set. Similarly, the higher the given degree of accuracy requirement for the accuracy estimates, the larger the sample size needed. The relationship between the sample size, the value σ, and the given degree of accuracy required needs to be established.

If the distribution is normal, the discussion is simpler. However, as discussed in Section 8.1, the distribution of DTM errors is not necessarily normal and, therefore, a new random variable with approximate normal distribution needs to be selected for further discussion. Let ΔH be the random variable of height differences $e(x, y)$ in discrete space; μ be the mean of a random sample of size n from a particular distribution; and M be the true value of the random variable. Then, the ratio

$$Y = \frac{\mu - M}{\sigma/\sqrt{n}} \tag{8.8}$$

is a standardized variable and has approximately the normal distribution $N(0, 1)$, even though the underlying distribution is not normal, as long as n is large enough (Hogg and Tanis 1977). Suppose the σ of a distribution is known but the M is unknown, then for the probability r and for a sufficiently large value of n, a value Z can be found from the statistical table for $N(0, 1)$ distribution, such that the probability that Y will be within the range from $-Z$ to Z is approximately equal to r, or approximately,

$$P(-Z \leq y \leq Z) \approx r \tag{8.9}$$

The closeness of the approximate probability r to the exact probability depends on both the underlying distribution and the sample size. If the distribution is unimodal (with only one mode) and continuous, the approximation is usually quite good for even a small value of n (e.g., 5). If the distribution is "less normal" (i.e., badly skewed or discrete), a large sample size is required (e.g., 20 to 30 points).

Substituting Equation (8.8) into Equation (8.9) and rearranging it, the following expression can be obtained:

$$P\left(\mu - \frac{Z\sigma}{\sqrt{2}} \leq y \leq \mu + \frac{Z\sigma}{\sqrt{2}}\right) \approx r \tag{8.10}$$

For a given constant S, the percentage of the probability, $(100r)\%$, of the random interval $\mu \pm S$ including M is called the confidence interval, where S is the specified degree of accuracy for the mean estimate, μ in this case. In general, if the required confidence interval $(100r)\% = 100(1-\alpha)\%$, then the sample size n can be expressed as follows:

$$n = \frac{Z_r^2 \times \sigma^2}{S^2} = Z_r^2 \times \left(\frac{\sigma}{S}\right)^2 \tag{8.11}$$

where Z_r is the limit value within which the values of the random variable Y will fall with probability r. Its value can be found in the statistical table for the $N(0, 1)$ distribution. The mathematical expression is as follows:

$$\Phi(Z) = 1 - \alpha/2 \tag{8.12}$$

and the commonly used values are as follows:

$$Z_{r=0.95} = 1.960, \quad Z_{r=0.98} = 2.326, \quad Z_{r=0.99} = 2.576$$

For example, if the accuracy required for the mean estimate is 10% of the standard deviation of the DTM errors (i.e., σ), and the confidence level is 95%, then the required sample size is

$$n = Z_r^2 \times \left(\frac{\sigma}{S}\right)^2 = 1.96^2 \times \left(\frac{100}{10}\right)^2 = 384$$

Similarly, there is also a relationship between the accuracy specified for the standard deviation estimate σ and the required sample size. According to Burington and May (1970), the variance of the standard deviation estimate from a sample can be expressed as follows:

$$\sigma_\sigma^2 = \frac{\sigma^2}{2(n-1)} \tag{8.13}$$

that is,

$$n = \frac{\sigma^2}{2\sigma_\sigma^2} + 1 \tag{8.14}$$

For example, if the accuracy σ_σ required for the standard deviation estimate σ is 10% of σ, then the required sample size is 51.

The variation of DTM accuracy estimate values with the number of checkpoints used has been intensively tested by Li (1991). The number of checkpoints was reduced systematically from 100 to 1% to produce a number of new sets of checkpoints. These new sets of checkpoints were then used to assess the DTM accuracy and produce new sets of DTM accuracy estimates. The test results confirm the relationships expressed by Equations (8.11) and (8.14).

Equations (8.11) and (8.14) can be used to estimate the number of checkpoints required. In such calculations, it is implicit that the checkpoints are free of errors. However, this is not the case in practice. If the accuracy of the set of checkpoints is lower than the expected DTM accuracy, then the result of the DTM accuracy estimated from the height differences is meaningless. This means that the relationship between the required accuracy of checkpoints and the given degree of accuracy for the DTM accuracy estimate should be established. In this discussion, the accuracies are discussed in terms of the standard deviations.

Let ΔH_2 be the error involved in the checkpoints and ΔH_1 the true height difference. Then,

$$\Delta H = \Delta H_1 + \Delta H_2 \tag{8.15}$$

By applying the error propagation law to Equation (8.15), the following expression can be obtained:

$$\sigma^2 = \sigma_{\Delta H_1}^2 + \sigma_{\Delta H_2}^2 \tag{8.16}$$

The value of σ itself is not of interest but the value of $\sigma_{\Delta H_2}$ is. The attempt is made here to find a critical value for $\sigma_{\Delta H_2}$ so that the σ is still acceptable as being the

ACCURACY OF DIGITAL TERRAIN MODELS

representative of $\sigma_{\Delta H_1}$. As expressed in Equation (8.13), the standard deviation of $\sigma_{\Delta H_1}$ has a variance approximately as follows:

$$\sigma^2_{\sigma_{\Delta H_1}} = \frac{\sigma^2_{\Delta H_1}}{2(n-1)} \tag{8.17}$$

Therefore, the acceptable range for σ to deviate from $\sigma_{\Delta H_1}$ can be expressed as follows:

$$\sigma_{\Delta H_1} - \frac{\sigma_{\Delta H_1}}{\sqrt{2(n-1)}} \leq \sigma \leq \sigma_{\Delta H_1} + \frac{\sigma_{\Delta H_1}}{\sqrt{2(n-1)}} \tag{8.18}$$

It is much more convenient to use a single value, so the square root of these two terms is used as the representative value because they are independent. Then, the following equation can be obtained:

$$\sigma^2 = \sigma^2_{\Delta H_1} + \frac{\sigma^2_{\Delta H_1}}{2(n-1)} = \frac{(2n-1)\sigma^2_{\Delta H_1}}{2(n-1)} \tag{8.19}$$

Combining Equation (8.19) with Equation (8.16), the following expression can be obtained:

$$\sigma^2_{\Delta H_2} = \frac{\sigma^2}{2n-1} \tag{8.20}$$

or

$$\sigma_{\Delta H_2} = \frac{\sigma}{\sqrt{2n-1}} = \frac{1}{\sqrt{2n-1}} \times \sigma \tag{8.21}$$

For example, if the sample size of the set of checkpoints is 51, then the required accuracy of the checkpoints in terms of the standard deviation is 10% of the standard deviation (σ) of the DTM errors. In mapping sciences, the accuracy of checkpoints is usually specified in terms of RMSE, then RMSE might be used to replace σ in Equation (8.20).

The last consideration is the distribution of the checkpoints. An intensive test as to whether random distribution is as good as even distribution (e.g., in grid form) was conducted by Li (1991). Two test areas (see Section 8.3) were used. The numbers of checkpoints for the areas were 1892 and 2314. From each set of checkpoints, 15 subsets of checkpoints, each with 500 points, were randomly generated. The randomness of selection was achieved by using a set of random numbers from a uniform distribution generated by a computer subroutine for random numbers. In the generation of random numbers, the range was determined by the total number of points in the original set of checkpoints. After this, those checkpoints with the same numbering as the generated random numbers were taken from the original set to form the sample. As expected, there were differences among the 15 accuracy estimates. However, the variation was very small and well within the acceptable range. Therefore, it might be assumed that the random selection of checkpoints is acceptable if the selection is over the whole test area.

8.3 EMPIRICAL MODELS FOR THE ACCURACY OF THE DTM DERIVED FROM GRID DATA

From the literature it can be seen that many experimental investigations into the accuracy of DTM have been conducted by many researchers. The best known investigation was the international test organized by the International Society for Photogrammetry and Remote Sensing (ISPRS) in the early 1980s (Torlegard et al. 1986). A number of institutions all over the world participated in the acquisition of DTM source data by using the photogrammetric method. Six areas with different types of terrain were tested. However, this international test failed to produce any empirical model for DTM accuracy. In the early 1990s, a systematic investigation into the relationship between sampling intervals and DTM accuracy was conducted by Li (1990, 1992a, 1994) using three sets of the ISPRS test data. Through this testing, an empirical model for DTM accuracy prediction was produced. Cases both with and without terrain features were considered and a different model for each was produced. Recently, in the community of geo-information, similar tests have also been conducted (e.g., Gong et al. 2000; Tang 2000). This section is based mainly on the tests by Li (1990, 1992a, 1994).

8.3.1 Three ISPRS Test Data Sets

The three ISPRS data sets used were for the Uppland, Sohnstetten, and Spitze areas. The basic characteristics of these test areas are described in Table 8.4. A set of photogrammetrically measured contour data, a set of square-grid data, and a set of F-S data for each of these areas were used. Some information about the test data is given in Table 8.5. The checkpoints were measured from much larger-scale aerial photographs and therefore have much higher accuracy then the test data points. Some information about these checkpoints is given in Table 8.6.

Figure 8.2 shows the contour maps of these areas. The corresponding F-S data are superimposed onto each of these maps. The Uppland area is relatively flat, with a few mounds. In the Sohnstetten area, a valley runs through the middle of the area, so most of the F-S points are along the edges and ravines. In the Spitze area, a road junction cuts through the right side of the area, so the F-S points are those along the break lines caused by these roads.

8.3.2 Empirical Models for the Relationship between DTM Accuracy and Sampling Intervals

A triangulation-based modeling system was used in this experiment and linear interpolation was used to avoid any misleading fluctuation on the constructed surface.

Table 8.4 Description of the ISPRS Test Areas

Test Area	Terrain Description	Height Range (m)	Mean Slope (°)
Uppland	Farmland and forest	7–53	6
Sohnstetten	Hills with moderate height	538–647	15
Spitze	Smooth terrain	202–242	7

ACCURACY OF DIGITAL TERRAIN MODELS

Table 8.5 Description of Test Data

Parameter	Uppland	Sohnstetten	Spitze
Photo scale	1:30,000	1:10,000	1:4000
Flying height (H) (m)	4500	1500	600
Grid interval (m)	40	20	10
Grid data accuracy[a] (m)	±0.67	±0.16	±0.08
CI (m)	5	5	1
Average planimetric CI[b] (m)	48	9	8
Contour point interval (m)	10.4–22.5	3.7–19.8	5.4–9.2
Contour data accuracy[a] (m)	±1.35	±0.45	±0.18

[a] Accuracy is represented in terms of RMSE.
[b] The mean planimetric CI is equal to CI cot α, where α is the mean slope angle.

Table 8.6 Description of Checkpoints

Test Area	Photo Scale	Flying Height (m)	Number of Points	RMSE (m)	Largest Error (m)
Uppland	1:6000	900	2314	±0.090	0.20
Sohnstetten	1:5000	750	1892	±0.054	0.07
Spitze	1:1500	230	2115	±0.025	0.05

Through the comparison of heights interpolated from DTM surfaces and the corresponding checkpoints, an error for each checkpoint is obtained, from which the accuracy estimates of the DTM surfaces can be computed.

To test the accuracy of DTM with sampling intervals (i.e., the grid intervals in this case were due to regular grid sampling), a number of new data sets with grid intervals larger than the interval of the original grid were produced by simple resampling without interpolation, as discussed in Chapter 4. The test results are shown in Table 8.7, which lists the variation of DTM accuracy with grid interval and changes in accuracy after F-S data are added.

The results for the Uppland and Sohnstetten areas are plotted in Figure 8.3. It is clear that

1. If the F-S points are sampled, the relationship between DTM accuracy and sampling intervals (grid intervals in this particular case) is quite linear.
2. If the F-S points are not sampled, the relationship between DTM accuracy and sampling intervals (grid intervals in this particular case) is a quadratic curve.

By regression, empirical models for DTM accuracy could be obtained. The general form is:

With F-S data:
$$\sigma_{\text{DTM-c}} = k_1 \times \sigma_{\text{Data}} + k_2 \times d \tag{8.22}$$

With no F-S data:
$$\sigma_{\text{DTM-g}} = k_1 \times \sigma_{\text{Data}} + k_2 \times d + k_3 \times d^2 \tag{8.23}$$

where d is the sample interval, that is, the grid interval for grid-based sampling.

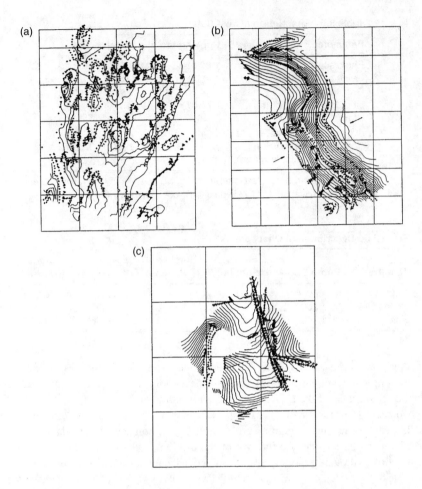

Figure 8.2 Contour maps of the test areas (photogrammetrically measured) superimposed with feature-specific points: (a) Uppland area (CI = 5 m); (b) Sohnstetten area (CI = 5 m); and (c) Spitze area (CI = 1 m), where the large blank area was not measured due to difficulties.

8.3.3 Empirical Models for DTM Accuracy Improvement with the Addition of Feature Data

From Equation (8.23) it can be found that the difference between the DTM accuracy with and without additional feature points is the second-order term. Therefore, the difference could be expressed as follows:

$$\Delta\sigma = \sigma_{\text{DTM-g}} - \sigma_{\text{DTM-c}} = a + b \times d^2 \tag{8.24}$$

The regression result is shown in Figure 8.4. It is clear that the curves fit the experimental data very well. In fact, in Figure 8.4, the ratio of the grid interval to its smallest interval (d/d_0) is used instead of the absolute value of d. The regression results also

ACCURACY OF DIGITAL TERRAIN MODELS

Table 8.7 The Relationship between the Accuracy of DTM and Grid Intervals

Test Area	Grid Interval (m)	Standard Error (σ) (m)		Difference in σ Value (m)	Ratio in Grid Interval
		No F-S Data	With F-S Data		
Uppland	$\sqrt{2} \times 20$	0.63	0.59	0.04	1
	40	0.76	0.66	0.10	$\sqrt{2}$
	$\sqrt{2} \times 40$	0.93	0.70	0.23	2
	80	1.18	0.80	0.38	$2\sqrt{2}$
Sohnstetten	20	0.56	0.40	0.16	1
	$\sqrt{2} \times 20$	0.87	0.55	0.32	$\sqrt{2}$
	40	1.44	0.77	0.67	2
	$\sqrt{2} \times 40$	2.40	1.08	1.32	$2\sqrt{2}$
Spitze	10	0.21	0.14	0.07	1
	$\sqrt{2} \times 10$	0.28	0.15	0.13	1
	20	0.36	0.16	0.20	$2\sqrt{2}$

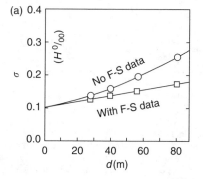
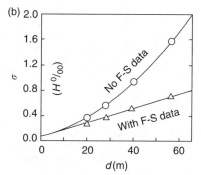

Figure 8.3 Variation of DTM accuracy with sampling interval (grid interval in this case) (Reprinted from Li 1994, with permission from Elsevier): (a) for Uppland area and (b) for Sohnstetten area.

reveal that the constant a in Equation (8.24) is close to 0; therefore, Equation (8.24) can be rewritten as

$$\frac{\Delta \sigma_2}{\Delta \sigma_1} = \left(\frac{d_2}{d_1}\right)^2 \quad (8.25)$$

where $\Delta \sigma_1$ and $\Delta \sigma_2$ represent the difference in value corresponding to d_1 and d_2.

8.4 THEORETICAL MODELS OF DTM ACCURACY BASED ON SLOPE AND SAMPLING INTERVAL*

Since the early 1970s, attempts have been made to establish a mathematical model for the prediction of DTM accuracy through experimental analysis. A number of such

---

* The materials included in this section were first published in Photogrammetric Record (Li 1993a, 1993b).

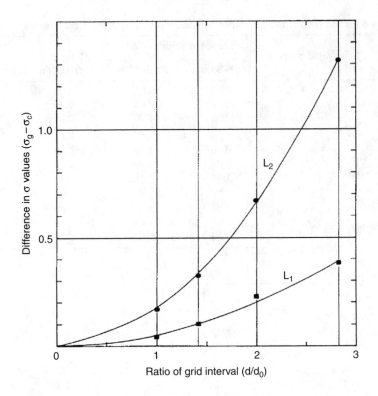

Figure 8.4 Relationship between the difference in DTM accuracy values (with and without F-S points) and the ratio of grid interval. The dot and square points represent the test result; the continuous curves are for regression results. L_1 and L_2 are for Uppland and Sohnstetten, respectively. d/d_0 is the ratio of the grid interval d to the smallest grid interval d_0 (Li 1994).

models have been developed. Most of them are either not reliable or not practical enough. In this section, the theories behind these models will be outlined. The model developed by Li (1990, 1993b) will be presented in detail, because it is similar to traditional map accuracy specification, that is, making use of slope and sampling interval.

8.4.1 Theoretical Models for DTM Accuracy: An Overview

It is understandable that a terrain profile can be expanded by a Fourier series. Through the analysis of these individual sine and cosine waves, the accuracy loss due to sampling and surface reconstruction from sinusoidal functions could then be estimated (Makarovic 1972). The fidelity of the reconstructed surface is represented by the ratio of the mean value of the magnitude of the linearly constructed sinusoidal waves to the amplitude of the input waves, as shown in Figure 8.5. In this figure, the profile ABCDEF, reconstructed by linear interpolation, is an approximation to the sinusoidal input; Δx is the sampling interval; and δy is the height error at X_i,

ACCURACY OF DIGITAL TERRAIN MODELS

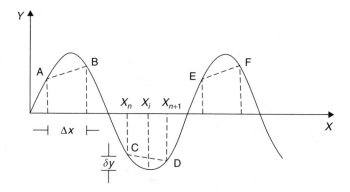

Figure 8.5 Sampling from a sine wave and reconstruction.

which is the height difference between the sine wave and the reconstructed profile. Suppose a is the original amplitude of the sine wave and m is the statistical mean error level over a sufficient length of the sine wave, then F in the expression

$$F = \frac{a - m}{a} \qquad (8.26)$$

represents the fidelity of the reconstructed data. Transfer functions can also be derived for different interpolation techniques. These fidelity figures may also be converted into standard deviation values. In this way, the accuracies of DTM surfaces can also be compared for different types of terrain surface.

> In principle, this theory is complete ... The task remains to investigate the frequency distribution of different terrain types and to relate the corresponding theoretical and empirical accuracy results (Ackermann 1979).

Covariance and variogram are two of the measures for terrain roughness and therefore can be used to estimate the accuracy loss due to sampling and reconstruction, thus to estimate DTM accuracy (Kubik and Botman 1976). First, covariance values are computed for different point intervals; then, these covariance values are approximated by either the exponential or the Gaussian function. In a similar way, the variogram can also be used as a terrain descriptor for DTM accuracy estimation (Frederiksen et al. 1986). Actually, they are all inter-related. Therefore, only the model based on the variogram is described here.

The values of semivariogram for different point intervals can be computed by Equation (2.9). After that, these semivariogram values are approximated by the following function:

$$2\gamma(d) = Ad^b \qquad (8.27)$$

where A and b are two constants.

The values of these two constants will depend on the type of terrain modeled. Figure 8.6 shows the semivariograms of the three ISPRS test data sets, described

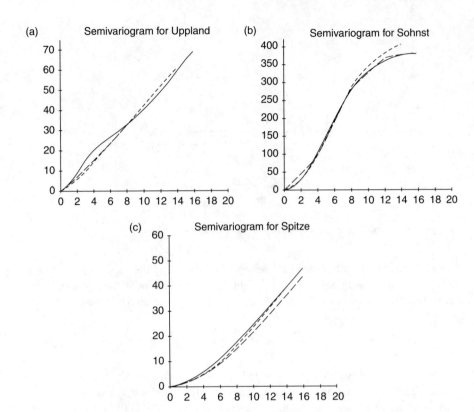

Figure 8.6 Semivariograms for the three ISPRS test areas, computed from data sets with various intervals.

in Section 8.3.1. The mathematical expression of an accuracy model based on the variogram is then expressed as follows (Frederiksen et al. 1986):

$$\sigma_{int}^2 = A \times \left(\frac{D}{L}\right)^b \left(-\frac{1}{6} + \frac{2}{(b+1)(b+2)}\right) \quad (8.28)$$

where σ_{int}^2 denotes the DTM accuracy in terms of error variance without taking the errors in the raw data into consideration; D is the sampling interval for the raw data; and L is the sampling interval of the profiles that were used to compute the two parameters A and b. The final expression is:

$$\sigma_{DTM}^2 = \sigma_{Data}^2 + \sigma_{int}^2 \quad (8.29)$$

where σ_{DTM} is the accuracy of the resulting DTM; σ_{Data} is the accuracy of measured raw data; and σ_{int}^2 is the accuracy loss due to sampling and reconstruction. All these values are expressed in terms of standard deviation.

ACCURACY OF DIGITAL TERRAIN MODELS

Table 8.8 Comparison of Experimental Results with the Predicted Accuracies by Variogram-Based Model

Test Area	Grid Interval (m)	Predicted Accuracy (m)	Grid Data		Composite Data	
			Tested Result (m)	Difference (m)	Tested Result (m)	Difference (m)
Uppland	40	1.04	0.76	0.28	0.66	0.38
	40√2	1.18	0.93	0.25	0.70	0.48
	80	1.38	1.18	0.20	0.80	0.58
Sohnstetten	20	0.74	0.56	0.18	0.43	0.31
	20√2	0.98	0.87	0.11	0.56	0.42
	40	1.38	1.45	−0.07	0.78	0.60
	20√2	1.77	2.40	−0.63	1.08	0.69
Spitze	10	0.29	0.21	0.08	0.16	0.13
	10√2	0.37	0.28	0.09	0.17	0.20
	20	0.48	0.35	0.13	0.18	0.30

It is interesting to note that accuracy predictions from this model are closer to the actual results obtained from the grid data sets (Li 1993a) in spite of large differences, whereas they are in very poor agreement with the results obtained from the composite data sets. This might be due to the fact that the values of the variogram used in this model were computed from grid data sets only and not from the composite data sets (Table 8.8), since it is complicated and difficult to compute variograms from nongrid data sets.

The parameters of the model based on variogram analysis were estimated from the whole set of data points. In practice, it is impossible to do this with confidence since the DTM accuracy for a given sampling interval needs to be predicted before the actual measurement of the data points can be carried out.

It is also understandable that the high-frequency part of the terrain surface is difficult to model and the accuracy loss in the process of terrain surface reconstruction can be determined by the summation of Fourier spectra of terrain profiles in their high-frequency part (Frederiksen 1980). In other words, those regions higher than $1/(2D)$, where D is the sampling interval form the error component. The mathematical expression of this model is as follows:

$$\sigma_{DTM}^2 = \sigma_{Data}^2 + \sum_{\lambda=2D}^{0} P_\lambda \tag{8.30}$$

where P_λ is the spectral value corresponding to the wavelength λ; D is the sampling interval; σ_{Data} is the accuracy of the source data; and σ_{DTM} is the accuracy of the final DTM.

Experimental results (Li 1993a) show that the results predicted by this model are very different from the experimental results, but in the case of the composite data, the difference is much smaller. This model always produces too optimistic a prediction (Table 8.9).

Table 8.9 Comparison of Test Results with the Predicted Accuracy by Frequency-Based Model

Test area	Predicted Accuracy (m)	Grid Data		Composite Data	
		Tested Result (m)	Difference (m)	Tested Result (m)	Difference (m)
Sohnstetten	0.26	0.46	−0.20	0.35	−0.09
Spitze	0.10	0.31	−0.21	0.20	−0.10
Drivdalen	1.25	1.57	−0.32	1.47	−0.22

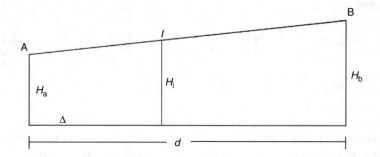

Figure 8.7 Linear interpolation of point I between points A and B.

As these models are not reliable and not conventional, from this section on, the mathematical model for DTM accuracy prediction-based slope and sampling intervals (Li 1990, 1993b) will be presented.

8.4.2 Propagation of Errors from DTM Source Data to the DTM Surface

As discussed previously, linear surfaces are the least misleading, thus the most reliable. The linear modeling of the square grid means representing terrain surfaces by continuous bilinear facets. The height of a desired position is then interpolated from the bilinear surface.

When discussing error propagation in linear modeling, error propagation in a profile should be considered first. Suppose points A and B in Figure 8.7 are two grid nodes with the interval of d; point I, between A and B, is to be interpolated. If the horizontal distance between points I and A is Δ, then:

$$H_i = \frac{d - \Delta}{d} H_a + \frac{\Delta}{d} H_b \tag{8.31}$$

Here H_a and H_b are the heights of points A and B, respectively, and H_i is the interpolated height of point I. If points A and B are measured with an accuracy σ_{nod}, then the accuracy of point I, σ_{nod}, which is propagated purely from the two grid nodes,

ACCURACY OF DIGITAL TERRAIN MODELS

can be expressed as follows:

$$\sigma_i^2 = \left(\frac{d-\Delta}{d}\right)^2 \sigma_{\text{nod}}^2 + \left(\frac{\Delta}{d}\right)^2 \sigma_{\text{nod}}^2 \tag{8.32}$$

Equation (8.32) is an expression of the accuracy (in terms of standard deviation) of a particular point located along one side of a surface. However, what is of interest here is the overall average value for all possible points along the line AB, which is a representative value for the DTM profile. In this case, the horizontal distance from these points to point A in Figure 8.7 (Δ in Equation [8.32]) should be regarded as a variable that takes a value from 0 (at point A) to d (at point B). Therefore, the average variance of all points between A and B can be computed as follows:

$$\sigma_S^2 = \frac{1}{d}\int_0^d \left(\left(\frac{d-\Delta}{d}\right)^2 \sigma_{\text{nod}}^2 + \left(\frac{\Delta}{d}\right)^2 \sigma_{\text{nod}}^2\right) d\Delta = \frac{2}{3}\sigma_{\text{nod}}^2 \tag{8.33}$$

where σ_S^2 refers to the overall average value of error variance of all points along the whole profile with a grid interval of d, but only with respect to errors propagated from the source data (i.e., grid nodes).

For the overall accuracy of the points along a profile, another term concerning accuracy loss due to the linear representation of the terrain surface should be added, thus giving the following formula:

$$\sigma_{\text{Pr}}^2 = \sigma_S^2 + \sigma_T^2 = \frac{2}{3}\sigma_{\text{nod}}^2 + \sigma_T^2 \tag{8.34}$$

where σ_T^2 denotes the accuracy loss caused by the linear representation of terrain surfaces in the form of variance (which will be discussed later); σ_{nod}^2 is the variance of errors at grid points; and σ_{Pr}^2 is the overall accuracy of DTM points along the profile with an interval of d, also in terms of variance.

In the case of bilinear surfaces, the interpolation of a point takes place in two perpendicular directions. Suppose A, B, C, and D are the four nodes and point E is the point to be interpolated on the bilinear surface (Figure 8.8). The interpolation can take place initially along AB and CD, using Equation (8.31). Thus, point I can be interpolated from A and B and similarly point J can be interpolated from D and C. The next step takes place between points I and J, that is,

$$H_e = \frac{d-\varepsilon}{d}H_i + \frac{\varepsilon}{d}H_j \tag{8.35}$$

where ε is the horizontal distance from point E to point I and H_e, H_i and H_j are the heights of points E, I, and J, respectively.

Thus, Equation (8.35) again expresses the linear interpolation along a profile with an interval of d. Fundamentally, it is identical to Equation (8.31). Therefore, the same development as for Equation (8.31) can be obtained. However, the accuracy of points I and J in Figure 8.8, as for point I in Figure 8.7, is different from that of points A, B, C, and D; and the actual accuracy value varies with the positions of I and J between the two nodes and the characteristics of the terrain surface. Therefore, the average

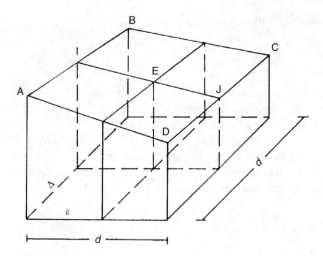

Figure 8.8 Bilinear interpolation of point E by use of four nodes (A, B, C, and D).

value expressed by Equation (8.34), σ_{Pr}^2, should be used as the representative for points I and J in Figure 8.8. Again, there is an accuracy loss (σ_T^2) due to the linear representation for profile IJ. Thus, an analog to Equation (8.34) can be obtained for the accuracy of the points interpolated from a bilinear surface as follows:

$$\sigma_{Surf}^2 = \frac{2}{3}\sigma_{Pr}^2 + \sigma_T^2 \tag{8.36}$$

By substituting Equation (8.34) into Equation (8.36), the following expression can be obtained:

$$\sigma_{Surf}^2 = \frac{2}{3}\left(\frac{2}{3}\sigma_{nod}^2 + \sigma_T^2\right) + \sigma_T^2 = \frac{4}{9}\sigma_{nod}^2 + \frac{5}{3}\sigma_T^2 \tag{8.37}$$

where σ_{Surf}^2 is the average value for the accuracy of the points on a bilinear surface; σ_{nod}^2 is the accuracy of nodes; and σ_T^2 is the accuracy loss due to linear representation of terrain profiles.

8.4.3 Accuracy Loss Due to Linear Representation of Terrain Surface

So far, the general form of the accuracy model of DTM surfaces has already been derived and is expressed by Equation (8.37). In this connection, two important problems needed to be solved:

1. the accuracy of grid points (σ_{nod}^2)
2. the accuracy loss caused by linear representation of terrain surfaces (σ_T^2).

The first problem was addressed in Section 8.1.4. Therefore, the remaining problem is to obtain a good estimate for σ_T^2.

8.4.3.1 Strategy for Determining σ_T^2

The value of σ_T^2 varies with the roughness of the terrain surface, which varies from place to place. Therefore, it is impossible to depict its inflexions using an analytical method, especially for small local deviations. These characteristics can only be handled by using statistical methods.

In linear modeling of a terrain surface, σ_T represents the standard deviation of all height differences (ΔH) between terrain surfaces and the resulting linear facets (the DTM surfaces) constructed from nodes free of errors. In this case, ΔH, that is, the $e(x, y)$ in Equation (8.5), is a random variable. According to the discussion in Section 8.1.3, for a given random variable, regardless of its distribution, its σ value (σ_T here) gives a strong indication of its dispersion. Mathematically,

$$P(|\Delta H - \mu| \leq K\sigma_T) \geq f(K) \qquad (8.38)$$

where μ is the mean value; K is a constant; and $f(K)$ is a function of K with its value ranging from 0 to 1. Suppose ΔH has a normal distribution; if K takes a value of 3, then $f(K)$ is equal to 99.73%. This means that for normal distribution, with the probability of 99.7%, ΔH will have a value (if sampled) from $-3\sigma + \mu$ to $3\sigma + \mu$. This probability is so large that in error theory, 3σ is regarded as the possible maximum error and any error larger than this value is regarded as gross error. Taking an analog from the practice of error theory, the following expression seems appropriate:

$$\sigma_T = \frac{E_{max}}{K} \qquad (8.39)$$

where σ_T is the accuracy loss due to linear representation of the terrain profile; E_{max} is the possible maximum error (which will be specifically discussed later); and K is the same constant as given in Equation (8.38) and its value is dependent on the distribution of ΔH. As DTM errors are not normally distributed, as discussed in Section 8.1.2, the value of K must be quite different from 3, which is the value for normal distribution. On the other hand, as can be seen from Table 8.1, experimental tests reveal that the probability of DTM errors larger than 4σ is approximately 0.3%. Therefore, it seems appropriate to take

$$K = 4$$

for Equation (8.39). As a consequence, the only task left here is to obtain a reliable estimate for the E_{max} in Equation (8.39).

8.4.3.2 Extreme Error (E_{max}) Due to Linear Representation

To analyze the possible extreme values of ΔH, it is necessary to consider some possible outlines of terrain profiles in extreme situations. Since only extreme cases are being examined, some of the analyses may seem unrealistic.

Figure 8.9(a) and Figure 8.9(b) illustrate the maximum possible errors at point C, for the same terrain feature but with different locations of nodes due to a fault or other geological structure giving rise to a steep change in slope. If information giving

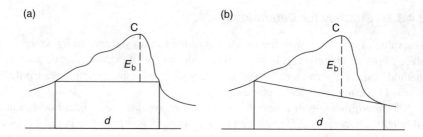

Figure 8.9 The possible maximum errors of linear representation, due to faults or breaks, with different locations of grids.

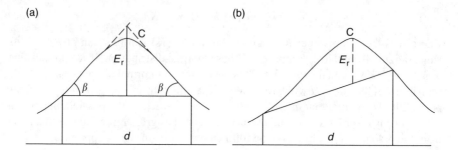

Figure 8.10 The possible maximum errors of representation using grid nodes only, with different locations of grids. (a) The maximum error occurring when a grid contains local maxima or minima. (b) E_r varies with the location of the grids.

a full description of this structure has not been collected, a huge error may result. The value of such an error, E_b here, varies with the characteristics of the terrain itself. Therefore, these values can only be measured directly, but not estimated analytically.

Figure 8.10(a) and Figure 8.10(b) show the possible positive maximum error at point C for different locations of nodes when only points that are located on regular grid nodes are sampled (in other words without F-S points). As shown in Figure 8.10(a), the possible maximum error of E_r arises when C lies in the middle of the grid, giving the following formula:

$$E_{r,\max} = \frac{1}{2}d \tan \beta \tag{8.40}$$

where $E_{r,\max}$ is the possible maximum error in such a case. Similarly, the possible negative error can also be estimated.

Figure 8.11(a) shows the possible errors that may occur for grid data with some F-S points for a convex slope. This figure can be justified because it is not practical to include all convex and concave points, even for the case where pure selective sampling has been carried out on a stereo model (in a photogrammetric system). Figure 8.11(b) is exaggerated from Figure 8.11(a) for the convenience of obtaining a numerical estimate. Point C in this diagram shows an extreme case of convex slope.

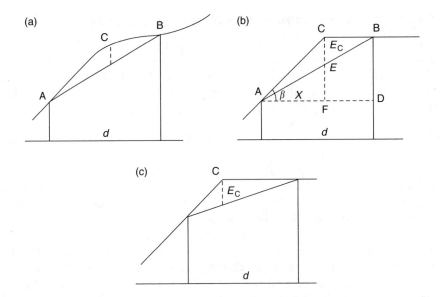

Figure 8.11 Possible maximum errors of the linear representation of ordinary terrain slopes: (a) a convex slope; (b) exaggeration of (a); and (c) variation of E_c with location of grids.

Line AB is the linearly constructed profile; $\angle CAD$ is the slope angle at point A (denoted as β); and line segment CE is the possible error at point C. Therefore:

$$CE = CF - EF = X \tan \beta - \frac{X^2 \tan \beta}{d} \quad (8.41)$$

Figure 8.11(c) shows that the value E_c varies with the location of the grid nodes. The next task is to find the maximum value for CE (Figure 8.11(b)) representing all possible locations of point C in terms of the horizontal distance from point A. If the first-order derivative of CE is considered to be equal to 0, then the location of C where the value of CE reaches its maximum can be determined as follows:

$$\frac{d(CE)}{dX} = \tan \beta - \frac{2X \tan \beta}{d} = 0 \quad (8.42)$$

From Equation (8.42) it can be seen that $X = d/2$. By substituting this value into Equation (8.41) and denoting CE with E_c,

$$E_{c,\max} = CB = \frac{1}{4} d \tan \beta \quad (8.43)$$

Therefore, it can be deduced that the value of possible extreme errors for the case of regular grid data only is double that for composite data. The maximum error due to linear representation is $E_{c,\max}$ for composite data whereas the situation is more complicated for grid data only.

8.4.3.3 A Practical Consideration Regarding E_{max} and σ

The three extreme values identified previously belong to three different types of distribution. E_b applies to grids taken across faults or break lines; E_r is related to grids taken across peaks, pits, ridges, and ravines; and E_c is used for ordinary terrain features and therefore for all the remaining grids. Suppose the proportions of grids that may contain E_c, E_r, and E_b are $P(c)$, $P(r)$, and $P(b)$, then:

$$P(c) + P(r) + P(b) = 1 \tag{8.44}$$

For composite data, both $P(r)$ and $P(b)$ are 0. It is only necessary to consider $P(r)$ and $P(b)$ for regular grid data. If there are no faults or break lines as shown in Figure 8.9, then $P(b)$ is 0. Otherwise, $P(b)$ can be estimated according to the height over the length and width of the faults or breaks.

Similarly, the estimation of $P(r)$ is not an easy task. For a smaller area, there is no better method than simply counting the number of grid cells across the ridge and ravine lines and then dividing by the total number of grids. For a large area, some alternatives may be used. The value of $P(r)$ is directly related to the wavelength of the terrain variation (Figure 8.12). However, the planimetric shape of a hill (expressed by contours) could be different from place to place. Even for the same hill, the wavelengths could be different if the profiles are taken along different directions. Therefore, even a rough estimate, such as an average value, could be valuable. The value of average wavelengths can be estimated as follows:

$$\lambda = 2H \cot \alpha \tag{8.45}$$

where H is the average relative height; α is the average slope angle; and λ is the average wavelength, all taken over the entire area to be modeled (Figure 8.12). In practice, the average value of local relief (half of the maximum minus minimum heights) can be used to represent H. Therefore,

$$\lambda = (H_{max} - H_{min}) \cot \alpha \tag{8.46}$$

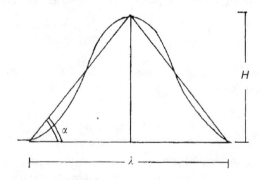

Figure 8.12 Approximate estimation of wavelength λ, where H is the average of height variation.

Once the estimation of λ is made, the value of $P(r)$ can then be estimated. Both the top and bottom of a spatial variation will occur over a single wavelength in one profile. Therefore, for a grid that has two profile directions perpendicular to each other, the frequency of E_r over a grid is as follows:

$$P(r) = \frac{4d}{\lambda} \qquad (8.47a)$$

where λ is the average wavelength; d is the grid interval; and $P(r)$ is the occurrence frequency of the extreme error E_r.

An idealized figure as shown in Figure 8.13 may help in understanding the estimation of $P(r)$. In this example, the total number of grid squares is $(1.5\lambda/d) \times (1.5\lambda/d)$. Suppose all profiles along both directions are identical to the one shown in Figure 8.13, then the total number of grid squares that may contain E_r is, as shown in Figure 8.13, approximately equal to $6(1.5\lambda/d)$. Thus, $P(r)$ is $4\lambda/d$, which is expressed in Equation (8.47a). However, a more important consideration is for the area with size $\lambda \times \lambda$. Figure 8.13 shows that in this unit area, $P(r) = 4d/\lambda = d(4\lambda/\lambda^2)$. Here, 4λ is the perimeter and λ^2 is the area. Therefore, the following formula may be more common and appropriate for the estimation of $P(r)$:

$$P(r) = \frac{\text{Perimeter of lowest contour}}{\text{Area enclosed by lowest contour}} \qquad (8.47b)$$

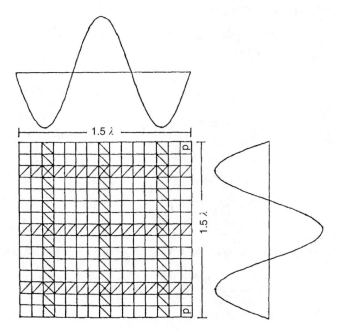

Figure 8.13 Estimation of ratio ($P(r)$) of grid nodes containing the local biggest and smallest points.

Equation (8.47b) could be very helpful for the estimation of $P(r)$ from existing contours, which from the map appear to be irregular in shape.

Therefore, for DTM surface linearly constructed from grid data only, the value of σ_T can be estimated as follows:

$$\sigma_T = \frac{P(r)E_{r,\max} + P(c)E_{c,\max} + P(b)E_{b,\max}}{K} \tag{8.48}$$

Such an averaging operation is not justified from a statistical point of view since E_b, E_r, and E_c belong to three different distributions. However, in the practice of DTM accuracy assessment, it is never possible to distinguish these three types of errors and estimates are made from a sample that contains all of them. Therefore, Equation (8.48) is an appropriate representation of DTM practice.

In practical terms, E_b rarely occurs and even if it does occur, it is normally sampled. Therefore, it is acceptable to neglect E_b in Equation (8.48), giving:

$$\sigma_T = \frac{P(r) \times E_{r,\max} + P(c) \times E_{c,\max}}{K} = \frac{P(r) \times E_{r,\max} + (1 - P(r)) \times E_{r,\max}}{K} \tag{8.49}$$

8.4.4 Mathematical Models of the Accuracy of DTMs Linearly Constructed from Grid Data

It has been shown that accuracy loss due to linear representation from measured grid data only can be rewritten as follows:

$$\begin{aligned}
\sigma_{T,r} &= \frac{E_{c,\max}}{K}(1 - P(r)) + \frac{E_{r,\max}}{K}P(r) \\
&= \frac{d \tan \alpha}{4K}(1 - P(r)) + \frac{d \tan \alpha}{2K}P(r) \\
&= \frac{d \tan \alpha}{4K}(1 + P(r)) \\
&= \frac{d \tan \alpha}{4K}\left(1 + \frac{4d}{\lambda}\right)
\end{aligned} \tag{8.50}$$

For a DTM constructed from composite data, the accuracy loss formula is as follows:

$$\sigma_{T,c} = \frac{E_{c,\max}}{K} = \frac{d \tan \alpha}{4K} \tag{8.51}$$

Substituting Equations (8.51) and (8.50) into Equation (8.37), the accuracy models of the DTM linearly constructed from composite data and grid data only, respectively, are as follows:

$$\sigma^2_{Surf/c} = \frac{4}{9}\sigma^2_{nod} + \frac{5}{48K^2}(d\tan\alpha)^2$$

$$= \frac{4}{9}\sigma^2_{nod} + \frac{5}{48 \times 4^2}(d\tan\alpha)^2$$

$$= \frac{4}{9}\sigma^2_{nod} + \frac{5}{768}(d\tan\alpha)^2 \quad (8.52a)$$

$$\sigma^2_{Surf/r} = \frac{4}{9}\sigma^2_{nod} + \frac{5}{48K^2}(1+P(r))^2(d\tan\alpha)^2$$

$$= \frac{4}{9}\sigma^2_{nod} + \frac{5}{768}\left(1+\frac{4d}{\lambda}\right)(d\tan\alpha)^2 \quad (8.52b)$$

where $\sigma^2_{Surf/c}$ and $\sigma^2_{Surf/r}$ denote the accuracies of DTM linearly constructed from composite data and grid data only (both in terms of variance); σ^2_{nod} is the variance of errors at measured grid nodes; K is a constant (approximately equal to 4 depending on the characteristics of the terrain surface); α is the average slope angle of the area; $P(r)$ is the proportion of grids that may contain E_r and is expressed in Equation [8.47]); and λ is the wavelength of the profiles in the area, and is expressed by Equation (8.46).

At this point, the derivation of the formulae have all been completed. However, the discussion can be extended to provide an approximation of Equations (8.52) as follows:

$$\sigma_{Surf/c} = \frac{2}{3}\sigma_{nod} + \frac{\sqrt{5}}{\sqrt{48K}}(d\tan\alpha)$$

$$= \frac{2}{3}\sigma_{nod} + \frac{\sqrt{5}}{16\sqrt{3}}(d\tan\alpha) \quad (8.53a)$$

$$\sigma_{Surf/r} = \frac{2}{3}\sigma_{nod} + \frac{\sqrt{5}}{\sqrt{48K}}(1+P(r))(d\tan\alpha)$$

$$= \frac{2}{3}\sigma_{nod} + \frac{\sqrt{5}}{16\sqrt{3}}\left(1+\frac{4d}{\lambda}\right)(d\tan\alpha) \quad (8.53b)$$

These equations represent an analog to the Koppe Formulae, which are widely used for specifying map accuracy in middle European countries. Equation (8.53) proves to be a very good approximation of Equation (8.52) in the case where grid intervals are relatively small and is more convenient to use in practice.

The experimental test results obtained in Section 8.3 were used to evaluate this model (Li 1990, 1993b). The basic facts about these areas were given in Table 8.5. The wavelengths for these three areas are estimated as 470 m (computed by Equation [8.46]), 214 m (i.e., the width of the test area checked), and 300 m (the width of the test area checked). The results are given in Table 8.10. These results reveal that this set of mathematical models produce reasonably reliable estimates of DTM accuracy.

Table 8.10 Comparison of the Predicted Accuracy and the Test Results

Test Area	Grid Interval (m)	Grid Data Only			Composite Data		
		Predicted Accuracy (m)	Tested (m)	Difference (m)	Predicted Accuracy (m)	Tested (m)	Difference (m)
Uppland	$20\sqrt{2}$	0.54	0.63	−0.09	0.51	0.59	−0.08
	40	0.64	0.76	−0.13	0.56	0.66	−0.10
	$40\sqrt{2}$	0.85	0.93	−0.08	0.66	0.70	−0.04
	80	1.24	1.18	0.06	0.81	0.80	0.01
Sohnstetten	20	0.63	0.56	0.07	0.45	0.43	0.02
	$20\sqrt{2}$	0.97	0.87	0.10	0.63	0.56	0.07
	40	1.56	1.45	0.11	0.87	0.78	0.09
	$40\sqrt{2}$	2.58	2.40	0.18	1.23	1.08	0.15
Spitze	10	0.17	0.21	−0.04	0.12	0.16	−0.04
	$10\sqrt{2}$	0.25	0.28	−0.03	0.15	0.17	−0.02
	20	0.38	0.35	0.03	0.20	0.18	0.02

For theoretical evaluation, the following set of parameters can be used as standards: (a) accuracy, (b) descriptive realism, (c) precision, (d) robustness, (e) generality, (f) fruitfulness (Meyer 1995), and (g) simplicity (Li 1990, 1993b).

8.5 EMPIRICAL MODEL FOR THE RELATIONSHIP BETWEEN GRID AND CONTOUR INTERVALS

In Section 8.4, theoretical models for the accuracy of DTMs derived from grid data were presented. In this section, the accuracy of DTMs derived from contour data will be discussed and then the relationship between contour and grid intervals will be addressed.

8.5.1 Empirical Model for the Accuracy of DTMs Constructed from Contour Data

According to contour map accuracy specifications, the accuracy of a DTM derived from contour data can be expressed by the following expression:

$$\sigma_{\text{DTM, cont}}^2 = \frac{\sigma_{\text{DCD}}^2}{C} + \left(\frac{\text{CI}}{K}\right)^2 \qquad (8.105)$$

where σ_{DCD} is the standard deviation of errors in digital contour data; CI is the contour interval; K and C are two constants; and $\sigma_{\text{DTM,cont}}$ is the standard deviation of errors at the DTM derived from the digital contour data.

As a general guide, the value of $\sigma_{\text{DTM,cont}}$ is about $\frac{1}{3}$CI. Experimental results confirm such a conclusion. Table 8.11 lists some results obtained from the three ISPRS test areas (Li 1990, 1994). In each area, there is a set of photogrammetrically

ACCURACY OF DIGITAL TERRAIN MODELS

Table 8.11 The Improvement of DTM Accuracy with the Addition of F-S Data into Contour Data

Parameter	Uppland With F-S Data	Uppland Without F-S Data	Sohnstetten With F-S Data	Sohnstetten Without F-S Data	Spitze With F-S Data	Spitze Without F-S Data
RMSE (m)	0.93	1.74	0.35	0.91	0.17	0.27
μ (m)	0.47	1.05	0.11	0.22	0.09	0.10
σ (m)	0.80	1.39	0.35	0.88	0.15	0.24
$+E_{max}$ (m)	3.25	5.91	1.73	4.52	0.75	0.94
$-E_{max}$ (m)	−5.18	−5.18	−2.48	−3.01	0.95	−0.95
CI/σ	6.25	3.60	4.29	5.68	6.67	4.17
Reduction in σ (%)	42.45		60.23		37.50	

measured contour data. Like the case of square grid data, two types of data could be generated by a combination of contour data with F-S points: (a) contour data only and (b) contour data with F-S data. The improvement of DTM accuracy with the inclusion of F-S data is noticeable.

The exact value of C in Equation (8.54) is difficult to estimate as each triangle may have a different shape. However, a value of 3 seems appropriate since three points have been used for interpolation within a linear facet. The value of K in Equation (8.54) ranges from 4.5 to 5.9. This implies that the error budget deriving from the loss in fidelity of terrain topography that is selectively represented by only contours is in the range from CI/4 to CI/6, depending on the characteristics of the terrain topography. When F-S data are included, the K in Equation (8.54) ranges from 10 to 30. In terms of standard deviation, the level has been reduced to CI/6 to CI/15 from the original level of CI/3 to CI/5. The improvement is about 40 to 60%. As a result, Equation (8.54) may be written as follows:

$$\sigma^2_{DTM,cont} = \frac{\sigma^2_{DCD}}{3} + \left(\frac{CI}{K}\right)^2, \quad K = \begin{cases} 4\text{–}6, & \text{if no F-S data} \\ 10\text{–}30, & \text{if with F-S data} \end{cases} \quad (8.106)$$

This is based on the photogrammetrically measured contour data. However, if it is digitized from contour maps, the accuracy would be poorer than the one expressed by Equation (8.55).

8.5.2 Empirical Model for the Relationship between Contour and Grid Intervals

As has been discussed previously, photogrammetrically measured data and contour data are the two major types of data for digital terrain modeling. Among the various strategies for photogrammetric sampling for DTM data acquisition, regular grid sampling is the most popular. Therefore, it would be of interest to discuss the relationship between grid-based and contour-based terrain representations, in terms of fidelity, accuracy, or quality.

To compare the quality of these two types of DTM, it is helpful to have a comparison of data density between them. For grid data, the density is represented by grid interval (d) while for contour data the density is expressed by (vertical) contour intervals (CI). The average value of the planimetric contour interval for a contour interval of CI can be expressed as follows:

$$D = \text{CI} \times \cot \alpha \qquad (8.107)$$

where α is the average slope angle of the terrain surface, CI is the vertical contour interval; and D is the planimetric contour interval.

Theoretically speaking, the accuracy of the DTM derived from contour data with an interval of CI should be approximately equal to that derived from grid data with intervals of D as expressed in Equation (8.56) if the terrain is homogenous; the effect of errors involved in source data on the final results is the same.

Experimental results (Li 1990, 1994) show that the relationship is not that straightforward and a factor, K, needs to be included in Equation (8.56), that is,

$$d = K \times D = K \times \text{CI} \times \cot \alpha \qquad (8.108)$$

where

$$K = \begin{cases} 1.2\text{–}2.0, & \text{if no F-S data} \\ 1.0\text{–}1.5, & \text{if with F-S data} \end{cases} \qquad (8.109)$$

CHAPTER 9

Multi-Scale Representations of Digital Terrain Models

The previous two chapters addressed the issues on the quality of DTMs. The representations of DTM data will be discussed in the next three chapters. This chapter will look at some issues on multi-scale representation.

9.1 MULTI-SCALE REPRESENTATIONS OF DTM: AN OVERVIEW

Before multi-scale representations of DTM data can be discussed, it is essential to clarify some concepts related to scale, to understand what is to be addressed in this chapter.

9.1.1 Scale as an Important Issue in Digital Terrain Modeling

"Scale is a confusing concept, often misunderstood, and meaning different things depending on the context and disciplinary perspective" (Quattrochi and Goodchild 1997, p. 395). Scale is an old issue in geosciences such as cartography and geography.

In cartography, maps are produced at different scales, for example, 1:10,000 and 1:100,000. A ground area with a fixed size will be mapped into a bigger map space at a larger scale. For example, a ground area of 10 km × 10 km will be a map area of 1 m × 1 m at a scale of 1:10,000. However, at 1:100,000, it is an area of only 0.1 m × 0.1 m on map. Due to the reduction in size (from 1 m × 1 m to 0.1 m × 0.1 m), the same level of detail (LOD) as represented on the larger-scale map cannot be represented on the smaller-scale map. This means that the representations of the same feature in the same area will be different when the maps are at different scales. Therefore, there is a multi-scale issue in cartography, that is, how to derive small-scale maps from large-scale maps through operations such as simplification and aggregation. This issue is called *generalization*.

In geography, there is a similar issue. Normally, geographical data are sampled from enumeration units. The size of the enumeration unit is a scale indicator. In some geographical applications, a larger unit is needed as the basis for analysis. The sampled data need to be aggregated into a larger enumeration unit from the original unit. However, the statistical results may be different when the analysis is carried out based on different sizes (i.e., different scales) of the enumeration unit. Therefore, the issue is how to aggregate data from small enumeration units to larger units for processing. This issue is called the "modifiable areal unit" issue (Openshaw 1994). Indeed, there are similar issues in all geosciences, such as geomorphology, oceanography, soil science, biology, biophysics, social sciences, hydrology, environmental sciences, and so on.

In digital terrain modeling, there is also a similar issue. Currently, DTM data at a national level are produced at various scales, for example, 1:10,000, 1:50,000, and 1:100,000, for different applications. They are mainly derived from contour maps at the corresponding scales. It would be desirable to maintain and update the DTM data at the largest scale only and to derive DTM data at any smaller scale by a generalization process, that is, DTM generalization.

In computer graphics and games, landscape is often used as a background in various situations such as driving and flight simulations. In such cases, the LOD of the terrain surface may appear to be different if viewed from different distances. Therefore, in a scene, some parts may be represented in more detail and other parts in less detail. This is also a multi-scale issue and is simply termed LOD, which is also an important issue in DTM visualization.

As will be discussed later, LOD is an issue quite different from generalization. In this chapter, the emphasis is on generalization although the basic ideas on LOD will also be outlined.

9.1.2 Transformation in Scale: An Irreversible Process in Geographical Space

Here, the term *geographical space* means the real world. However, the term *real world* is still confusing because different disciplines study different aspects of it. Nuclear physics studies particles at the sub-molecular level in units of nanometers. This is an extreme at a micro-scale. At the other extreme, astro-physics studies the planets at an intergalactic level in units of light-years (the distance travelled by light in the period of a year). Such studies are at a macro-scale. Between these two extremes, many scientific disciplines study the planet earth, such as geology, geography, geomatics, geomorphology, geophysics, which are collectively called *geosciences*. Here, *real world* refers to the world studied by the geosciences. Such studies are at a scale called *geo-scale*. By an analogy to electromagnetic (EM) spectrum, the scale range, from micro-scale to geo-scale to macro-scale, is termed the *scale spectrum* (Figure 9.1). Like the visible light band in the EM spectrum, geo-scale is also a small band in the scale spectrum (Li 2003). Digital terrain modeling is a branch of geoscience.

The transformation in the scale of geographical space is much more complicated than that in Euclidean space. In Euclidean space, any object has an integer dimension, that is, 0 dimension for a point, 1 for a line, 2 for a plane, and 3 for a volume.

MULTI-SCALE REPRESENTATIONS OF DIGITAL TERRAIN MODELS

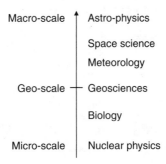

Figure 9.1 Geo-scale in the scale spectrum.

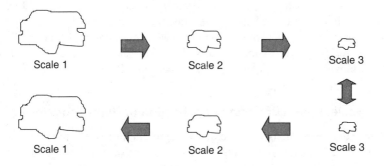

Figure 9.2 Scale change in 2-D Euclidean space: a reversible process.

An increase (or decrease) in scale will cause an increase or (or decrease) in length in a 2-D space and in volume in a 3-D space. However, the shape of the object remains unchanged. Figure 9.2 illustrates the scale reduction in a 2-D Euclidean space. Scale 2 is a two-time reduction of scale 1 and scale 3 is a four-time reduction of scale 1. In such a transformation process, the length of the perimeter is reduced by two and four times, respectively, and the area of the object is reduced by 2^2 and 4^2 times, respectively. When the object at scale 3 is increased four times, the shape of the object is identical to the original one shown at scale 1. That is, the transformations are reversible.

However, in geographical space, the dimensions are not integers. The concept of fractal dimensions was introduced by Mandelbrot (1967). In such a space, a value between 1.0 and 2.0 is the dimension of a line and a value between 2.0 and 3.0 is the dimension of a surface. In fractal geographical space, it was discovered long ago that different lengths will be obtained for a coastline represented on maps at different scales. The length measured from smaller-scale maps will be shorter if the same unit size (at map scale) is used for measurement. This is because different levels of reality (i.e., the Earth's surface with different degrees of abstraction) have been measured. Indeed, on maps at a smaller scale, the level of complexity of an object is reduced to suit the representation at that scale. But when the representation at a smaller scale is enlarged back to the original size, the level of complexity cannot be recovered.

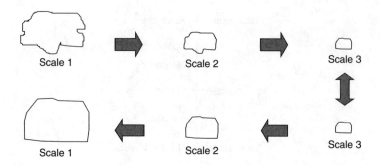

Figure 9.3 Scale change in 2-D geographical space: an irreversible process.

Figure 9.3 illustrates that such a transformation is not reversible. Li (1996, 1999) regarded such kind of transformation as "transformation in scale dimension."

This means that the generalization of a DTM from a large-scale to a smaller-scale representation is an irreversible process. It will be shown in Section 9.3 that this process follows a natural principle (Li and Openshaw 1993).

9.1.3 Scale, Resolution and Simplification of Representations

The size of the basic unit for measurement or representation is referred to as *resolution*. If the data are in raster format, the size of the raster of pixels is referred to as the resolution. The larger the pixel size, the lower the resolution. In the case of grid DTM, the grid interval is usually regarded as the resolution.

Normally, the scale and resolution of spatial data are tightly packed. With the resolution of one's eyes fixed, when one views an object more closely, the images formed in one's eyes are larger, thus the images have a higher resolution or are at a larger scale. That is, in normal cases, the resolution is also good indicator of scale for DTM data. This is because resolution means the *level of detail* and scale means the *level of abstraction*. However, scale is not equal to resolution. Scale could refer to the *ratio of distances* as well as the *relative size of interest*. Figure 9.4 shows four images at the same scale but with four different resolutions. Similarly, digital maps can be plotted at any scale one wishes, but the resolution is fixed.

With the introduction of resolution, it is easier to explain the difference between scale reduction in Euclidean and geographical spaces. In Euclidean space, reduction in the size of an object does not cause a change in its complexity. This can be understood with the following line of thought. When the scale is reduced, the size of the object is also reduced, but at the same time, the basic resolution of the observation instrument is also refined by the same magnification. This is implied in Euclidean geometry. For example, if the scale is reduced by two times, the size is reduced by two times and the resolution is two times finer (i.e., higher). On the other hand, in geographical space, when the scale is reduced, the size of an object is also reduced, but the basic resolution of the observation instrument remains unchanged. That is, this change of complexity is achieved by changing the relationship between the size of the object and the resolution of observation.

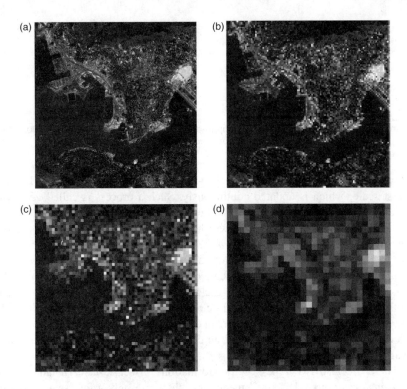

Figure 9.4 Four images of Hong Kong with the same scale but different resolutions. The color plate can be viewed at http://www.crcpress.com/e_products/downloads/download.asp?cat_no=TF1732.

Table 9.1 The Cause and Effect of Scale Reduction in Euclidean and Geographical Space

	Effect		Cause	
Space	Relative Complexity	Absolute Complexity	Instrumentation Resolution	Observer's Resolution
Euclidean space	Increased	Unchanged	Reduced	Unchanged
Geographical space	Unchanged	Decreased	Unchanged	Unchanged

There are ways in which to achieve this result (Table 9.1). The first is to change the size of the object but, at the same time, to retain the basic resolution of the observational instrument. The second is (a) to retain the size of the object but change the basic resolution of the observation instrument and then (b) to change the size of the observed objects by simple reduction in Euclidean space.

9.1.4 Approaches for Multi-Scale Representations

It becomes clear that there are two different types of multi-scale representations. The first one is map-like, emphasizing the metric quality, and is thus useful for measurement. The multi-scale issue on the DTM is related to how to automatically

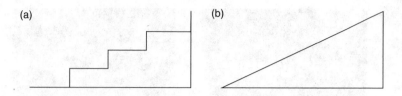

Figure 9.5 Steps and linear slope compared to discrete and continuous transformations: (a) discrete (steps) and (b) continuous (linear slope).

derive DTM data suitable for any smaller-scale representation from the DTM at the largest scale, which is updated continuously. Such a process is called *generalization* and it is applied uniformly across the whole area covered by the DTM data so that all data points within the area will have a uniform accuracy. This is referred to as *metric multi-scale representation* here. The other type, for visual impression only (e.g., for computer graphics and games), is called *visual multi-scale representation*. On the same image, the scale of the image pixels produced from DTM data is not the same over the whole image and is a function of viewing distance. In other words, the LOD represented on an image varies from place to place. This kind of approach is simply called *LOD* in computer graphics. In some literature, it is also called *view-dependent LOD* and, in contrast, metric multi-scale representation is called *view-independent LOD*.

There are also two types of transformations in scale, *discrete* and *continuous transformations*. "Discrete" means that there are only a few scales available, for example, 1:10,000, 1:100,000, and 1:1,000,000. The transformation jumps from 1:10,000 to 1:100,000, then to 1:1,000,000. Discrete transformation is like fixed steps in a staircase (Figure 9.5a). "Continuous" means that transformation can be to any scale, for example, 1:50,000 or 1:56,999, although in practice some scales are not used (e.g., 1:56,999). Continuous transformation is like a linear slope (Figure 9.5b).

From the above discussions, the approaches for multi-scale representation of DTM data can be summarized as in Figure 9.6.

9.2 HIERARCHICAL REPRESENTATION OF DTM AT DISCRETE SCALES

The hierarchical representation seems to be popular for the representation of DTM data at discrete scales, particularly in computer graphics and games and terrain visualization, in order to speed up the data processing. Both grid and triangular networks could be represented in hierarchy (de Berg and Dobrindt 1998; de Floriani 1989).

9.2.1 Pyramidal Structure for Hierarchical Representation

The pyramid is the most commonly used hierarchical representation of DTM data. Figure 9.7 shows three-level pyramid structures of square grid and triangular grids. That is, four squares (or triangles) at the third level form a larger square quadrilateral

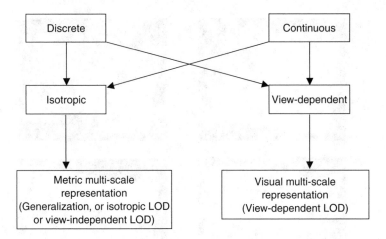

Figure 9.6 Alternative approaches for multi-scale representation of DTM data.

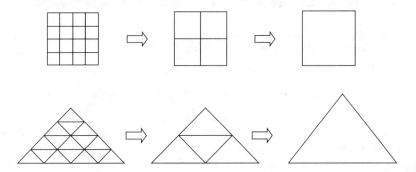

Figure 9.7 Pyramid representations of grid and triangular networks.

(or triangle) at the second level. Similarly, four squares (or triangles) at the second level form the largest square (or triangle) at the first level. The number of squares (triangles) at the nth level is 4^{n-1}.

The sizes of the squares at the same level of the pyramid structure are identical. Figure 9.8 is an example of the grid pyramid representation of DTM data, where the original DTM is represented in three hierarchical levels. In the four-to-one aggregation process, simple averaging is adopted to compute the height value of the new grids. For example, if the heights of the four grid nodes at the fourth level are 5, 6, 4.5, and 5.5 m, then the average height value of these, that is, $(5+6+4.5+5.5)$ m/4 = 5.25 m, is used as the height for the new grid node at the third level.

The hierarchical concept of a simple pyramid emphasizes the level of grid sizes, that is, different resolutions, to represent the terrain surface at different scales. This is also the simplest LOD for visualization of DTM data. However, it does not take into consideration certain terrain features and thus usually produces relatively obvious visual distortions due to the loss of important surface characteristics and discontinuity at the boundaries between grid cells.

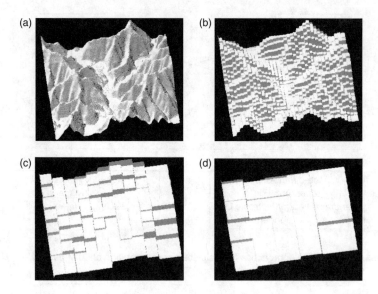

Figure 9.8 Pyramid representation of grid DEM: (a) original DEM at 1:20,000; (b) the second level; (c) the third level; and (d) the fourth level.

9.2.2 Quadtree Structure for Hierarchical Representation

A major shortcoming of the simple pyramid structure is that the grid intervals among the grid cells are identical at any hierarchical level irrespective of whether the terrain surface is complicated or simple. This may cause problems if some parts of an area are complicated while other parts are simple. In this case, a hierarchical structure with varying grid sizes would be more desirable. The complicated parts could be represented with grid cells with finer resolution (i.e., smaller grid interval) and the simpler parts could be represented by grid cells with coarser resolution that is, (larger grid interval). *Area quadtree*, or simply *quadtree*, is such a kind of grid in common use. Figure 9.9 is an illustration of quadtree structure. Similarly, triangular cells also can be represented by quadtree. Figure 9.10 is an example of triangular quadtree.

The aggregation of four cells into one is similar to the simple pyramid. The only difference is that in a quadtree, some criteria must be set so as to determine whether aggregation should take place for the four given cells. For example, if the height differences are larger than the threshold, then no aggregation should take place, otherwise they are aggregated into one. Figure 9.11 is an example of a hierarchical representation by a quadtree.

In visualization with such a representation, when a more detailed level is desired, the next level with smaller grid intervals will be displayed, only in those parts with complicated terrain variations. To speed up the visualization process, one would like to generate an LOD for in as many levels as possible. However, in normal practice, only three to five LODs are produced and stored. Further levels are generated in real time by algorithms.

MULTI-SCALE REPRESENTATIONS OF DIGITAL TERRAIN MODELS

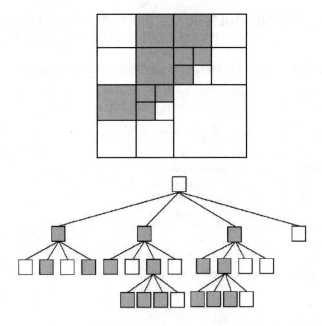

Figure 9.9 Hierarchical representation of grid by quadtree structure.

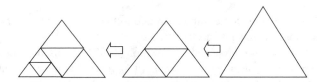

Figure 9.10 Hierarchical description of terrain triangular network by use of quadtree.

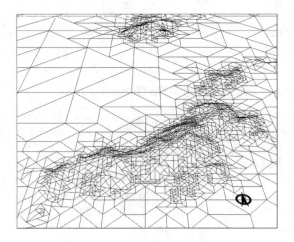

Figure 9.11 Quadtree representation of DTM (Cheng 2000).

9.3 METRIC MULTI-SCALE REPRESENTATION OF DTM AT CONTINUOUS SCALES: GENERALIZATION

9.3.1 Requirements for Metric Multi-Scale Representation of DTM

Since the later 1970s, metric multi-scale representation of DTM data has been a research topic. A list of six criteria were proposed by Weibel (1987) for the evaluation of the methodology used:

1. to run as automatically as possible
2. to perform a broad range of scale changes
3. to be adaptable to the given relief characters
4. to work directly on the basis of the DTM
5. to enable an analysis of the results
6. to provide the opportunity for feature displacement based on the recognition of the major topographic features and individual landforms (major scale reduction).

Three approaches have been proposed for metric multi-scale representation of DTM data.

1. filtering methods (e.g., Loon 1978)
2. generalization of structure lines (e.g., Wu 1997)
3. a hybrid (Weibel 1987) of the above two.

However, these methods do not directly relate to the filtering process to scales and, therefore, are not true generalization methods.

If the set of criteria proposed by Weibel (1987) are used to evaluate these pyramidal representations, the results are not too good. The most serious shortcoming is that only a fixed number of scales are produced, at least for hierarchical grid networks, that is, 2-time, 4-time, 8-time, 16-time, ... scale reduction. Therefore, this approach is only convenient for data structures with which visualization of DTMs could be speeded up, but it cannot be used to produce smaller-scale DTMs from larger-scale ones. In this section, an approach based on a natural principle will be introduced.

9.3.2 A Natural Principle for DTM Generalization

The natural principle formalized by Li and Openshaw (1993) mimics the generalization process of the Nature. The example they used is the Earth's surface viewed from different heights. If one views the Earth from the Moon, all terrain variations disappear and the Earth appears like a blue ball. If one views it from a satellite, then the terrain surface becomes very smooth. These phenomena can easily be checked by forming a stereo model from a pair of satellite images such as SPOT images with 10-m resolution or Spacelab Metric Camera photography. Such a stereo model is at a very small scale. When one views the terrain surface from an airplane, small details are still not visible but the main characteristics of the terrain variations become clear. It is a commonplace to photogrammetrists that stereo models formed from high-altitude photography are more generalized than those formed from low-altitude photography. These illustrate the generalization process, or the transformation in scale.

This is also due to the limitation of the human eye's resolution. One is not able to see objects beyond this resolution. When the viewpoint is higher, the ground area corresponding to the human eye's resolution becomes larger, thus the ground surface appears more abstract. In the case of stereo models formed from images, it is due to the resolution of the images. That is, all information within the image resolution (e.g., 10 m per pixel in the case of SPOT images) disappears. These examples underline a universal principle, a natural principle as described by Li and Openshaw (1993), which states:

> for a given scale of interest, all details about the spatial variations of geographical objects beyond certain limitation are unable to be presented and can thus be neglected.

It follows, therefore, that a simple corollary to this process can be used as a basis for the transformations in scale dimension. The corollary can be stated as follows:

> By using a criterion similar to the limitation of the human eye's resolution and neglecting all the information about the spatial variation of spatial objects beyond this limitation, zooming (or generalization) effects can be achieved.

Li and Openshaw (1992) also term this limitation as the smallest visible object (SVO) but it will be called the smallest visible size (SVS) in this context. Figure 9.12

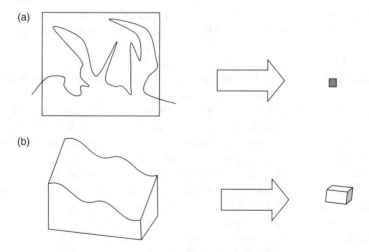

Figure 9.12 The natural principle given by Li and Openshaw (1993): a point or a cell can be used to represent the spatial variations within a certain limitation. (a) In 2-D: spatial variations within this area can be completely neglected and then represented by a point (or a raster cell). (b) In 3-D: spatial variation within this volume can be completely neglected and then represented by a point (or a voxel cell).

illustrates the natural principle for a 2-D (Figure 9.12a) and a 3-D representation (Figure 9.12b).

Now the critical question is how to compute the size of the SVS. Through intensive testing, Li and Openshaw (1992) found that a value from 0.5 to 0.7 mm on the target (or output) map will enable them to produce line generalization results similar to that done by human beings. Let k be the SVS on the map and K be the SVS on the ground, then

$$K = k \times S_T \tag{9.1}$$

where S_T is the scale factor of the target map.

However, there is a problem associated with this formula. That is, the value of K is the same no matter how large the scale S_S of the source (input) map is. To solve this problem, Li and Openshaw (1992, 1993) modified Equation (9.1) into

$$K = k \times S_T \times \left(1 - \frac{S_S}{S_T}\right) \tag{9.2}$$

This equation seems more reasonable. When the difference between S_S and S_T is small, the K value is also small. This means that little needs to be changed. In extreme, when $S_T = S_S$, $K = 0$, no generalization will be done.

9.3.3 DTM Generalization Based on the Natural Principle

This principle was successfully implemented for digital map generalization (Li and Openshaw 1992). It has also been demonstrated that this natural principle is equally applicable for 3-D representations. Figure 9.13 illustrates the generalization of a 3-D surface. In this figure, one views the terrain from two different heights, that is, level A and level B. Suppose the viewpoint at each level is the same and the viewing resolution is also identical, then the surface could be generalized easily, as shown in Figure 9.13(b) and Figure 9.13(c), respectively.

Figure 9.13 illustrates the generalization based on a central projection and the degree of generalization will vary with the viewing distance. But if one considers only a small area of the surface just below the viewpoint each time and moves the viewpoint little by little, then an orthogonal projection is approximated by the central projection and the DTM surface will be generalized more uniformly.

The process of applying the natural principle to DTM generalization is illustrated in detail in Figure 9.14. It works like a convolution process, carried out cell by cell on the input DTM. Each time, a template with a size equal to the SVS is placed onto a cell of the input DTM, all the cells within this template will be used to estimate the height of this cell in the output DTM. The computation process is illustrated in Figure 9.15. Two methods were used (Li and Li 1999). One is the average of all cells within the template and the other is the average of the cells along the edges of the template. Indeed, all the point-based interpolation methods described in Chapter 6 can be used for this purpose.

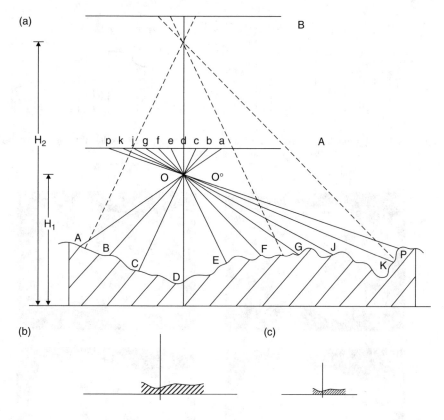

Figure 9.13 Generalization of a 3-D representation based on the natural principle (Li and Openshaw 1993).

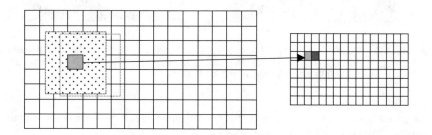

Figure 9.14 The movement of an SVS template over the input DTM pixel by pixel, within which all spatial variations are ignored, for example, nine cells (16 nodes) are aggregated into one in this example.

Figure 9.16 shows a set of DTMs generalized based on the method described in this section. The 1:20,000-scale DTM, as shown in Figure 9.8, was then generalized to produce DTMs at 1:50,000, 1:100,000, and 1:200,000. It is clear that the surface becomes smoother as the scale becomes smaller. The contour plots of these DTMs

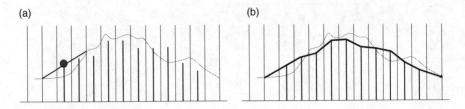

Figure 9.15 Aggregation of pixels within an SVS template (4 × 4 pixels) (the average of the boundary pixels only is used in this example).

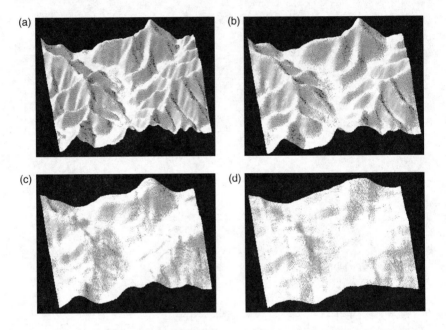

Figure 9.16 Generalization of DTM from 1:10,000 to 1:200,000, based on the natural principle: (a) original DTM at 1:20,000; (b) generalized for 1:50,000; (c) generalized for 1:100,000; and (d) generalized for 1:200,000.

are given in Figure 9.17. It is clear that contour lines are smoother when the scale is smaller. This is in accordance with the natural generalization.

It must be noted here that this generalization method is similar to low-pass filtering but they differ in many aspects. First, the template can be moved gradually pixel by pixel or jumped from one pixel to another by a distance smaller than template size. In an extreme case, there may be no overlap between templates, leading to the simple pyramidal structure. Second, the size of the template is computed from the values of the source and target scales.

In this method, the degree of generalization is the only concern but the resulting grid size of the final DTM is not. If desirable, a quadtree structure may be used to represent (or compress) the generalized DTM.

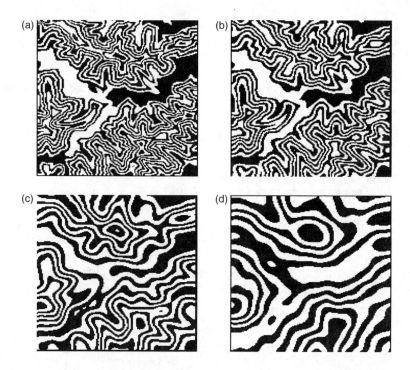

Figure 9.17 Contour representation of DEMs at different scales: (a) 1:20,000; (b) 1:50,000; (c) 1:100,000; and (d) 1:200,000.

This method is equally applicable to triangular networks. In such a process, the height of each triangular vertex will be modified.

9.4 VISUAL MULTI-SCALE REPRESENTATION OF DTM AT CONTINUOUS SCALES: VIEW-DEPENDENT LOD

As discussed previously, the LOD technique is used for computer graphics and DTM visualization.

9.4.1 Principles for View-Dependent LOD

The basic idea of LOD is simple. That is, more detail is used for scenes or objects closer to the viewpoint and less detail for scenes or objects further from the from viewpoint.

If the DTM is in grid form, then coarser grid cells will be used for the representation of the scenes or objects that are more distant from the viewpoint. If the DTM is triangular, then coarser triangular cells will be used for the representation of the scenes or objects that are more distant from the viewpoint. Coarser cells, either grid or triangular, are generated by a number of operations such as *collapse* and *removal*. Figure 9.18 illustrates four of the basic operations for the simplification of a triangular network for LOD purposes:

1. *Vertex removal*: A vertex in the triangular network is removed and new triangles are formed (Figure 9.18a).
2. *Triangle removal*: A complete triangle with three vertices is removed and new triangles are formed (Figure 9.18b).
3. *Edge collapse*: An edge with two vertices is collapsed to a point and new triangles are formed (Figure 9.18c).
4. *Triangle collapse*: A complete triangle with three vertices is collapsed to a point and new triangles are formed (Figure 9.18d).

Now the question arising is when to use these operations. This is about the constraints of simplification. Recalling the VIP selection discussed in Chapter 4, two constraints were used: (a) the number of VIPs to be retained and (b) the allowable accuracy loss. These two constraints can also be used to simplify the DTM data for the generation of view-dependent LOD, leading to two distinct approaches. However,

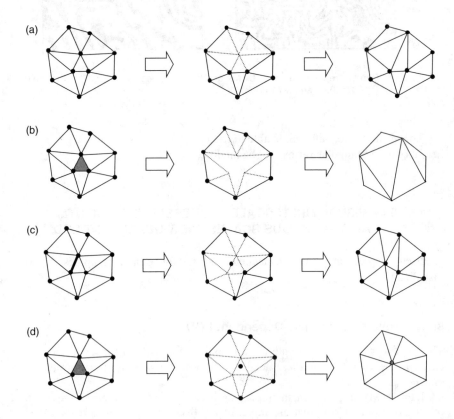

Figure 9.18 Basic geometric operations for simplification of the triangular network: (a) vertex removal; (b) triangle removal; (c) edge collapse; and (d) triangle collapse.

the number of triangles should be used in LOD instead of the number of VIPs. The method making use of allowable error as the constraint is also called *fidelity-based simplification* and that making use of the number of triangles as the constraint is also called *budget-based simplification* (Luebke et al. 2003).

9.4.2 Typical Algorithms for View-Dependent LOD for DTM Data

One of the first real-time continuous LOD algorithms for terrain grids was the early work of Lindstrom et al. (1996) as cited by Luebke et al. (2003). The simplification scheme involves a vertex removal approach in which a pair of triangles is reduced to a single triangle. Figure 9.19 illustrates the principle. In this figure, $\triangle ABC$ and $\triangle BCD$ are the two triangles considered. If vertex C is removed, then there will be a vertical error (δ) from C to line $\overline{AD}$. If δ on the screen is smaller than a threshold, then vertex C can be removed. This is the working principle of a fidelity-based simplification.

An extremely popular continuous LOD algorithm is the real-time optimally adapting meshes (ROAM) algorithm developed by Duchaineau et al. (1997). A continuous mesh is produced by applying a series of split and merge operations on a binary triangle tree (Figure 9.20). Again, screen-based geometry error is used as a threshold for split and merge operations. Figure 9.21 is an example of the LOD of a terrain surface determined by this method.

The ROAM algorithm includes a number of other interesting features and optimizations, including an incremental mechanism to build triangle strips. Real-time display of complex surfaces is provided by dynamically computing a multi-resolution triangular mesh for each view. The meshes minimize geometric distortions on the screen while maintaining a fixed triangle count. Pop-ups are minimized in several ways, and efficient mesh corrections ensure that selected lines of sight or object proximity are correctly represented. An incremental priority-queue algorithm uses frame-to-frame coherence to quickly compute these optimal meshes (Duchaineau et al. 1997).

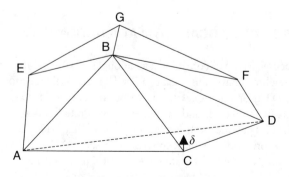

Figure 9.19 Fidelity-based LOD by Lindstrom et al. (1996).

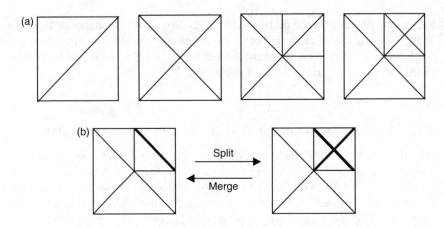

Figure 9.20 Split–merge of binary triangle tree in ROAM algorithm. (a) Binary triangle tree. (b) Split and merge of binary triangle tree structure.

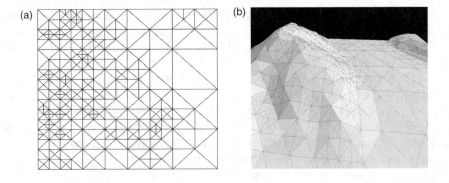

Figure 9.21 LOD of DTM by ROAM algorithm; (a) Binary triangle tree structure; (b) Perspective display.

9.5 MULTI-SCALE DTM AT A NATIONAL LEVEL

DTMs have become a type of core data in a national spatial data infrastructure (NSDI) and thus most countries have multi-scale DTM data sets. In this section, two examples of multi-scale DTM at a national level are briefly described. One is China, from the eastern part of the world, and the other is the United States, from the western part of the world. They are both large countries so that DTMs at many scales are available. The information for the Chinese multi-scale DTMs is extracted from the home page of the National Geomatics Center of China and that for the American multi-scale DTMs is extracted from the home page of the U.S. Geological Survey (USGS).

9.5.1 Multi-Scale DTM in China

So far, China has built its national DTM database at three scales, that is, 1:1,000,000, 1:250,000, 1:50,000, covering the whole country. For some important regions, such as river basins of China's seven major rivers, DTMs at 1:10,000 have also been produced. Larger-scale DTM databases are produced by provincial or metropolitan bureaus of surveying and mapping.

The national 1:1,000,000 DTM database was produced from over 10,000 sheets of 1:50,000 and 1:10,000 topographic maps. The grid intervals are $28''.125 \times 18''.750$ (longitude × latitude). There are a total of 25,000,000 grid nodes all over the country.

The national DTM database of China at 1:250,000-scale was produced from the 1:250,000-scale topographic maps. There are two sets of national DTM data at this scale, one stored in the national (grid) coordinate system with intervals of 100 m × 100 m and the other stored in the geographical coordinator system with intervals of $3'' \times 3''$.

The national DTM database of China at 1:50,000-scale was produced from topographic maps at three different scales, that is, 1:10,000, 1:50,000, and 1:100,000. The grid interval is 25 m.

DTMs at 1:10,000-scale for the river basin areas of all China's seven major rivers have been produced by the National Geomatics Center of China. The grid size is 12.5 m × 12.5 m. The production of DEMs at this scale is done at the provincial level and is still in progress.

9.5.2 Multi-Scale DTM in the United States

The United States has many scales of DTM as well, that is, 1-degree, 30 minutes, 15 minutes, and 7.5 minutes. The 1-Degree DTM corresponds to the 1:250,000-scale USGS topographic map series, and is available for all of the contiguous United States and most of Alaska. It is expressed in geographic coordinates (latitude and longitude). The grid interval is $3'' \times 3''$.

The 30-Minute DTM corresponds to the 1:100,000-scale topographic maps. It is available for the coterminous United States and Hawaii. It is expressed in geographic coordinates (latitude and longitude). The grid interval is $2'' \times 2''$.

The 15-Minute Alaska DTM corresponds to the USGS 1:63,360-scale topographic map series for Alaska. It is expressed in geographic coordinates (latitude and longitude). The grid interval is $2'' \times 3''$ (latitude/longitude).

The 7.5-Minute DTM corresponds to the USGS 1:24,000- and 1:25,000-scale topographic quadrangle maps, and is available for all of the United States and its territories. Most files will have a grid interval of 30 m but 10-m grids are also available for some locations. However, for Alaska, the grid interval is $1'' \times 2''$ (latitude/longitude).

CHAPTER 10

Management of DTM Data

In Chapter 9, it was discussed that DTM data have become part of an NSDI and one usually produced at the national level with multi-scales. For large countries like Brazil, Canada, China, India, and the United States, the volume of DTM data could be huge. Therefore, efficient management of DTM data in a computerized system is an important task at national or provincial geospatial information centers. Therefore, this chapter is devoted to management of DTM data.

10.1 STRATEGIES FOR MANAGEMENT OF DTM DATA

Spatial data, including DTM data, must be managed efficiently and database technology plays an important role. There are different strategies to deal with the problems in the management of DTM data.

10.1.1 Strategy for Making DTM Data Management Operational

To make the management of spatial data operational, spatial data sets are partitioned according to five attributes, horizontal or vertical positions, time, theme, and scale. In the management of DTM data, scale and horizontal positions are used. The use of scale was discussed in Chapter 9 and, therefore, only the use of horizontal position will be described in this section.

If the area to be modeled is large such as a nation or a province, one is concerned with the arrangement of the huge volume of DTM data. Questions such as "should distributed or centralized databases be used," or "how can the data of the whole area be split into small pieces so that they can be managed efficiently" are the concern here.

As contour maps have been widely used for DTM production, DTMs at a national scale are usually arranged in a way similar to map sheets. Figure 10.1 shows the arrangement for the 1:1,000,000-scale topographic maps of China. Table 10.1 shows the size of each map sheet at different scales, ranging from 1:1,000,000 to

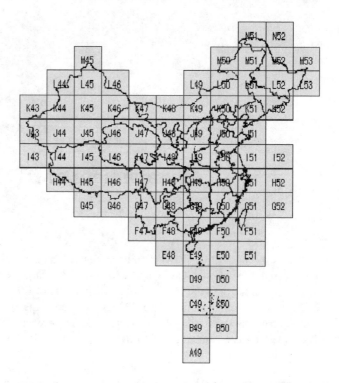

Figure 10.1 The tiling system of China's map sheets at 1:1,000,000 scale. (Courtesy of the National Geomatics Center of China.)

Table 10.1 The Sizes of China's Map Sheets From 1:1,000,000 to 1:10,000 Scales

Scale	1:1,000,000	1:500,000	1:250,000	1:100,000	1:50,000	1:25,000	1:10,000
Size (long/lat)	6° × 4°	3° × 2°	1.5° × 1°	30′ × 20′	15′ × 10′	7.5′ × 5′	3′45″ × 2′30″

1:10,000. Taking the DTM of China at 1:1,000,000 as an example, it is in a grid form and there are a total of 25,000,000 data points (at grid nodes). The heights of these points are divided into tiles, which follow the 1:500,000-scale topographic maps (http://nfgis.nsdi.gov.cn/). In other words, each tile covers an area of 3° × 2° (longitude/latitude). This kind of partition is the *operational strategy* for DTM data management. Such a strategy is equally applicable for any project with a relatively large area to be modeled.

10.1.2 Strategy for Using Databases for DTM Data Management

The second strategy is about the use of databases to store DTM data. The traditional database is good at managing of event (or attribute) data but it is not good for geometric data. On the other hand, all spatial data, including DTM data, have two components,

MANAGEMENT OF DTM DATA

geometric and attribute, and therefore are quite different from ordinary event data, which have only one component. Therefore, special arrangements for spatial data must be made according to the characteristics of these two components. Currently, the mainstream practice is to use files to store geometric data and to use relational tables to store attribute data (and relational data if any) in a traditional relational database. The files for geometric data are then managed by a computer system. The geometric and attribute data are then linked by pointers. This is also common for DTM data management. Files are cataloged and indexed so that efficient retrieval is possible. This is helped by metadata, or "data about data." Metadata contain the information describing the contents, quality, status, and other characteristics. Metadata can also be indexed using files. However, if complicated, metadata can also be managed by databases. In this way, databases do now come into the area of geometric data management, but indirectly.

Current development is toward object-relational databases. In such databases, geometric data (mainly the coordinates) are also organized into tables and stored and managed by the relational database management system. This has become popular for the management of large-volume DTM data. In practice, when data volume is not very large, a file system is still commonly used due to its convenience and the high cost of object-relational databases. Purely object-oriented databases have also been under development. However, there is still a long way to go before they will be commonly used.

10.2 MANAGEMENT OF DTM DATA WITH FILES

In the previous section, it was discussed that file systems are still commonly used for the management of DTM data. The structure of such files will be discussed in this section.

10.2.1 File Structure for Grid DTM

When the DTM is in a grid form, it can be represented by point matrix (Figure 10.2), or raster format. The topology between a grid point and its adjacent grid points is implicitly built in the rows and columns of the matrix.

The coordinates of a grid node can be computed based on the coordinates (x_0, y_0) of the origin of the area and the square grid intervals d. Suppose the lower-left corner

$$\begin{pmatrix} 56 & 58 & \ldots & 60 & 56 \\ 57 & 59 & \ldots & 63 & 58 \\ \ldots & \ldots & \ldots & \ldots & \ldots \\ 64 & 70 & \ldots & 68 & 66 \\ 62 & 68 & \ldots & 66 & 62 \end{pmatrix}$$

Figure 10.2 Matrix representation of grid DTM data.

Table 10.2 Typical File Structure for Grid DTM

File Components	Contents	Comments
Header	Coordinates of the origin; coordinate data type; height range; height data type; grid interval; numbers of rows and columns; order of rows and columns, position (in the file) where the body starts; position (in the file) where the footer starts; use of compression or not; etc.	The information in the footer can also be recorded here
Body	Height values of grid nodes	Row by row and column by column in blocks
Footer	Data describing the general characteristics of DTM, e.g., name, boundary, producer, projection parameters, version, accuracy, date of production, date of revision, linage, etc.	Metadata

point of a matrix (m, n) is used as the origin, then the coordinates of the grid node at (i, j) are:

$$\begin{aligned} x_{i,j} &= (j-1) \times d + x_0, & j &= (0, n-1) \\ y_{i,j} &= (i-1) \times d + y_0, & i &= (0, m-1) \end{aligned} \quad (10.1)$$

In other words, this elevation matrix records the heights at DTM grid nodes. However, some additional information is required to tell users how to read the height information. The location of the origin and the grid interval are necessary for the computation of coordinates, and information about the sequence of the height values is also needed so that each grid node can be assigned a height value. In a typical file of raster data, such additional information is recorded as the header and the matrix is the file body. In the body, heights are recorded row by row and then column by column, or column by column and then row by row, or block by block. Some other relevant information may also be recorded, either in the header or in a footer. Therefore, the typical file structure for a grid DTM is as shown in Table 10.2.

10.2.2 File Structure for TIN DTM

The TIN model represents a surface comprising a series of contiguous triangles, hence triangulated. A triangle has three vertices, which can be arbitrarily located, here irregular in shape. This contrasts with the grid model where points are spaced regularly in a lattice. The big difference between the management of TINs and grids is that, for the TIN model, apart from elevation values, the coordinates (x_i, y_i) of each vertex (say ith) and the information describing the topological relations among the three vertices need to be recorded. The topological relationship between triangles also needs to be recorded in most cases.

MANAGEMENT OF DTM DATA

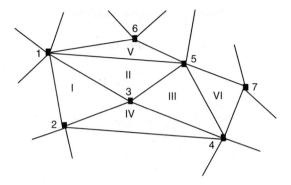

Figure 10.3 A triangulated irregular network (TIN).

Table 10.3 A List of Coordinates of Points

No.	X	Y	Z
1	429.1	269.6	57.5
2	437.3	200.3	60.2
3	504.7	234.1	55.3
4	607.2	190.5	56.1
5	555.4	265.8	50.2
6	506.7	280.3	52.5
7	621.2	251.4	53.8
⋮	⋮	⋮	⋮

Table 10.4 A List of Triangles

No.	Vertex 1	Vertex 2	Vertex 3
I	1	2	3
II	1	3	5
III	3	4	5
IV	2	4	3
V	1	5	6
VI	4	7	5
⋮			

The recording of geometric information is illustrated in Figure 10.3 and Table 10.3 and Table 10.4, that is, a table of points containing all their coordinates and a table of triangles with their corresponding three vertices. Apart from geometric information, the topological information is recorded for efficient retrieval of data. Table 10.5 lists the adjacent relations between these triangles.

The file structure for a TIN DTM is simply the list of points with their coordinates, with some metadata also included in the header. The file structure is like that given in Table 10.6. The topological information about these triangles is stored either in a database or in a file. Table 10.7 illustrates a possible arrangement of such topological information in a file.

Table 10.5 Adjacent Relations of Triangles

Triangle	Edge Neighbors		
I	—	IV	II
II	I	III	V
III	IV	VI	II
IV	—	III	I
V	II	—	—
VI	—	—	III
⋮			

Table 10.6 Typical File Structure for TIN Point Coordinates

File Components	Contents	Comments
Header	The coordinates of the points on the boundary (convex hull); ranges of X, Y, and Z coordinates; coordinate data type; data types; numbers of points; position (in the file) where the body starts; position (in the file) where the footer starts; use of compression or not; etc.	The information in the footer can also be recorded here
Body	X, Y, and Z coordinates of points in sequence	May also be in blocks
Footer	Data describing the general characteristics of DTM, e.g., name, producer, projection parameters, version, accuracy, date of production, date of revision, linage, the null points code, etc.	Metadata

Table 10.7 Typical File Structure for TIN Topology

File Components	Contents	Comments
Header	Number of triangles, the bytes of data for Table 10.4 or Table 10.5, data types, etc.	The information in the footer can also be recorded here
Body	Information in Table 10.4 or information in Table 10.5	Adjacent triangular topology is not always necessary
Footer	Other relevant information	Metadata

10.2.3 File Structure for Additional Terrain Feature Data

As discussed in Chapter 4, a hybrid DTM network may be generated if data from composite sampling (i.e., grid plus feature points and lines) are used. In normal practice, the grid and feature data are stored in separate files. When modeling or interpolation is needed, grids are split into triangles and feature points and lines are added to the regular triangular network to update local triangles (Figure 10.4).

MANAGEMENT OF DTM DATA

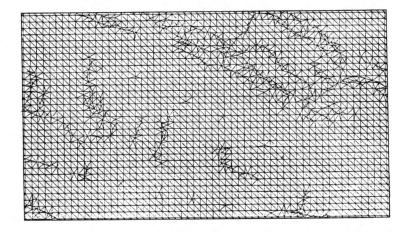

Figure 10.4 Hybrid of regular grid and TIN.

$$1(\text{line ID}), N_1 \text{ (No.of points on line 1)}$$
$$X_{11}, Y_{11}, Z_{11}$$
$$X_{12}, Y_{12}, Z_{12}$$
$$\ldots$$
$$X_{1N_1}, Y_{1N_1}, Z_{1N_1}$$
$$2(\text{line ID}), N_2 \text{ (No.of points on line 2)}$$
$$X_{21}, Y_{21}, Z_{21}$$
$$X_{22}, Y_{22}, Z_{22}$$
$$\ldots$$
$$X_{2N_2}, Y_{2N_2}, Z_{2N_2}$$
$$\ldots$$
$$\ldots$$

Figure 10.5 The body of vector file structure for terrain feature data.

Feature data may be stored in one or two files, one for points and the other for lines. The file structure for terrain feature points is similar to that for the points of TINs. However, for lines, it is slightly different. In the header, the number of lines is specified and in the body the data could be organized as shown in Figure 10.5.

10.3 MANAGEMENT OF DTM DATA WITH SPATIAL DATABASES

In the previous section, the file structures for both grid and TIN DTMs were discussed. These files are managed using an indexing system, which can be organized into files or into tables and managed by a database if the indexing is rather complex. In this case, an ordinary relational database will serve for the purpose. On the other hand, the DTM data can also be organized into tables in an object-relational database, in which DTM data are stored in block as a field.

10.3.1 Organization of Tables for Grid DTM Data

A large area (e.g., a country) may be divided into a number of smaller regions (e.g., provinces) and each region can be further divided into a number of smaller units called tiles. Each tile may also be further divided into a number of small units. This is a hierarchical structure and can be indexed efficiently for the management of DTM in a grid form. Figure 10.6 shows an indexing system for the hierarchy in three levels, region, tile, and block. It is not necessary to have rectangular shapes for the tiles. For example, the boundaries will be irregular if the DTM data of a nation is managed based on drainage area or administrative region.

In some commercial systems, the block is the basic unit for access and retrieval. Each block comprises many rows and columns. Through the structural index for "region–tile–block–row–column," the height of any location within the database can be uniquely determined. The spatial index formed by the region–block–unit hierarchy ensures fast retrieval of and seamless access to DTM data. The arrangement of tables for a regional DTM in an object-relational database is shown in Table 10.8, Table 10.9, and Table 10.10, which are created by the authors for illustration purposes only.

The above data organization method may also apply to TINs for large areas. As the TIN boundary of each region is irregular, to avoid the edge-matching problem between adjacent blocks, a certain degree of overlapping is necessary in block partitioning.

Suppose each region is organized into a database. There are only four fields in a record. This is illustrated in Table 10.8.

In Table 10.8, the field Region-table-name is the name of the table containing DTM data (see Table 10.9); the field Region-DTM-info is an abstract data type using database BLOB field (variable length), that is, a data stream, and has a

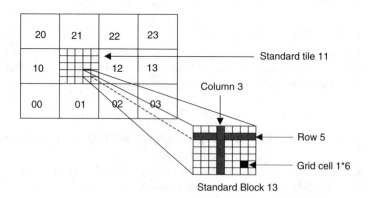

Figure 10.6 Hierarchical structure based on region–tile–block (Modified from ESRI, 1992).

MANAGEMENT OF DTM DATA

Table 10.8 An Index Table for a Regional DTM

Region-ID	Region-Table-Name	Region-DTM-info	Range-of-region
1	GridDEM50000_1	GridDTM50000_1_INFO	GridDTM50000_1_ENVELOPE
2	GridDEM50000_2	GridDTM50000_2_INFO	GridDTM50000_2_ENVELOPE
⋮	⋮	⋮	⋮

structure that contains the information about the region. For example, the structure GridDTM50000_1_INFO contains all the tile and block information about this region. The following is an example:

```
* GridDTM50000_1_INFO
{
    int XtilesNum;          //number of tiles in column
                              direction, e.g., four in
                              Figure 10.6//
    int YTilesNum;          //number of tiles in row
                              direction, e.g., three in
                              Figure 10.6//
    int XBlocksNum;         //number of blocks in each tile,
                              in column direction, e.g.,
                              five in Figure 10.6//
    int YBlocksNum;         //number of blocks in each tile,
                              in row direction, e.g., five
                              in Figure 10.6//
    int BlockRow;           //number of rows in each block,
                              e.g., seven in Figure 10.6//
    int BlockColumn;        //number of columns in each
                              block, e.g., eight in
                              Figure 10.6//
    float BlockCellSize;    //interval of DTM cells,
                              e.g., 25.0 for 25.0 m//
    int Scale;              //scale factor of the DTM,
                              e.g., 50,000 for 1:50,000//
    BOOL bOriDataLayer;     //whether it is original or
                              updated, e.g., TRUE
                              if original//
    BOOL bCompressed;       //whether or not data
                              compression is
                              used, e.g., FALSE if
                              no compression//
};
```

In Table 10.8, the field Range-of-region is also the abstract data type BLOB, that is, a pointer to a structure that contains the coordinates of the four corners of the region. For example, the structure GridDTM50000_1_ENVELOPE may contain:

```
GridDTM50000_1_ENVELOPE
{
    float XMin;     //the smallest X coordinates, e.g.,
                      850,000 for 850,000 m//
    float XMax;     //the largest X coordinates, e.g.,
                      860,000 for 860,000 m//
    float YMin;     //the smallest Y coordinates, e.g.,
                      810,000 for 810,000 m//
    float XMax;     //the largest Y coordinates, e.g.,
                      830,000 for 830,000 m//
};
```

In Table 10.8, the index table of the DTM at region level is set for the logical structure of the region–block–tile–block hierarchy. The actual heights at grid nodes are arranged in blocks and stored in a table with a name given under the field Region-table-name. In other words, the height data are stored block by block. Three different ways have been used to organize data in blocks, which are shown in Tables 10.9 and Table 10.10.

In Table 10.9, the Block-ID is the main key, which is unique to each block. Each Block-ID consists of four numbers. The first two indicate the location of the corresponding tile (which contains this block) in the region, one for the numbering

Table 10.9 Organization of DTM Height Data for Region GridDEM50000_1 in Block

Block-ID	Bytes-of-Block	Block-Data
0000	112	$h_{0,0}\ h_{0,1} \ldots h_{0,6}\ h_{1,0} \ldots h_{6,7}$ (of Block 0000)
⋮	⋮	⋮
1113	112	$h_{0,0}\ h_{0,1} \ldots h_{0,6}\ h_{1,0} \ldots h_{6,7}$ (of Block 1133)
⋮	⋮	⋮
2344	112	$h_{0,0}\ h_{0,1} \ldots h_{0,6}\ h_{1,0} \ldots h_{6,7}$ (of Block 2344)

Table 10.10 Organization of DTM Height Data for Region GridDEM50000_1 in Tiles

FILE-ID	DTM-Info	DTM-Data
00	DTMINFO00	Heights at tile 00
01	DTMINFO01	Heights at tile 01
⋮	⋮	⋮
23	*DTMINFO23	Heights at tile 23

of the tile in row direction and the other in column direction. The other two indicate the location of the current block in the corresponding tile, one for block numbering in row direction and the other in column direction. For example, the block labeled in Figure 10.6 is assigned an ID of 1113. The first two digits, that is, "11" indicate a location of the second tile in row direction and second tile in column directions in the region. The last two digits, that is, "13," indicate the location of the block within tile 11, that is, the second in row and the fourth in column directions. The data type for block-data is BLOB. In Figure 10.8, the size of each block is a 7×8 matrix. The 7×8 height for each block is then recorded in this field. In this table, as the number of bytes for each block is 112, a float (or integer) of two bytes is used for each height value. $h_{0,0}\ h_{0,1} \ldots h_{0,6}\ h_{1,0} \ldots h_{6,7}$ are the heights of the respective 7×8 data blocks.

In fact, it is also possible to organize the DTM data file using tile as the basic unit. That is, one file is used for each tile. In this case, the file format of each tile may not necessarily be the same. Table 10.10 shows the organization of tables using tile as the basic unit.

The file-ID in this table is the tile number. The data types for the fields DTM-Info and DTM-Data are both BLOB, which is a data stream. The former is a structure as follows (using DTMINFO00 as an example):

```
DTMINFO00
{
    Char Filename;          //file name of file for DTM data of
                              tile 00//
    Float ENVELOPE2D;       //the area covered by tile 00, i.e.,
                              the coordinates of the
                              four corners//
    Int Data-Bytes;         //the size of data file in terms of
                              bytes//
};
```

In the DTM-Info field, a file name is included to refer to the height data block in the DTM-Data field of the corresponding record. This is for the convenience of retrieval. The height values in each tile form a data stream and are stored in the field DTM-data. The data may be stored in separate files (i.e., not as a field in the table). In this case, Table 10.9 simply stores the logical information for the management of DTM files.

10.3.2 Organization of Tables for TIN DTM Data

As was done for the region–tile–block hierarchical structure of grid DTM, similar arrangements can also be made for TIN DTM. A region can be divided into a number of (e.g., $M \times N$) blocks for the TIN. Table 10.11 shows the indexing of TIN blocks.

Table 10.11 The TIN Block Indexing Table

Region-ID	Region-Table-Name	Region-Info	Range-of-Region
0	TIN50000_0	TIN50000_0_INFO	TIN50000_0_ENVELOPE
1	TIN50000_1	TIN50000_1_INFO	TIN50000_1_ENVELOPE
⋮	⋮	⋮	⋮
K	TIN50000_K	TIN50000_K_INFO	TIN50000_M_ENVELOPE

The fields of Table 10.11 are defined as the ID of the region, the name of the region, other information about the region, and the envelope (range) of the region, respectively. For the ith region, the structure in the Region-Info field is called TIN50000_i_INFO, defined as follows:

```
TIN50000_ i_INFO
{
    float Xblocksize;       //the size of each block in the
                              region, in X direction//
    float Yblocksize;       //the size of each block in the
                              region, in Y direction//
    int XBlocksNum;         //the number of blocks in the
                              region, in X direction//
    int YBlocksNum;         //the number of blocks in the
                              region, in Y direction//
    int NPoint;             //the total points number in the
                              region//
    int Scale;              //scale factor of DTM data, e.g.,
                              50,000 for 1 : 50,000//
    BOOL bCompressed;       //whether or not the data are
                              compressed//
};
```

TIN50000_i_ENVELOPE is the defined area coverage of the ith region, which is similar to the structure GridDTM50000_i_ENVELOPE discussed previously.

A triangle is the basic unit in a TIN. There may be a number of triangles in each region or block. The data structure for a triangle is shown in Figure 10.7. Three tables are required for this structure, one for point coordinates, one for the relationship between a triangle and its three vertices, and one for the relationship between a triangle and its three edge neighbors, for example the tables given in Section 10.2.2, Table 10.3, Table 10.4, and Table 10.5. These three tables can also be stored in a database table. Table 10.12 illustrates the table formats in a spatial database.

In Table 10.12, BLOCK-ID is the main key of the data table (i.e., ID), Both Triangle-List and Point-list are data streams (i.e., type BLOB). Triangle-List contains the data for Table 10.4 and Table 10.5. The Point-list contains the coordinates of points in Table 10.2.

MANAGEMENT OF DTM DATA

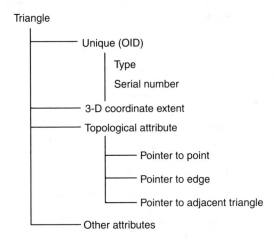

Figure 10.7 Structure of a triangle entity in a TIN.

Table 10.12 The TIN Topology Data Table

BLOCK-ID	Triangle-List	Point-List
00	Triangle list for block 00	Point coordinates of Block 00
01	Triangle list for block 01	Point coordinates of Block 01
⋮	⋮	⋮
$M-1, N-1$	Triangle list for block $M-1, N-1$	Point coordinates of Block $M-1, N-1$

Vertex is the basic entity of the TIN in a database. The topological relationship, including the links between a triangle and its three vertices, between a node and its adjacent nodes, and between a triangle and its adjacent triangles, can also be set up in the database using a pointer system for vertices. This structure is given in Figure 10.8.

10.3.3 Organization of Tables for Additional Terrain Feature Data

As has been discussed in Section 10.2.3, if additional terrain feature data are available, they need to be stored in a separate file (if file system is used) or table (if spatial database is used) from the grid DTM data. The arrangement of line data in a vector file is given in Section 10.2.3. In this section, the organization of such vector data into tables in spatial databases is given. The structure of such vector lines is given in Figure 10.9. The data format is given in Table 10.13.

In this table, an integer is used for line types, for example, 1 for ridge lines, 2 for break lines, 3 for rivers, and so on. A data stream (i.e., BLOB) is used to store the coordinates of all the points on a line. In fact, the terrain features could be points, line, and areas. If there is more than one type of terrain feature, an indexing table can be used to manage them. Table 10.14 is an example.

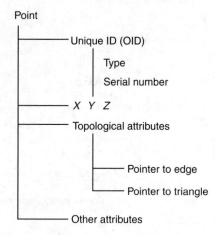

Figure 10.8 Pointer structure of a point in a TIN database.

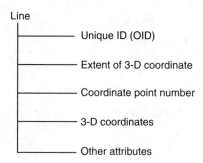

Figure 10.9 Line structure of linear entity.

Table 10.13 The Linear Entity Data Table

Line-ID	Line-Type	Number-of-points	Coordinates-of-the-Line
1	1	N_1	$X_{11}, Y_{11}, Z_{11}\ X_{12}, Y_{12}, Z_{12} \ldots X_{1N_1}, Y_{1N_1}, Z_{1N_1}$
2	3	N_2	$X_{21}, Y_{21}, Z_{21}\ X_{22}, Y_{22}, Z_{22} \ldots X_{2N_2}, Y_{2N_2}, Z_{2N_2}$
⋮	⋮	⋮	⋮

Table 10.14 An Indexing Table for Terrain Features

Region-ID	Region-table-name	Feature-Info	Range-of-region
0	Feature50000_0	Feature50000_0_INFO	Feature50000_0_ENVELOPE
1	Feature50000_1	Feature50000_1_INFO	Feature50000_1_ENVELOPE
⋮	⋮	⋮	⋮
K	Feature50000_K	Feature50000_K_INFO	Feature50000_K_ENVELOPE

The meanings of the fields in this table are similar to those in Table 10.11. The structure of Feature50000_*i*_INFO is defined as:

```
Feature50000_i_INFO
{
        INT  LRID;         //ID of feature table//
    Char TABLENAME;       //name of table//
    BLOB TABLEFLAG;       //flag of table, such as 1 for point
                            table, 2 for line table, 3 for area
                            table, etc.//
};
```

10.3.4 Organization of Tables for Metadata

"Metadata" is "data about data." Metadata describes the content, quality, status, and other features of data. Metadata also help to locate and understand the data. Metadata are an important basis for sharing spatial data. Through inquiring about and browsing through the metadata, users obtain general information about the kind of data available, which data are of interest, and where such data are kept. A major part of the interface used in the clearinghouse of a national spatial data infrastructure is to provide interactive queries on metadata. Through metadata it is convenient to obtain descriptions of spatial data (e.g., DTM) and the data themselves. Metadata have four fundamental functions:

1. *Availability*: to indicate whether a data set exists for a certain geographic area.
2. *Fitness for use*: to evaluate whether the data set is applicable.
3. *Access*: to determine the means of acquiring verified data.
4. *Transfer*: to successfully handle (e.g., transfer) and utilize the data set.

Metadata generally contain the following:

1. *Basic identifiers*: the fundamental information about data sets, for example, title, geographic extent, currency (updatedness), rules of access or utilization.
2. *Quality*: quality evaluation of data sets, including accuracy of location and attributes, completeness, consistency, information sources, producing methods.
3. *Data organization*: the mechanism for representing spatial information in data sets, for example, whether the spatial location is represented directly by a raster or a vector or indirectly by a street address or zip code.
4. *Spatial reference*: description of the coordinate systems of data sets, including projection, parameters, benchmark of the plane, and elevation.
5. *Entity and attribute*: the content of data sets, including type, attributes, domain.
6. *Issuance*: obtaining of datasets, for example, contact information, format, and how to get information about data and prices from the Web and physical media.
7. *Metadata reference*: description information about the currency (updatedness) of the metadata and its producers, etc.

Table 10.15 An Example of a Metadata Set in a Table

Table-ID	Table-Name	Institution-Name	Product	Updating-Date	Scale	...
100,000	PubMetadata	GeomaticsCenter	DTM	10-12-2004	50,000	
⋮	⋮	⋮	⋮	⋮	⋮	⋮

Table 10.16 Standards of Metadata

Name of Standards	Institution or Organization
Content Standard for Digital Geospatial Metadata (CSDGM)	U.S. Federal Geographic Data Committee (FGDC)
Directory Interchange Format (DIF)	U.S. National Aeronautics and Space Administration (NASA)
Government Information Locator Service (GILS)	U.S. Federal Government
CEN Geographic Information — Data Description — Metadata	CEN (European Committee for Standardization) TC287
Geographical Data Description Directory (GDDD)	Multipurpose European Ground Related Information Network (MEGRIN)
Geomatic Data Sets Cataloguing Rules	Canadian General Standards Board (CGSB) and Committee on Geomatics (CoG)
ISO Geographic Information — Metadata	ISO (International Organization for Standardization)/TC211

As metadata files are descriptive documents with text and numeric data types, generally a relational database is sufficient. Two tables are required for metadata management, one for publishing metadata for the public (as shown in Table 10.15), and the other for internal applications. In fact, the structures of these two tables are the same, but with different fields.

The importance of metadata is well recognized and international efforts have been made on the standardization of metadata. Table 10.16 provides a list of such standards.

10.4 COMPRESSION OF DTM DATA

As discussed in the previous section, in the structure for DTM data, one variable states whether the data have been compressed. If compressed, then decompression needs to be applied before using the data. In this section, a brief discussion on DTM data compression is given.

10.4.1 Concepts and Approaches for DTM Data Compression

There are two basic ideas behind data compression, leading to two different approaches:

1. *Lossless compression*: Some data provide no extra information, or are redundant. Redundancy can be minimized by more efficient coding methods. After decoding, the data can be recovered with 100% fidelity.

MANAGEMENT OF DTM DATA

2. *Lossy compression*: Information or accuracy loss is still acceptable even though changes to some of the data points are made. However, the information or accuracy lost during compression can never be recovered.

In digital terrain modeling, data redundancy is inevitable, especially when image-based techniques such as automated photogrammetric systems, LIDAR and InSAR are used for data acquisition (see Chapter 3 for a more detailed discussion).

Data redundancy exists in a number of different guises. It is more than just height values being constant. Rather the inefficiency of using fixed-length data coding and also the correlation between neighboring data values are the big concerns.

In fact, both lossless compression and lossy compression techniques are used in digital terrain modeling. The selection of VIPs discussed in Chapter 4 is a lossy compression technique. Therefore, this section contains only a brief discussion on lossless compression for grid DTM. In fact, the quadtree structure discussed in Chapter 9 is a very efficient method of data compression. Other traditional methods such as run-length and block encoding, and general-purpose methods such as gzip or bzip2 can also be used to reduce file sizes. In this section, the widely used Huffman coding is first described, then another simple method is introduced.

10.4.2 Huffman Coding

The basic idea behind the Huffman coding is that some values occur more frequently than others. If shorter codes are used for more frequent values and longer codes for less frequent values, then the overall storage required will be significantly reduced. Table 10.17 shows a set of grid DTM, that is, heights, in a matrix. The heights range from 212 to 216. Therefore, an 8-bit space (i.e., a byte) is required to store one height value and a total of 8×36 bits (288 bits) are required for this data set.

Table 10.18 shows that height 213 occurs most frequently and 216 least frequently. The idea is to devise a coding system to let the value with higher frequency have a shorter code and the lower one have a longer code. In this example, the value 213 should be assigned the shortest code and 216 the longest code. The Huffman coding does this and is illustrated in the right part of Table 10.18.

First, the values are put in an order of frequency, with the most frequent value at the top. This process is called source reduction. In this process, two basic principles are involved:

1. The last two frequencies are always summed to form one value in each round.
2. The frequency values are always sorted in order, with the largest frequency value at the top.

Table 10.17 A Set of DTM Data in a Grid

213	213	213	212	212	213
216	212	213	212	212	216
214	215	215	213	216	215
212	213	213	213	214	214
212	213	215	214	213	212
212	213	213	213	213	212

Table 10.18 Huffman Coding for Data Compression

	Original Source				Source Reduction				
Value	Occurrence	Frequency	First	Code 3	Second	Code 2	Third	Code 1	Final Code
213	15	0.42	0.42	1	0.42	1	0.58	0	1
212	10	0.28	0.28	01	0.30	00	0.42	1	01
214	4	0.11	0.19	000	0.28	01			001
215	4	0.11	0.11	001					0000
216	3	0.08							0001

After the number of frequencies is reduced to two, codes can be assigned to each of them. In the code assignment process, three principles are involved:

1. The code is binary, that is, always 0 and 1.
2. Codes at higher levels are propagated into lower levels.
3. Codes are assigned by tracing back the source reduction process.

In the example given in Table 10.18, three rounds of reduction are required for the data set with five values. The last two frequency values (0.58 and 0.42 under the column head "Third") are assigned 0 and 1, respectively. As 0.42 is the frequency for value 213, code 1 is assigned to value 213. On the other hand, in the third round, the frequency 0.58 was combined from the frequencies 0.30 and 0.28 in the second round; therefore, the code 0 for frequency 0.58 will be propagated to frequencies 0.30 and 0.28. In this way, frequencies 0.30 and 0.28 are assigned 00 and 01, respectively. Again, the frequency 0.30 in the second round was combined from the frequencies 0.19 and 0.11 in the first round; therefore, the code 00 for frequency 0.30 will be propagated to frequencies 0.19 and 0.11. Thus, frequencies 0.19 and 0.11 are assigned 000 and 001, respectively. Again, the frequency 0.19 in the first round was combined from the original frequencies 0.11 and 0.08; therefore, the code 000 for frequency 0.19 will be propagated to frequencies 0.11 and 0.08. Thus, frequencies 0.11 and 0.08 now have codes 0000 and 0001, respectively. Therefore, the final codes for values 3, 2, 4, 5, and 6 are 1, 01, 001, 0000, and 0001. The total number of bits required for this coding is $1 \times 15 + 2 \times 10 + 3 \times 4 + 4 \times 4 + 4 \times 3 = 80$. As a result, the compression ratio is $288/80 = 3.6$. Generally, a maximum ratio of 5 is achievable with lossless compression.

10.4.3 Differencing Followed by Coding

It was discussed previously that there is a higher correlation between close neighbors. This means that the differences in heights between neighbor DTM cells must be smaller than the original height values. Therefore, it is natural to make use of the differences and then encode these differences (Kidner and Smith 2003; Wessel 2003).

Kidner and Smith (2003) suggested using the optimal linear predictor to compute DTM heights, then computing the height differences between the original and the predicted DTM points, and last encoding these differences with a Huffman coding. Figure 10.10 illustrates the principle.

MANAGEMENT OF DTM DATA

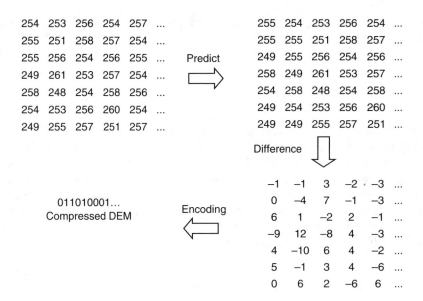

Figure 10.10 DTM data compression based on linear prediction.

The prediction is based on the three-point linear predictor as follows:

$$Z = a - b + c \tag{10.2}$$

If the coefficients can be optimized, then a much higher compression ratio may be reached. Kidner and Smith (2003) suggested the use of the following predictor:

$$Z = \text{INT}(w_1 \times a + w_2 \times b + w_3 \times c) \tag{10.3}$$

where INT refers to take the nearest integer and w_1, w_2, and w_3 are weights whose sum is equal to 1. When $w_1 = 1, w_2 = (-1), w_3 = 1$, the predictor is equivalent to the three-point predictor expressed by Equation (10.2).

After this differencing process, Huffman coding can be applied to the differences. Kidner and Smith (2003) revealed that differencing followed by Huffman coding is capable of producing significant reduction. They showed that all the USGS 1:250,000-scale DTMs can be successfully compressed into a CD-ROM. The test results revealed that this lossless DTM compression method offers a typical storage saving of 90% compared with the traditional 2-byte ASCII or binary representations. The compressed files are less than half the size of GZIP-encoded DTMs.

10.5 STANDARDS FOR DTM DATA FORMAT

DTM data are a type of fundamental data in a national spatial data infrastructure. Like other data sets, there must be some standards for them, including accuracy and format. In this chapter, only format is discussed as this chapter deals with data management.

10.5.1 Concepts and Principles of DTM Data Standards

Data are important and expensive. It is therefore important for data to be shared. DTM data often have different formats. Each software producer sets a specific format for the data in its digital terrain modeling system. So one system may need to provide a large number of programs under the import/export menu in order to read other formats. This is the case of data exchange between any two systems. If there are n data formats, one needs to write C_n^2 programs. For example, if there are ten different formats, one needs to write $C_{10}^2 = 10!/2 = 10 \times 9/2 = 45$ programs. As illustrated in Figure 10.11(a), this is very inefficient.

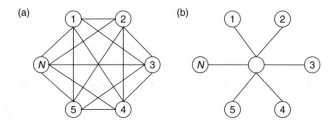

Figure 10.11 DTM data exchange standards: (a) exchange between any two and (b) exchange via a neutral format.

Table 10.19 U.S. Standards for DTM Data Format

Element	Comments
Filter	Blank fill
Origin code	Free format mapping origin code
DTM level	1 = DEM-1; 2 = DEM-2, 3 = DEM-3; 4 = DEM-5
Pattern	1 = regular; 2 = random; reserved for future use
Coordinate system	0 = geographic; 1 = UTM; 2 = state plane
Zone	UTM coordinate zone
Map projection	Specify the type and parameters of map projections
Unit for planimetry	0 = radius; 1 = feet; 2 = meter; 3 = arc-second
Unit for height	1 = feet; 2 = meter
Number of bounding polygons	Set to $n = 4$
Corner coordinates	Four corners of the quadrangle corners, from lower-left corner, clockwise
Minimum and maximum heights	In the same unit as for unit for height
Axis orientation	Zero of the same as easting and northing, or geographic system
Accuracy code	0 = unknown; 1 = recorded
Resolutions	In X and Y, as well as Z
Row and column	Number of rows and columns of the height matrix

MANAGEMENT OF DTM DATA

Table 10.20 Chinese Standards for DTM Data Exchange

Element	Description
DataMark	Geospatial data exchange format of China — the tag of DTM data exchange format (CNSDTF-DTM)
Version	Version number of the exchange format, e.g., 1.0
Unit	Coordinate unit, K for kilometer, M for meter, D for trapeze with degree as unit, S for trapeze expressed by degree-minute-second (i.e., DDDMMSS.SSSS)
Alpha	Directional angle
Compress	Compression method, e.g., 0 for noncompression, 1 for run-length encoding
Xo	X coordinate of the original point on the top-left corner
Yo	Y coordinate of the original point on the top-left corner
DX	Interval in X direction
DY	Interval in Y direction
Row	Number of rows
Col	Number of columns
ValueType	Type of elevation value
HZoom	Magnification rate, i.e., the number used to make elevation data stored in integer, e.g., 100 is used to make 213.56 become 21356
Coordinate	Coordinate system; G for geodetic coordinate system; M for mathematical coordinate system; M is the default value
Projection	Projection type (optional)
Spheroid	Reference spheroid (optional)
Parameters	Projection parameters
MinV	Minimum value of the grid height
MaxV	Maximum value of the grid height

An alternative solution is to develop a neutral format and have all software system support this format. In this way, exchange is more efficient than the exchange between two systems. Figure 10.11(b) illustrates this.

There are many format standards available such as the international standardization agreements STANAG 3809 published by NATO, the DTED (digital terrain elevation data) level 1 and level 2 files specified by the Department of Defense of the United States. However, in this section, only the ones by US and China are briefly described in this section, as did for the multi-scale DTM data in Chapter 9.

10.5.2 Standards for DTM Data Exchange of the United States

Intended to facilitate the interchange and use of DEM data, The National Mapping Division of the USGS specified the logical ASCII format for DEM data sets in its Standards for Digital Elevation Models (USGS 1998), as listed in Table 10.19.

These basic elements are contained in the old format although there is a new version containing additional information.

10.5.3 Standards for DTM Data Exchange of China

Similar to the U.S. Standards, China's Standards also list a few essential elements for the description of a DTM data set (SBQTS 1999). Table 10.20 is an extraction from the standard.

CHAPTER **11**

Contouring from Digital Terrain Models

Contour production has been one of the traditional applications of DTMs. This chapter is devoted to the production of contours from DTM.

11.1 APPROACHES FOR CONTOURING FROM DTM

Contour lines could be produced from either grid-based or triangulation-based DTM, either by a vector-based or by a raster-based method. Thus, theoretically speaking, four combinations are available for contouring from DTM:

1. vector-based contouring from a grid-based DTM
2. vector-based contouring from a triangulation-based DTM
3. raster-based contouring from a grid-based DTM
4. raster-based contouring from a triangulation-based DTM.

The first two approaches are widely used. Raster-based contouring from a grid-based DTM is not widely used although it is not difficult. Raster-based contouring from a triangulation-based DTM is not practical because the triangulations are normally for irregularly distributed data and thus it is not possible to form a raster structure without interpolation. It is possible to produce a stereomate for a contour map so that a stereo model of terrain can be seen if viewed using a stereoscope. This results in various approaches for contouring from DTM, as shown in Figure 11.1.

11.2 VECTOR-BASED CONTOURING FROM GRID DTM

There are many ways to produce contour lines from a grid DTM. The basic principles are similar although the procedures may differ. All methods in vector-based

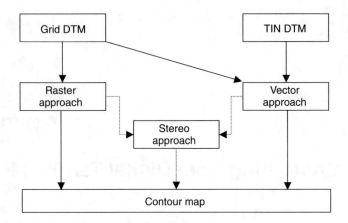

Figure 11.1 Approaches for contouring from DTMs.

approaches try to solve the following problems:

1. to search for the starting point of each contour line
2. to interpolate the contour points by computing the coordinates of the intersection points between each contour line and grid edges
3. to trace contour lines
4. to smooth traced contour lines, if desirable.

11.2.1 Searching for Contour Points

To obtain all contour lines in the area of concern, first the heights of the lowest ($h_{\min}$) and highest ($h_{\max}$) contours are computed from the heights of the lowest ($Z_{\min}$) and highest ($Z_{\max}$) points in the area:

$$h_{\min} = (Z_{\min}//\Delta h + 1) \times \Delta h$$
$$h_{\max} = (Z_{\max}//\Delta h) \times \Delta h \quad (11.1)$$

where Δh is the desirable contour interval for the area and // represents the integer division, that is, taking only the integer part of the division result, for example, $5//2 = 2$.

The next step is to search for all the contours systematically and then to trace each one. There are two methods to search for contour points on grid edges. The first starts with a given height and checks grid edges one by one to see whether or not the contour passes through any grid cells. The other starts with a grid edge and then checks all the possible contour heights to see whether this grid cell contains any contours. They are basically the same and the only difference is the searching sequence.

Usually, the search for grid edges starts from a corner, for example, the upper-left, and ends until a contour point is found for the given height or until all grid edges have been checked. Open contours are closed to the boundaries of the area and therefore their starting points should be found on the grid edges along the boundaries. Closed

CONTOURING FROM DIGITAL TERRAIN MODELS

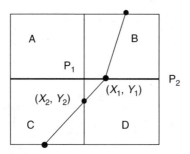

Figure 11.2 Interpolation of contour points and tracking of contour lines.

contours are inside the DTM area. An open contour may go through as few as two grid cells and a closed contour may occupy as few as four cells. Therefore, searching needs to be conducted on each grid cell to ensure that no contour is missed.

Whether or not a grid edge (e.g., $\overline{P_1P_2}$ in Figure 11.2) contains one or more contour points can be judged by the following equations:

$$
\begin{array}{lll}
Z_{P_1} > h > Z_{P_2}, \text{ or} & Z_{P_1} < h < Z_{P_2} & \Rightarrow \text{within } \overline{P_1P_2} \\
Z_{P_1} = h, & \text{or } h = Z_{P_2} & \Rightarrow \text{through node } P_1 \text{ or } P_2 \\
\text{otherwise,} & & \Rightarrow \text{not contained in this edge}
\end{array}
\quad (11.2)
$$

where h is the contour of concern, P_1 and P_2 are the two nodes of a cell edge; and Z_{P_1} and Z_{P_2} are the heights of the two nodes. Equation (11.2) can be simplified to:

$$(Z_{P_1} - h)(Z_{P_2} - h) \leq 0 \quad (11.3)$$

In Equation (11.3), if the result is equal to 0, then the contour will passes through a grid node. This may cause difficulties in determining the next direction. To avoid this, it is a normal practice to add or subtract a small value from the height of the node that is equal to h (height of the contour).

11.2.2 Interpolation of Contour Points

If the cell edge $\overline{P_1P_2}$ does contain a point of the contour with height h, then the exact location of this point on $\overline{P_1P_2}$ needs to be interpolated. As in traditional mapping, linear interpolation is usually adopted, that is,

$$
\begin{aligned}
X_h &= X_{P_1} + \frac{(h - Z_{P_1})}{(Z_{P_2} - Z_{P_1})} \times (X_{P_2} - X_{P_1}) \\
Y_h &= Y_{P_1} + \frac{(h - Z_{P_1})}{(Z_{P_2} - Z_{P_1})} \times (Y_{P_2} - Y_{P_1})
\end{aligned}
\quad (11.4)
$$

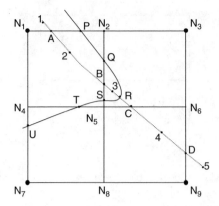

Figure 11.3 Incorporation of contour lines when terrain feature lines are sampled.

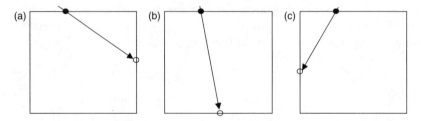

Figure 11.4 Three possible exits for a given point on a grid edge.

Figure 11.2. shows simple interpolation by using grid nodes. If terrain feature points and lines have been sampled, such as the line $\overline{1, 2, 3, 4, 5}$ in Figure 11.3, then they have to be considered in the interpolation of contour lines, as follows:

1. To interpolate the intersections (i.e., A, B, C, and D in Figure 11.3) between a feature line and the grid cells to obtain the X, Y, and Z coordinates of these intersections.
2. To interpolate contour points by using these intersections and grid nodes. For example, points A and N_2 are used to interpolate contour point P; and points 3 and C are used to interpolate contour point R.
3. To obtain the final contour points (i.e., P, Q, R, S, T, and U in Figure 11.3) by contour tracing and interpolation.

11.2.3 Tracing Contour Lines

Once the starting point of a contour is located, the next step is to trace the contour through the grids. For a given point on a grid edge, there are three possible exits within this grid, as shown in Figure 11.4. Therefore, tracing a contour line means to determine the exit for the current point and forward this exit through the grid cells, until the contour closes or reaches the boundary of the DTM area. Figure 11.2 also illustrates the tracing of a contour line from cells B to D and C.

The basic principle of contour tracing is that the exit edge of the current grid cell is naturally the entrance edge of the adjacent grid cell. Figure 11.2 also shows the

CONTOURING FROM DIGITAL TERRAIN MODELS

threading process, where the contour line enters cell D from cell B. In this case, cell edge $\overline{P_1P_2}$ is the exit edge of cell B and the entrance edge of cell D at the same time.

It is likely that there is more than one contour line of the same height in the DTM area, therefore it is still necessary to search for other possible contours with the same height. After all contours with one height are interpolated, the search for contours with a new height is carried out. This process is repeated until all contour lines are interpolated. Figure 11.5 shows an example of a contour map.

It is also possible that a contour has more than one exit, causing an ambiguity in contour line direction. Figure 11.6 shows ambiguity of line trend with five possible cases. The third case is impossible. One solution to this problem is to add a central point, whose height is the average of the heights of the four grid nodes. An alternative is to arbitrarily set a priority criterion (e.g., the right side is higher).

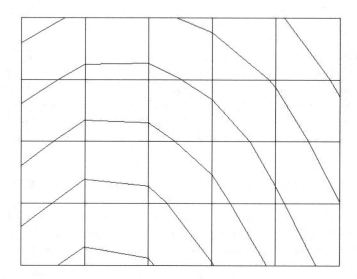

Figure 11.5 Contour lines interpolated from grid DTM by using the vector approach.

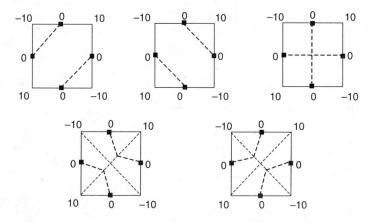

Figure 11.6 Ambiguity of contour line direction (modified from Petrie 1990a).

11.2.4 Smoothing Contour Lines

Figure 11.5 shows that contour lines interpolated from grids may be not as smooth as required for practical applications. Therefore, smoothing processes need to be applied. There are two solutions available:

1. to subdivide a grid cell into a number (e.g., nine) of sub-cells, so as to obtain smooth contour lines
2. to apply a smoothing technique to the interpolated contour lines.

For the first solution, a polynomial surface is fitted to a grid cell (or a few grid cells) so that nodes of finer grid cells are interpolated from the polynomial surface. Figure 11.7 illustrates the subdivision of nine cells from the original cell to obtain smoother contour lines.

For the other solution, many kinds of smoothing techniques are available. Figure 11.8 classifies these techniques. In contouring practice, curve fitting is the most widely used, including least-square curve fitting and splines (e.g., Bezier curves, B-splines, tension splines, and cubic splines). Figure 11.9 shows two examples. As discussion on splines and least squares for surface modeling was provided in Chapter 6 and line smoothing is 1-D surface modeling, no further discussion on these topics will be given here. Splines are more widely used for contour smoothing because each data point might be honored by this technique.

11.3 RASTER-BASED CONTOURING FROM GRID DTM

It was discussed in the previous section that a set of processes (i.e., searching, tracing, interpolation, and smoothing) are needed to produce contour in vector-based contouring. It will be seen later that a raster-based approach is easier and more intuitive

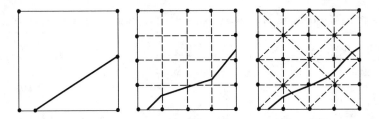

Figure 11.7 Smooth contours obtained from finer grid cells.

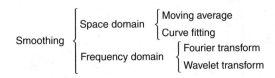

Figure 11.8 A classification of smoothing techniques.

CONTOURING FROM DIGITAL TERRAIN MODELS

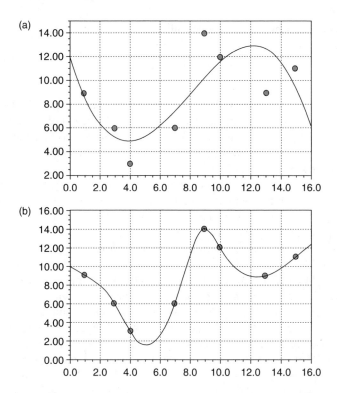

Figure 11.9 Curve fitting techniques for smoothing: (a) cubic curve fitting by least squares and (b) cubic curve fitting by spline.

if grid DTM with high density is available, where fully digital photogrammetric workstations powered with image-matching techniques are in use. In raster mode, techniques developed in digital image processing can easily be used to produce contours, such as slicing technique for binary contouring and edge detection for contour line tracing from binary contour to form edge contour (Eyton 1984). In this section, a number of such contouring techniques in raster mode will be presented.

11.3.1 Binary and Edge Contouring

Image slicing is used to produce a number of slices (classes) with a given height range, which is the CI. Then, height classes are alternately assigned black and white colors. The boundaries between contrasting classes define the contour lines. To derive edge contours, edge tracing is then applied.

The principle of applying slicing technique to binary contouring is simple. First Equation (11.1) is applied to compute the highest and lowest contours. The total number of contours is then computed as follows:

$$N_T = \frac{h_{max} - h_{min}}{\Delta h} \quad (11.5)$$

The number of classes for a grid node with height Z can then be computed as follows:

$$N = \frac{Z - h_{min}}{\Delta h} + 1 \qquad (11.6)$$

Class 0 is reserved for the heights below h_{min}. For example, the height range in an area is from 3 to 16 m and the CI is 5 m. Then, the height of the lowest contour is $h_{min} = 5$. All points with height Z in the range $5 \leq Z < 10$ are classed as 1; all with height Z in the range $10 \leq Z < 15$ are classed as 2; and all with height Z equal to 15 or above are classed as 3.

After slicing, a color is assigned to each class. In binary contouring, only two colors, black and white, are used. A simple rule for color assignment could be: black for odd classes and white for even classes. Figure 9.17 shows some examples of such a contour produced from the DTM shown in Figure 9.16.

This method can also be implemented with an on-line plotter (Eyton 1984). A row of DTM data are read into memory, then assigned a class number according to Equation (11.6). If N is an odd number, a black pixel is drawn at the corresponding position and if N is an even number, then a white pixel is drawn.

To extract the boundaries of height classes as contour lines, an edge tracing procedure is needed. Many operators have been designed in image processing. These are usually expressed in the form of a template for convenience in discrete mode. The Sobel operator is one of the most popular techniques for edge detection. The two templates for this operator are shown in Figure 11.10, one for vertical and the other for horizontal edges. For an edge in an arbitrary direction, the root square of the results (or the sum of the absolute values) for these two directions is used.

Figure 11.11 shows an example of edge detection by a Sobel operator. However, as the pixel values in the classified height image are homogenous, the search for

	(a)			(b)		
	1	2	1	−1	0	1
	0	0	0	−2	0	2
	−1	−2	−1	−1	0	1

Figure 11.10 Sobel operator for edge detection: (a) for horizontal edge and (b) for vertical edge.

(a)
```
2 2 2 2 8 8 8 8
2 2 2 2 8 8 8 8
2 2 2 2 8 8 8 8
2 2 2 2 8 8 8 8
2 2 2 2 2 2 2 2
2 2 2 2 2 2 2 2
2 2 2 2 2 2 2 2
2 2 2 2 2 2 2 2
```

(b)
```
—  —  —  —  —  —  —  —
—  0  0  24 24 0  0  —
—  0  0  24 24 0  0  —
—  0  0  19 25 24 24 —
—  0  0  8  19 24 24 —
—  0  0  0  0  0  0  —
—  0  0  0  0  0  0  —
—  —  —  —  —  —  —  —
```

Figure 11.11 Edge detection by a Sobel operator: (a) the original image and (b) the edge detected.

CONTOURING FROM DIGITAL TERRAIN MODELS

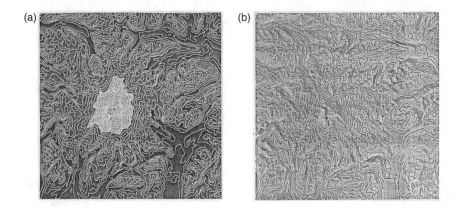

Figure 11.12 Gray-scale display of contour map (Reprinted from Eyton 1984, with permission from Elsevier): (a) Contours over a gray background. (b) Illuminated contour map.

boundary points between two adjacent height classes may be simplified. For example, in Figure 11.11(a), if code number 2 is used to represent the class with height range [5,10) and code number 8 to represent the class with height range [10,15), then all pixels with code 8 adjacent to the pixels with code 2 together form the contour line with height 10, which is the boundary between these two classes. The edge contours are in fact ordinary contour lines and therefore there is no need to give an example here.

11.3.2 Gray-Tone Contouring

In fact, a variety of contour products could be produced in raster mode. For example, gray-scale contour display can easily be produced, such as

1. *three-tone contours*: that is, to make use of black, white, and gray
2. *contours over a gray background*: that is, to draw white contours over a gray background
3. *illuminated contours*: that is, to draw contours over a background of shading (see Chapter 13).

Figure 11.12 shows two contour plots with gray-tones. Figure 11.12(a) shows contours plotted over a gray background and Figure 11.12(b) shows a contour map plotted over a shading.

11.4 VECTOR-BASED CONTOURING FROM TRIANGULATED DTM

Contouring from grid data is straightforward and convenient for algorithm implementation. However, there might be an accuracy loss in the random-to-grid interpolation process if the sampled data are not in grid form. If the data points are irregularly distributed, a TIN (see Chapter 5) structure is often the solution.

The process for contour tracing in a TIN structure is similar to the one described in Section 11.2:

1. to search for the starting point of each contour line
2. to interpolate the contour points by computing the coordinates of the intersection points between each contour line and the triangle edges
3. to trace contour lines through the triangles
4. to smooth traced contour lines, if desirable.

The search for and interpolation of contour points in vector-based contouring from TINs are similar to contouring from grid data. On the other hand, tracing contour points is slightly different because the data structure is different. For a triangle, there are only two possible exits for a given contour point on the edge of a triangle (see Figure 11.13).

It is also the basic principle of contour tracing that the exit edge of the current triangle is the entrance edge of the adjacent triangle. Figure 11.14 shows the threading process, where the contour line enters triangle II from triangle I. In this case, triangle edge $\overline{AC}$ is both the exit edge of triangle I and the entrance edge of triangle II. Figure 11.15 is an example of contouring from a TIN-based DTM.

The contours in Figure 11.15 do not look nice. There is also a smoothing problem here. Similar to contouring from grid DTM, either of the following two methods can be used to solve the problem: curve fitting or subdivision of a triangle into a number of sub-triangles (e.g., nine sub-triangles).

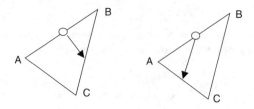

Figure 11.13 Two possible exits of a contour point on triangle edge $\overline{AB}$.

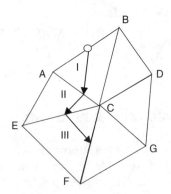

Figure 11.14 Contour threading on a TIN-based DTM.

CONTOURING FROM DIGITAL TERRAIN MODELS

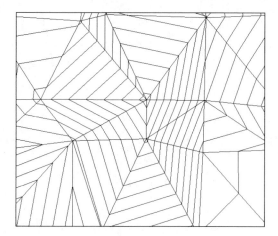

Figure 11.15 Contours produced from a TIN-based DTM.

Thus, triangulations produce unambiguous contour segments, and hence complete contour strings. A more interesting problem, as in grids, is how to extract all contour strings efficiently from the topological structure, as all cells must be checked for the presence of contours that may only be small closed loops. Various algorithms have dealt with this. Gold and Cormack (1987) showed that a single traversal of the triangulation may be used to maintain a sweepline that collects and maintains the partially discovered contour portions. van Kreveld (1996) develops more advanced data structures that improve the searching efficiency significantly. Nevertheless, aside from efficiency concerns, every cell in the grid or triangulation must be checked somehow to see if portions of contours pass through them, and the contour strings must be assembled either on an ongoing basis or by matching the individual segments after they have all been collected.

11.5 STEREO CONTOURING FROM GRID DTM

The natural world is 3-D. 2-D contours are not as attractive as a 3-D display of the terrain surface. However, on such a surface, the height information is not easily perceived or obtained when viewed. This stimulated the idea of producing stereo contour maps (Jensen 1980; Eyton 1984). One contour map is, as usual good for metric measurement, while the other is the stereomate only used for stereo viewing. This is an analog of orthoimage and its stereomate, which was introduced by Collins (1968). This is particularly useful in flat areas where the terrain variation is not great and thus it is difficult to recognize slope trends because the contours are sparsely distributed.

11.5.1 The Principle of Stereo Contouring

The human beings sense and receive 3-D information through binocular parallax, to form stereo models in their mind. Parallax plays a crucial role in the transformation of 2-D images to 3-D.

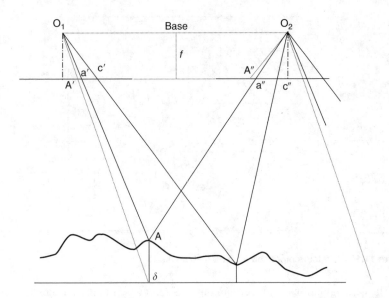

Figure 11.16 X-parallax and stereo measurement.

In orthoimage, the displacement of an image point caused by terrain height over a desired datum is corrected point by point so that the corrected image can be used as a plan, that is, measurable in planimetric positions. In this case, height measurement is not possible anymore. To make orthoimages measurable for height, the idea of stereomates was proposed by Collin (1968). The principle is illustrated in Figure 11.16. Ground point A is imaged on the left photo as a' and on the right photo as a''. The x-parallax of this image point is computed as follows:

$$p_a = x_{a'} - x_{a''} \tag{11.7}$$

where $x_{a'}$ is the x coordinate of a' on the left image and $x_{a''}$ is the x coordinate of a'' on the right.

If the left image is made into an orthoimage, then the image position of ground point A will be corrected to point A' from a'. Then, the difference of the x coordinates $\delta x_a = x_{a'} - x_{A'}$. If this orthoimage is used with the original right image to form a stereo model, then the height of A is changed because the x-parallax p_a is reduced by an amount of δx_a. In order to make the height of point A remain the same, the position of a'' needs to be modified, that is, to be shifted by the same amount as δx_a (from a'' to A'' in the figure).

In stereo contouring, the principle is the same. First, an ordinary contour is produced from the original DTM. Then, a procedure is applied to DTM data so as to make the x coordinate of each DTM point shift by an amount that is proportional to the height of this point. In this way, a new DTM, that is, the mate DTM, is created. The contour map produced from this mate DTM is then the stereomate of the contour map produced from the original DTM.

11.5.2 Generation of Stereomate for Contour Map

As a contour map is already measurable, there is no need for an exact of x-parallax for each point. Therefore, any function that is able to provide relatively correct parallaxes for DTM data points will serve the purpose. The simplest is a linear function (Eyton 1984), as follows:

$$\delta x_j = \frac{\delta x}{Z_{max} - Z_{min}} \times (Z_j - Z_{min}) \tag{11.8}$$

where Z_{max} is the greatest height in the area; Z_{min} is the lowest height; δx is the maximum amount of x-shift (i.e., the x-shift for Z_{max}); Z_j is the height of point j; and δx_j is the required x-shift for point j. The new x coordinate of each point is

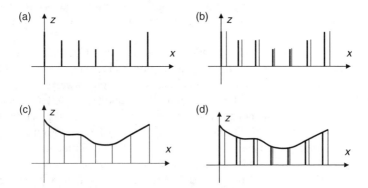

Figure 11.17 The process of generating a new grid DEM with the desired parallax. (a) One row of the original grid. (b) Each grid point shifted to the right proportionally. (c) Fitting a curve to the shifted points. (d) Interpolating a height for each node of the original grid.

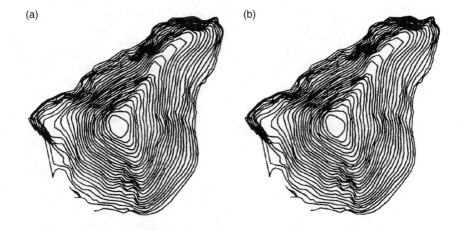

Figure 11.18 Stereo contour maps.

computed as follows:
$$X_{j,\text{new}} = X_{j,\text{old}} + \delta x_j \tag{11.9}$$
but the y coordinate and height of each point remain the same.

If the data are irregularly distributed, after an x-shift to every point, a triangulation procedure is applied to the new set of data and the mate contour map is then produced from the TIN-based DTM.

If the original data are in a regular grid form, then, the new data are not regular any more in X because an x-shift was introduced to every point according to the height of the point. There are two alternative approaches for the generation of the mate contour map:

1. to apply a triangulation procedure to build a TIN and then produce contours from the TIN-based DTM
2. to apply an interpolation procedure to interpolate the new grid data set, that is, to obtain new height value at the original location of each grid node.

Figure 11.17 illustrates the process for interpolation of the new grid data set. Figure 11.17(a) is a row in the original grid; Figure 11.17(b) represents the situation after applying an x-shift to each grid point, that is, each point is moved from the node position by an x-shift computed by Equations (11.8) and (11.9). After the shifts, the intervals between points are not equal anymore. That is, the original grid is broken. To obtain a new set of data in grid form, a new height for each of the original grid nodes needs to be computed from the x-shifted points. A polynomial function is fitted to the x-shifted points for interpolation, as shown in Figure 11.17(c). Figure 11.17(d) shows the new heights at the original grid nodes. A contour map, that is, the stereomate, can then be produced from these new data. Figure 11.18 shows an example of stereo contour maps.

CHAPTER 12

Visualization of Digital Terrain Models

It has been estimated that over 80% of information one obtains is through our visual systems and thus our visual systems are overloaded. From an other point of view, visualization is an important issue in all disciplines, including digital terrain modeling.

12.1 VISUALIZATION OF DIGITAL TERRAIN MODELS: AN OVERVIEW

DTM visualization is a natural extension of contour representation, which has been discussed in Chapter 11. In order to understand this, the basic concepts, that is, variables used at different stages, approaches, and basic principles, will be discussed here.

12.1.1 Variables for Visualization

Visual representation is an ancient communication tool and contouring is a graphic representation for visual communication. Here, communication means to present information (results) in graphic or other visual forms that are already understood. Six primary visual variables are available for such a presentation:

1. three geometric variables
   - shape
   - size
   - orientation
2. three color variables
   - hue
   - value or brightness
   - saturation or intensity

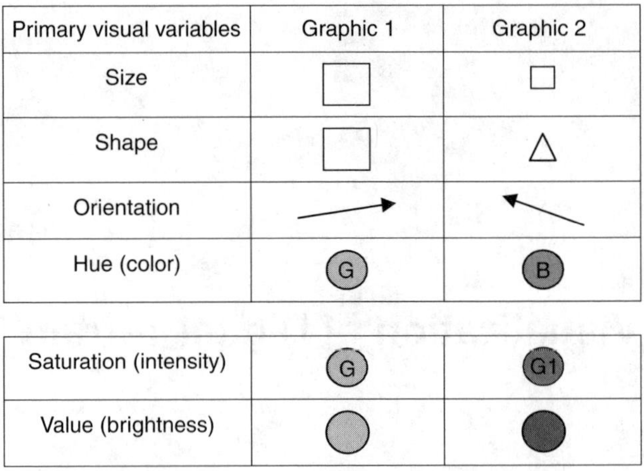

Primary visual variables	Graphic 1	Graphic 2
Size	□	▫
Shape	□	△
Orientation	→	↙
Hue (color)	G	B
Saturation (intensity)	G	G1
Value (brightness)	○	●

Figure 12.1 Six primary variables for visual communication. The color plate can be viewed at http://www.crcpress.com/e_products/downloads/download.asp?cat_no=TF1732.

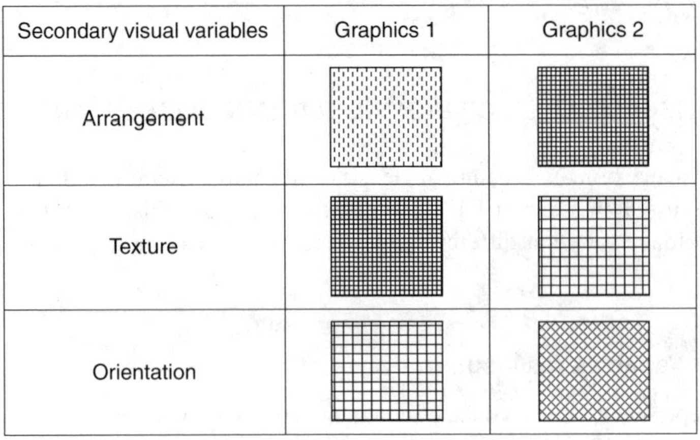

Secondary visual variables	Graphics 1	Graphics 2
Arrangement		
Texture		
Orientation		

Figure 12.2 Three secondary variables for visual communication.

Figure 12.1 shows these six variables graphically. In addition, three secondary visual variables (Figure 12.2) are available:

1. *Arrangement*: shape and configuration of components that make up the pattern.
2. *Texture*: size and spacing of components that make up a pattern.
3. *Orientation*: directional arrangement of parallel rows of marks.

Visualization is a natural extension of communication and goes into a domain called visual thinking (DiBiase 1990). Visualization emphasizes an intuitive representation of data to enable people to understand the nature of phenomena represented by the data. In other words, visualization is concerned with exploring data and information graphically — as a means of gaining understanding and insight into the data.

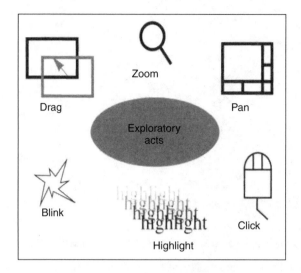

Figure 12.3 Exploratory acts for visual analysis (Reprinted from Jiang 1996 with permission).

Table 12.1 Variables at the Different Stages of Visualization

Stage	Variables in Use				
Paper graphics	Visual variables	—	—	—	—
Computer graphics	Visual variables	Screen variables	—	—	—
Visualization	Visual variables	Screen variables	Dynamic variables	Exploratory acts	—
Web-based visualization	Visual variables	Screen variables	Dynamic variables	Exploratory acts	Web variables

Thus, visualization has been compared to visual analysis, with an analogy to numerical analysis.

Visualization is a fusion of a number of scientific disciplines, such as computer graphics, user-interface methodology, image processing, system design, cognitive science, and so on. The major components are rendering and animation techniques. In visualization, in additional to the traditional visual variables, some other sets of variables are in use. One set, related to analysis, is called exploratory acts (Figure 12.3), which consists of drag, click, zoom, pan, blink, and highlight and so on (Jiang 1996). Theoretically, some variables particular to screen display such as blur, focus, and transparency (Kraak and Brown 2001) are also in use. In the era of Web-based visualization, more exploratory acts are in use, particularly the browse and plug-in. Table 12.1 lists the sets of variables in use at different stages.

The dynamic variables (DiBiase et al. 1992) are related to animation, including duration, rate of change, and order. These variables will be discussed in Section 12.5.

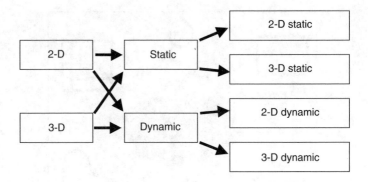

Figure 12.4 Approaches for graphic representation of DTM surface.

12.1.2 Approaches for the Visualization of DTM Data

Visualization of DTM data means to make use of these variables for visual presentation of the data so that the nature of the terrain surface could be better understood. In fact, in Chapter 1, a brief discussion on the representation of terrain surface was conducted and it was pointed out that terrain surfaces could be represented by either graphics or mathematical functions (Figure 1.4). This chapter focuses on graphic representations. It is understandable that there are 2-D and 3-D representations, both in static and dynamic modes. Figure 12.4 shows a classification of these visualization approaches.

This chapter gives a brief discussion of 2-D representation techniques and a few new developments in 3-D representations, as follows:

1. *Texture mapping*: This is to produce virtually real landscapes by mapping aerial photographs or satellite images onto the digital terrain model. This method can show the color and texture of all kinds of ground objects and artificial constructions, but the geometric texture of terrain relief cannot be clearly represented. Therefore, the method is often used to represent smooth areas where there are many ground objects and human activities, such as towns and traffic lines.
2. *Rendering*: This is like shading, but in 3-D representations. It makes use of illumination models to simulate the visual effect produced when lights shine on the terrain. This method can be used to simulate micro ground relief (geometric texture) and color using pure mathematical models. Terrain simulation based on fractal models is considered to be the most promising method.
3. *Animation*: This can be used to produce dynamic and interactive representations.

If all these techniques are compared, one would find that some are more abstract than others and some are more symbolic than others. Figure 12.5 summarize this.

12.2 IMAGE-BASED 2-D DTM VISUALIZATION

In two dimensions, contouring is the most popular technique. A detailed description of contouring was given in Chapter 11. This section presents some image-based

VISUALIZATION OF DIGITAL TERRAIN MODELS

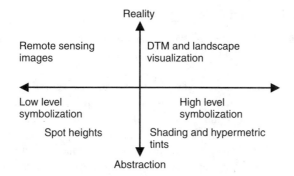

Figure 12.5 A comparison of various techniques for terrain visualization.

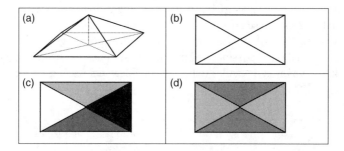

Figure 12.6 Shading of terrain surface: (a) a pyramid-like object; (b) the orthogonal view; (c) hill shading; and (d) slope shading.

techniques. It is possible to make the 2-D representation dynamic through animation; however, it is not common to do so, therefore 2-D dynamic representation will not be discussed here.

12.2.1 Slope Shading and Hill Shading

Among these image-based techniques, shading is still widely used. Two types are available, hill (or oblique) and slope (or vertical) shading.

Slope shading assigns a gray value to each pixel according to its slope value. The steeper the slope, the darker the image. Figure 12.6(a) is pyramid consisting of four triangular facets and a base. Figure 12.6(b) is the orthogonal view of Figure 12.6(a). Figure 12.6(d) is the result of slope shading. It can be found that the two facets with identical slope angles are assigned the same gray shade.

Figure 12.6(c) is the result of hill shading. The idea is to portray the terrain variations with different brightness by illuminating the pyramid so that shadow effects are produced, thus leading to the stereoscopic sense, which is produced by the readers' experience (but not by perception on a physical level). In hill shading, a light source is assumed, normally from the northwest. The facet facing the light is brightest and the facet facing away the darkest.

12.2.2 Height-Based Coloring

Here, the term *height-based coloring* means to assign a color to each image pixel based on the heights of the DTM data. Two approaches are in use, interval-based and continuous coloring.

Hypermetric tinting (color layers) is an interval-based coloring widely used. The basic principle is to use different colors for areas with different altitudes. Theoretically, one could use an infinite number of colors to represent heights. However, in practice, terrain surface is classified into a few intervals according to height and one color is assigned to each class. The commonly used colors are blue for water, green for lower altitude, yellow for medium, and brown or red for higher altitude. Figure 12.7(a) is an example.

Gray can also be used to produce an image similar to Figure 12.7(a). Figure 12.7(b) is an example. It is possible to use a continuous variation of gray tones to illustrate the variations of the terrain surface (instead of height ranges). In other words, gray levels from 0 to 255 are used to represent the heights of the terrain surface. A mapping process is needed to fit the terrain height variations into the gray range of [0,255]. Figure 12.8 shows some possible mappings. The simplest is linear stretching (if the range of heights is much smaller than 256) or linear depression (if the variation is

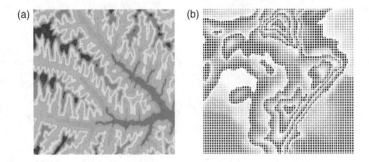

Figure 12.7 Interval-based coloring of terrain heights: (a) hypermetric tints (color layers) and (b) half toning (gray layers). The color plate can be viewed at http://www.crcpress.com/e_products/downloads/download.asp?cat_no=TF1732.

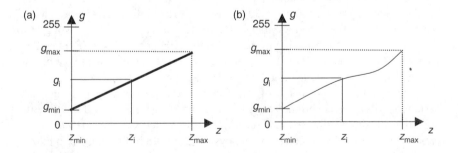

Figure 12.8 Height value to gray level mapping: (a) linear mapping and (b) nonlinear mapping.

VISUALIZATION OF DIGITAL TERRAIN MODELS

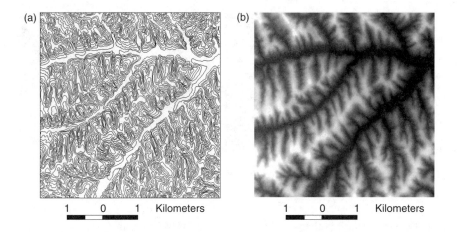

Figure 12.9 Representation of DTM by continuous gray image: (a) a contour map and (b) the gray image of the contour map.

outside the range of [0,255]). Equation (12.1) is the formula for a linear mapping.

$$g_i = g_{min} + \frac{g_{max} - g_{min}}{z_{max} - z_{min}}(z_i - z_{min}) \qquad (12.1)$$

where g_i is the gray value of height z_i; g_{min} is the desired minimum gray value, $0 \leq g_{min} < g_{max}$; g_{max} is the designed maximum gray value, $g_{min} < g_{max} \leq 255$; g_{min} is the lowest height in the area; and z_{max} is the largest height value in the area. In this way, the height range $[z_{min}, z_{max}]$ is mapped into a gray range $[z_{min}, z_{max}]$. Usually, the full gray range [0,255] is used and thus $z_{min} = 0$ and $z_{max} = 255$. Figure 12.9 is an example of the continuous gray image of a DTM, which clearly shows the shape of the landscape.

12.3 RENDERING TECHNIQUE FOR THREE-DIMENSIONAL DTM VISUALIZATION

With the development of computer graphics, 3-D visualization has become the mainstream of DTM visualization. The 3-D wire frame (Figure 12.10) is widely used, especially in computer-aided design. However, rendering, which employs some illumination models to produce a vivid representation of 3-D objects, has become a more popular technique for DTM visualization.

12.3.1 Basic Principles of Rendering

The basic idea of rendering is to produce vivid representations of 3-D objects. A surface is split into a finite number of polygons (or triangles in the case of TIN); all these polygons are projected onto the view plane of a given viewpoint; each visible pixel is assigned a gray value, which is computed based on an illumination model

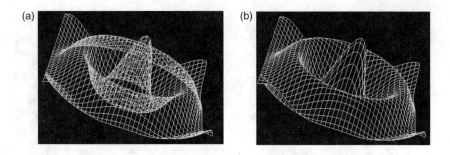

Figure 12.10 Three-dimensional wire frame of a surface: (a) hidden lines not removed and (b) hidden lines removed.

and the viewpoint. In other words, rendering of DTM is to transform a DTM surface from a 3-D to a 2-D plane. The rendering process follows these steps:

1. to divide the surface to be rendered into a set of contiguous triangular facets
2. to set a viewpoint, determine the observing direction, and transform the terrain surface into an image coordinate system
3. to identify the visible surfaces
4. to calculate the brightness (and color) of the visible surface according to an illumination model
5. to shade all the visible triangular pieces.

The first step is omitted here because triangulation was discussed in Chapters 4 and 5, and the subdivision of triangles was discussed in Chapter 9.

12.3.2 Graphic Transformations

What can be displayed on the screen is determined by the position of the observer (or viewpoint) and the direction of the sight line. Rendering begins with the transformation of the terrain surface from the ground coordinate system (GCS) O–XYZ to the viewpoint-centered eye-coordinate system (ECS) O_e–$X_e Y_e Z_e$ and then it projects the surface onto the display screen which is parallel to the O_e–$X_e Y_e$ plane. This series of transformations is called graphical transformations, which consists of shifting, rotating, scaling, and projection.

Both the GCS and the ECS are right-hand 3-D Cartesian coordinate systems. For the ECS, its origin is fixed on the viewpoint, and its axis Z_e is opposite the observing direction. Based on the characteristics of digital computation with a computer, a vector in 3-D space is described by three direction cosines. This simplifies the relationships between two 3-D coordinate systems and makes the computation of coordinate transformations more efficient. All subsequent processes, such as recognition of visible facets, projective transformation, and the shading process, will be carried out in the ECS. Figure 12.11 shows the relationship between the two coordinate systems.

Given the coordinates of the viewpoint in the GCS as $(X_{O_e}, Y_{O_e}, Z_{O_e})$ and an observing direction (azimuth angle α and pitch angle β), the direction cosine of each

VISUALIZATION OF DIGITAL TERRAIN MODELS

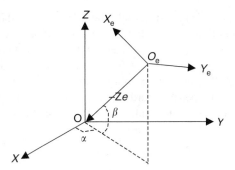

Figure 12.11 The ground coordinate and eye-coordinate systems.

eye-coordinate axis can be calculated. In order to simplify the calculation, the vector O_eO (from the viewpoint O_e to the origin of the GCS O) and the direction of the sight line are merged here. This joint direction will be considered as the future projection direction. This simplifies the problem. That is, when the direction of the sight line and the viewing distance D_S from O_e to O are known, then the coordinates of the viewpoint can be derived as follows:

$$\begin{bmatrix} X_{O_e} \\ Y_{O_e} \\ Z_{O_e} \end{bmatrix} = \begin{bmatrix} D_S \times \cos\beta \times \cos\alpha \\ D_S \times \cos\beta \times \sin\alpha \\ D_S \times \sin\beta \end{bmatrix} \quad (12.2)$$

The three direction cosines are the cosines of the angles between the vector from the origin to a point P and each of the coordinate axes (in the plane including the vector and the axis). If vector $\overrightarrow{OP}$ is of unit length, these direction cosines reduce to P_X, P_Y, and P_Z (usually called l, m, and n).

Let the direction cosines of O_eX_e, O_eY_e, O_eZ_e be represented by $(l_1\ l_2\ l_3)$, $(m_1\ m_2\ m_3)$, and $(n_1\ n_2\ n_3)$. Suppose O_eX_e is the horizontal axis, then

$$n_1 = \frac{X_{O_e}}{D_S}, \quad n_2 = \frac{Y_{O_e}}{D_S}, \quad n_3 = \frac{Z_{O_e}}{D_S} \quad (12.3)$$

$$l_1 = -\frac{n_2}{r}, \quad l_2 = -\frac{n_1}{r}, \quad l_3 = 0 \quad (12.4)$$

where $r = \sqrt{n_1^2 + n_2^2}$

$$m_1 = -n_3 l_2 = -\frac{n_1 n_3}{r}, \quad m_2 = n_3 l_1 = -\frac{n_2 n_3}{r}, \quad m_3 = r \quad (12.5)$$

And the relationship between the ground coordinate (X, Y, Z) and the eye-coordinate (X_e, Y_e, Z_e) is:

$$\begin{bmatrix} X_e \\ Y_e \\ Z_e \end{bmatrix} = \begin{bmatrix} l_1 & l_2 & l_3 \\ m_1 & m_2 & m_3 \\ n_1 & n_2 & n_3 \end{bmatrix} \begin{bmatrix} X - X_{O_e} \\ Y - Y_{O_e} \\ Z - Z_{O_e} \end{bmatrix} \quad (12.6)$$

To project the 3-D terrain surface onto the 2-D screen, either parallel or central (perspective) projection can be used. To obtain the visual effects consistent to the human eye and to produce perspective views with strong stereo sense and realism, the perspective projection is used in the field of computer graphics. Suppose a plane parallel to the O_e–$X_e Y_e$ plane and with a distance f to the viewpoint is used as a projection plane (screen), then the coordinates of a point in the ECS can be transformed into the coordinates (u, v) on the display screen by using the following formula:

$$u = \frac{X_e}{Z_e} \times f \qquad (12.7)$$

$$v = \frac{Y_e}{Z_e} \times f \qquad (12.8)$$

In these formulae, f is similar to the focus of the camera, expressing the distance between the projection plane (screen) and the observer. Experience shows that optimal visual effects can be obtained when f is three times the size of the screen.

12.3.3 Visible Surfaces Identification

The challenge in generating graphic images with a stereo sense is the removal of hidden surface, which is similar to the hidden line removal in the 3-D wire frame. This means that those facets that can be seen from the position of the current viewpoint need to be identified. Surface facets outside the view field are cut out, and those facets that are in the view field but are partially blocked by others have to be identified. This process is also called the recognition of the visible surface facets in the literature. Figure 12.12 shows these different surface facets.

All algorithms for visible surface recognition make use of a form of geometric classification to identify the visible and hidden surfaces. Visible surface recognition

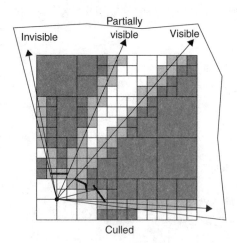

Figure 12.12 Different surface facets, completely hidden, partially visible, and visible.

can be carried out either in image or in object space. Image-based algorithms make a judgment through the examination of the projected images, while space-based algorithms directly examine the definition of the object. The commonly used algorithms are depth sorting (i.e., an object-based method), and Z-buffer (depth buffer), area subdivision, and scanning lines. These are image-based methods.

For n triangular facets producing N pixels, the computation complexity for image-based algorithms is $O(nN)$ as they examine the image pixel by pixel. By contrast, object-based methods compare each surface facet and thus the computation complexity is lower — $O(n^2)$. Experience shows that depth sorting is the most efficient method for situations where the number of triangles is less than 10,000; all methods except the depth buffer are significantly slow when the number of triangles is more than 10,000. It might be said that the depth sorting algorithm is more suitable for DTMs with a TIN structure and the depth buffer algorithm is more appropriate if the fractal subdivision of the grid DTM is employed.

In the depth sorting algorithm, first sort all the triangles based on the distance between the triangles and the viewpoint (called depth in the ECS), then process each triangle in sequence from far to near. This method is often called a painter algorithm, as it is similar to the painter's creation — first paint the background, then gradually add the foreground objects on the background. Obviously, the color of the close objects will cover the color of the objects behind, and finally the hidden parts are naturally removed. Since there are no intersections and no gaps between the TIN, the depth sorting algorithm is reliable.

The characteristic of the depth buffer algorithm is that it needs to reserve a 2-D array (Z-buffer) to access the depth (the value of Z_e) of the pixels currently in the computer frame buffer. The triangular facets are divided into parts as large as pixels, and the depth of each part (assumed to be fixed) is compared with that in the Z-buffer. If some part is closer than the current pixels, it will be written into the frame memory, and the Z-buffer will be updated with the new depth. The size of the Z-buffer is decided by the display resolution.

No matter what method is used to identify visible surfaces, the results from the processing are applicable only to the specific viewpoint and observing direction. As a result, real-time updating of graphics with change in viewpoint and view direction is restrained by the efficiency of the visible surface recognition (i.e., hidden surface removal). It is worth noting that in the ECS, the depths of all points have negative values.

12.3.4 The Selection of an Illumination Model

When a surface facet is identified as being visible, the next step is to assign different colors or gray values to different parts of the surface facet because when light illuminates the surface, the shading of each part is different. Therefore, to a large extent, the realism of a 3-D terrain display depends on the shading effect. To do so, the surface is decomposed into pixels and a color is assigned to each pixel. To produce vivid shading, illumination of the surface is the key element. There are two approaches to color assignment, that is, to make use of a model or to make use of the real texture of the object. In this section, only the use

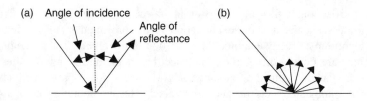

Figure 12.13 Reflectance of lights: (a) specular reflector and (b) diffuse reflector.

of an illumination model is discussed and the use of real texture is addressed in Section 12.4.

Visible light reflected by objects contains two types of information, spatial and spectral, which are the basis for interpretation. As different kinds of natural ground objects have different reflectance characteristics, and they may be illuminated by different light sources, it is impossible to simulate the illumination effect of natural scenery with 100% realism.

There are two types of reflection, diffuse and mirror reflections, as shown in Figure 12.13. Mirror reflection, or specular reflection, is in a single direction. Diffuse reflection is uniform in all directions. However, the real terrain surface is neither a pure diffuse reflector nor a pure specular reflector. Rather, most earth surfaces are somewhere between the two. Therefore, a combination of both models seems to be a realistic solution. Also, both reflected light and environmental light need to be considered.

An illumination model establishes the relationships between the reflecting intensity at any ground point, light source, and features on terrain. The Lambert cosine law describes the illumination model for diffuse reflection. As shown in Figure 12.13(b), if the incidence angle between the normal vector of point P on the ground and the vector directing to the light source from P is θ, then the intensity of diffuse reflection light on point P, I_d, is:

$$I_d = I_P \times K_d \times \cos\theta \tag{12.9}$$

where I_P is the intensity of the light source and $K_d \in (0, 1)$ is the coefficient of diffuse reflection on the ground. Since the light is diffused in all directions uniformly, the intensity of the diffuse reflection is independent of the viewpoint.

On the other hand, with specular reflection, the light reflected is in a single direction (Figure 12.13a), that is, the direction with an angle equal to the angle of incidence. However, since real terrain is usually not a complete specular reflector, its mirror reflection does not follow the reflection law strictly. After considering this, Phong (1975) developed his famous Phong model as follows:

$$I_S = I_P \times W(\theta) \times \cos^n \alpha \tag{12.10}$$

where α is the angle between the complete reflecting direction and the sight line, $W(\theta) \in (0, 1)$ is the surface reflection function for mirror reflection related to the characteristics of real terrain surface, which is usually simplified with a constant $K_s \in (0, 1)$; and n is the focus index of mirror reflection, the smoother the surface, the bigger the value of n.

VISUALIZATION OF DIGITAL TERRAIN MODELS

In most cases, to increase the realism, environmental light is also taken into consideration. The characteristics of environmental light are described by a diffusion model,

$$I_a = I_E \times K_a \tag{12.11}$$

where I_E and K_a are the intensity of environmental light and the coefficient of the terrain reflected environmental light, respectively. Since its effect on the scene is the same, generally it is also treated as a constant with its value equal to 0.02 to 0.2 times $I_P K_d$.

Combining the diffuse and mirror reflection models, the Phong model is as follows:

$$I = K_a \times I_E + \sum [K_d \times I_P \times \cos\theta + K_s \times I_P \times \cos^n \alpha] \tag{12.12}$$

Here, $\sum$ indicates the sum of all the light sources and $K_d + K_s = 1$. In practice, vivid results can be obtained by using only a point light source. In this way, the computation is simplified.

12.3.5 Gray Value Assignment for Graphics Generation

After the illumination model is presented, the gray level for any area of the surface facet can be estimated. The Gouraud (1971) shading is a simple but effective method for this purpose. In this method, the gray values of the three vertices are first estimated from the Phong model, then all pixels within this triangle are linearly interpolated from these three vertices. Figure 12.14 shows the principle. The formulae for this linear interpolation were given in Chapter 6.

The result of shading by the Gouraud model looks smooth, since the intensities change continually across the polygon edges. This approach is still used in today's hardware accelerated rendering pipelines (Zwicker and Gross 2000).

As discussed in Chapter 4, the problem with linear interpolation is that it is not smooth across the boundary of two linear facets. To solve this problem, Phong introduced a more realistic model that is able to simulate specular highlights. In this method, interpolation is carried out by using normals instead of intensities. Figure 12.15 is the perspective view of DTM shading produced by this method.

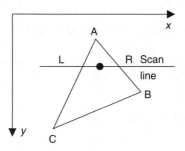

Figure 12.14 Scan line incremental method.

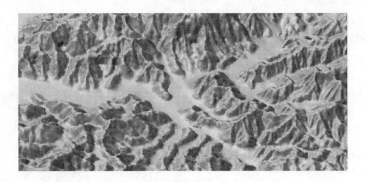

Figure 12.15 Shading of DTM.

Figure 12.16 Perspective view of DTM by altitude tinting. The color plate can be viewed at http://www.crcpress.com/e_products/downloads/download.asp?cat_no=TF1732.

To display terrain surfaces more realistically, apart from the gray levels, other colors with different intensities can also be used. Terrain with different altitudes may be represented by different colors, which makes the 3-D terrain image have the effect of hypermetric tints. Figure 12.16 is an example.

12.4 TEXTURE MAPPING FOR VIRTUAL LANDSCAPE GENERATION

This section discusses how to map texture and other attributes onto the terrain surface, so as to produce a more vivid view, called virtual landscape.

12.4.1 Mapping Texture onto DTM Surfaces

To improve the visual realism of images synthesized by rendering, a number of techniques have been developed. The basic idea is to add image-based information to the rendered primitives. The most commonly used technique is called texture mapping, that is, mapping a function of texture onto a 3-D surface. The function could

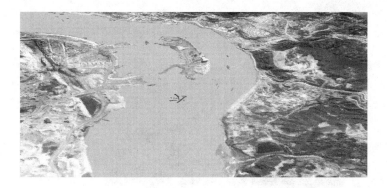

Figure 12.17 Mapping texture onto the surface of DTM. The color plate can be viewed at http://www.crcpress.com/e_products/downloads/download.asp?cat_no=TF1732.

be 1-D, 2-D, or 3-D and may be represented by discrete values in a matrix array or by a mathematical expression. Texture mapping enhances the visual richness of raster images while entailing only a relatively small increase in computation. Figure 12.17 is an example of such a product, showing part of the Yangtze River of China.

In this context, the texture is defined by a 2-D image array. The digital image data could be obtained from photographs or videos or generated by mathematical functions. As the data are in a discrete raster format, before texture mapping, a continuous texture function $f(U, V)$ in the texture space (U, V) has to be established by using these discrete data. The easiest method is to carry out an interpolation by using a bilinear function.

The first step in texture mapping is to map the texture onto the 3-D terrain surface; the second is to map the 3-D surface with texture onto the screen. To map from the texture space to the 3-D terrain, the most accurate method is to establish direct mapping between the texture coordinate system (U, V) and the 3-D ECS (X_e, Y_e, Z_e) based on central projective principles. The direct linear transformation (DLT) can be used for this purpose:

$$U = \frac{a_1 X_e + b_1 Y_e + c_1 Z_e}{a_3 X_e + b_3 Y_e + c_3 Z_e} \tag{12.13}$$

$$V = \frac{a_2 X_e + b_2 Y_e + c_2 Z_e}{a_3 X_e + b_3 Y_e + c_3 Z_e} \tag{12.14}$$

The computation required in this equation is heavy because it is a nonlinear function. In practice, a simple function similar to the affine function in 2-D can serve for this purpose:

$$U = a_1 X_e + b_1 Y_e + c_1 Z_e + d_1 \tag{12.15}$$

$$V = a_2 X_e + b_2 Y_e + c_2 Z_e + d_2 \tag{12.16}$$

At least four control points are required, whose texture and eye-coordinates are known. The control points in used photogrammetry or in DTM data may be used

Figure 12.18 Virtual landscape by mapping texture and other objects: (a) texture image and 2-D features mapped onto DTM and (b) texture image and 3-D features mapped onto DTM. The color plate can be viewed at http://www.crcpress.com/e_products/downloads/download.asp?cat_no=TF1732.

for this transformation. In digital photogrammetry, the texture coordinates and object space coordinates of all the DTM points are known.

12.4.2 Mapping Other Attributes onto DTM Surfaces

By mapping texture onto the DTM model one obtains vivid details of the terrain surface. In fact, the visual effect can be enhanced by adding other information onto the model, for example, designed roads, river, land use, vegetation, and images.

Aerial images can be mapped onto DTMs to produce realistic landscapes. In fact, images, vector data (lines), and 3-D objects on the ground (e.g., houses, trees), can also be mapped onto the DTM. Figure 12.18 shows such examples.

12.5 ANIMATION TECHNIQUES FOR DTM VISUALIZATION

In the previous sections, static techniques for 3-D visualization of DTM were discussed. However, these techniques can become dynamic by employing animation techniques.

12.5.1 Principles of Animation

The fundamental of animation is the page flipping technique, resulting in movies. First a number of frames of pictures are made and stored in computer memory, then they are displayed on screen in sequence. As mentioned in Section 12.1, three dynamic variables are available to control the animation process:

1. *Duration (time units for a scene)*: Normally, a frame duration of 1/30 sec (i.e., 30 frames per second) will produce a smooth animation. If the duration is too long, the action will be jerky.
2. *Rate of change (pace of animation or differences between two successive scenes)*: Figure 12.19 shows the animation of (up–down) vibration, with four frames. The differences between these frames are clear. If the rate is low, slow motion can be produced. On the other hand, fast motion is produced if the change rate is high.
3. *Order (the sequence of the frames)*: Frames could be arranged according to time, position, or attributes. The frame sequence in Figure 12.19 is arranged according to time. However, the frames in Figure 12.21 and Figure 12.22 are arranged according to the viewpoint.

In terrain visualization, "fly-through" and "walk-through" are commonly used. The animated image sequence is produced in an order of space, that is, by moving the viewpoint along a certain track. This type of animation is also called viewpoint animation.

There are two ways to access or display each picture frame, frame by frame or bit boundary block transfer (bitblt). Frame-based animation is full screen animation and page animation. First, a series of full screen images is produced and saved in a separate buffer, and then it is animated by displaying the pages in sequence. Frame animation is considered to be the best choice for complex and full shading. In bitblt, each frame is only a rectangular block of the full screen image. Less memory is required because only a small portion of the full screen display is manipulated each time. This can enhance the performance.

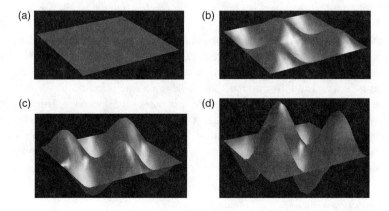

Figure 12.19 Four frames for animation of up–down vibration: (a) frame 1; (b) frame 2; (c) frame 3; and (d) frame 4.

For both kinds of animation, the image sequence has to be set up first. To obtain a fast speed, for example, 30 frames per second, all the frames are put into the memory. Therefore, both the number of frames and the capacity of each image are limited by computer memory. Various concepts for frame storage and display have been in use, such as RAM based, EMS/XMS based, and disk based. For example, the RAM-based method is usually used to produce smooth animation when a sequence is short and the amount of information is small (e.g., 30 frames, $160 \times 100 + 256$ colors).

12.5.2 Seamless Pan-View on DTM in a Large Area

With the development of computer graphics, it is possible to generate a seamless pan-view of the global DTM on a personal computer. On the other hand, the limitation of computers to real-time application of a large amount of DTM data is clear. Such limitations mainly rest in the size of memory, the volume of texture data, the precision of CPU floating points, the speed of display card for geometric shading and the speed of data transfer and access. With a given computer, the key to real-time display is (a) to reduce the computation required for rendering and (b) to speed up data access and display.

It is often the case that only a part of the terrain surface can be displayed at one time due to the large data volume, even when an LOD model as described in Chapter 9 is employed. To speed up the interactive real-time rendering of the terrain, usually only part of the data are selected for processing and the details in this part will also change dynamically with changes in viewpoint and sight line. An efficient mechanism for data organization and management is required to ensure the speedy dynamic triangular network updates required for scene changes with viewpoints. To manage the scenes, some parameters must be set to judge which part of the scene will be removed, updated, or accessed from the database and when to do so. That is, databases or data structures for DTM data storage must be able to support fast access to data.

To achieve real-time pan-views of a large area on a desktop PC, a common strategy is to apply multi-thread data paging based on subdividing the whole terrain into data blocks, as described in Chapter 10, double display buffers, and multi-thread process scheme. During panning, the data blocks in the current view field are selected according to the viewpoint and then different LODs are set according to the relationship between the data blocks, the viewpoint, and sight line. In this way, the number of models is reduced and the efficiency of scene rendering is increased.

The viewpoint is always located near the center of the data page. During panning, as the viewpoint moves, the data blocks on the data page need to be updated frequently. The moving direction of the viewpoint is judged by the offsets between the current position of the viewpoint (x_e, y_e) and the geometric center (x_c, y_c) of the data page, that is,

$$\Delta X = x_e - x_c \qquad (12.17)$$

$$\Delta Y = y_e - y_c \qquad (12.18)$$

VISUALIZATION OF DIGITAL TERRAIN MODELS

When ΔX is positive, the viewpoint moves toward the positive side of the x-axis, otherwise toward the opposite direction. If $|\Delta X| >$ BlockSize (the size of the data block) and $\Delta Y <$ BlockSize/2, a new column of data block in the moving direction is read into the data page; subsequently, the column of data block on the opposite side is deleted from the page, as shown in Figure 12.20.

Eight combinations of ΔX and ΔY are possible, up, down, left, right, upper-left, lower-left, upper-right, and lower-right, thus the forward direction of block movement could be in any of these eight directions. But, in each freshing, only one new row (or column) of a data block in the forward direction is added into the data page and one row (or column) in the backward direction is deleted.

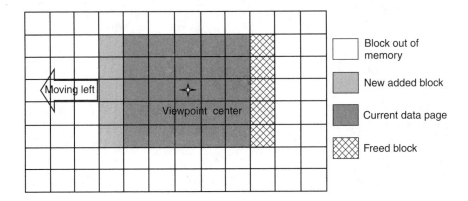

Figure 12.20 Dynamic data paging of data blocks.

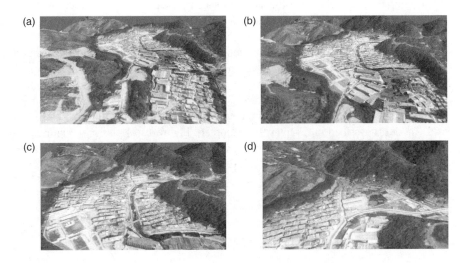

Figure 12.21 Four frames for fly-through animation: (a) frame 1; (b) frame 2; (c) frame 3; and (d) frame 4.

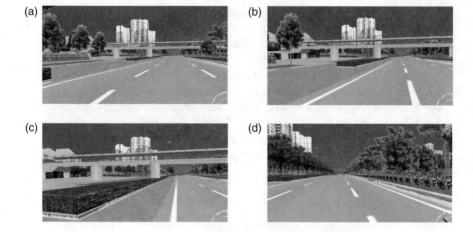

Figure 12.22 Four frames for walk-through animation: (a) frame 1; (b) frame 2; (c) frame 3; and (d) frame 4.

In this way, based on the offsets of the viewpoint and the geometric center of the data page, frequent updating of the data page is achieved and thus the real-time pan-view of a large area is realized.

12.5.3 "Fly-Through" and "Walk-Through" for DTM Visualization

Fly-through and *walk-through* are the two basic techniques used in terrain animation. They allow users to view a model from different angles. *Fly-through* provides a continuous bird's eye view to the landscape. That is, the viewpoint is far above the terrain surface. Therefore, the viewpoint can be moved in any direction in the 3-D space. *Walk-through* mimics the human view while walking. *Walk-through* can be considered as a special case of *fly-through*, that is, the viewpoint is low and its movement in vertical direction is restricted. The change in viewpoint for fly-through or walk-through can be controlled in various ways, such as using a mouse, keyboard, fixed route, or freedom to roam.

Similar to the pan-view of a large area, only the visible area is dynamically loaded and progressively rendered during the changes in the viewpoint. In most cases, an LOD model (described in Chapter 9) is adopted. Figure 12.21 shows the animation of a fly-through over a virtual landscape, with four frames. Figure 12.22 shows the animation of a walk-through the cityscape, again with four frames.

CHAPTER **13**

Interpretation of Digital Terrain Models

In Chapter 12, the visualization of DTM was discussed. Visualization can on the one hand be regarded as a representation and on the other hand compared to visual analysis. This chapter will cover DTM-based terrain analysis, or DTM interpretation.

13.1 DTM INTERPRETATION: AN OVERVIEW

To interpret a DTM means "to understand the terrain characteristics through the extraction/computation of the parameters." DTM interpretation is also called DTM-based terrain analysis.

The term *digital terrain analysis* means different things to people with different backgrounds because they emphasize different aspects. In some literature, a large part of digital terrain analysis is on interpolation methods for terrain surface modeling, which was discussed in Chapter 6; in some other literature a large part is on visualization of DTMs, which was the topic of Chapter 12; and for a third group, it means the derivation of attributes from terrain surfaces, which is the main content of this chapter.

It is understandable that people from different disciplines are interested in different sets of attributes of the terrain surface. A detailed discussion on all the possible attributes can be found in other literature (e.g., Moore et al. 1994; Wilson and Gallant 2000). This chapter considers the computation of commonly used attributes, such as slope and aspect, area and volume, roughness parameters, and hydrological parameters. In addition, the derivation of viewsheds and the analysis of inter-visibility between points on terrain surfaces are also presented.

13.2 GEOMETRIC TERRAIN PARAMETERS

This section discusses the computational models for geometric parameters, including surface area, projection area, and volume.

13.2.1 Surface and Projection Areas

The formula for the computation of the surface area of a triangle, S_Δ, is as follows:

$$S_\Delta = \sqrt{P(P - D_1)(P - D_2)(P - D_3)} \quad (13.1)$$

where D_i represents the length of the edge opposite the vertex I and is computed from Equation (13.2).

$$\begin{aligned} P &= \frac{1}{2}(D_1 + D_2 + D_3) \\ D_1 &= \sqrt{(x_3 - x_2)^2 + (y_3 - y_2)^2 + (z_3 - z_2)^2} \\ D_2 &= \sqrt{(x_3 - x_1)^2 + (y_3 - y_1)^2 + (z_3 - z_1)^2} \\ D_3 &= \sqrt{(x_1 - x_2)^2 + (y_1 - y_2)^2 + (z_1 - z_2)^2} \end{aligned} \quad (13.2)$$

The surface area of the whole DTM, S, is the sum of the surface areas of all triangles.

$$S = \sum_{i=1}^{N} S_{\Delta,i} \quad (13.3)$$

where N is the total number of triangles in the area. If the DTM is in a grid form, then each grid cell can be split into two triangles.

The area of the surface projected on the horizontal plane can also be computed from Equation (13.1). In this case, the heights for the three vertices of a triangle are set to 0. On the other hand, a more convenient method can be used for the computation of a horizontal area. Figure 13.1 shows the principle. In this figure, the three vertices are points 1, 2, and 3. If these three points are projected to the x-axis, then points 1′, 2′, and 3′ are obtained. Points 1 and 2, together with 1′ and 2′, form a trapezoid $\Delta 1, 2, 3$.

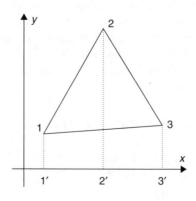

Figure 13.1 The area of $\Delta 1, 2, 3$ to be computed from three trapeziods.

Similarly, points 2 and 3, together with 2′ and 3′, form another trapezoid; and points 3 and 1, together with 3′ and 1′, form the third trapezoid. By adding the areas of the first two trapezoids together and subtracting the area of the third trapezoid, the area of the triangle $\Delta 1, 2, 3$ is obtained, that is,

$$A_{123} = |A_{122'1'}| + |A_{233'2'}| - |A_{311'3'}| \tag{13.4}$$

However, if the vertices are arranged clockwise and the areas are computed according to Equation (13.5), then the value of $A_{311'3'}$ will be negative and then Equation (13.4) could be written as Equation (13.6):

$$A_{122'1'} = \frac{y_1 + y_2}{2} \times (x_2 - x_1)$$

$$A_{233'2'} = \frac{y_2 + y_3}{2} \times (x_3 - x_2) \tag{13.5}$$

$$A_{311'3'} = \frac{y_3 + y_1}{2} \times (x_1 - x_3)$$

$$\begin{aligned}
A_{123} &= A_{122'1'} + A_{233'2'} + A_{311'3'} \\
&= \frac{1}{2}[(y_1 + y_2)(x_2 - x_1) + (y_2 + y_3)(x_3 - x_2) + (y_3 + y_1)(x_1 - x_3)] \\
&= \frac{1}{2}(y_1 x_2 + y_2 x_3 + y_3 x_1 - x_1 y_2 - x_2 y_3 - x_3 y_1) \\
&= \frac{1}{2}\begin{vmatrix} x_1 & y_1 & 1 \\ x_2 & y_2 & 1 \\ x_3 & y_3 & 1 \end{vmatrix}
\end{aligned} \tag{13.6}$$

In fact, Equation (13.6) can be extended to compute the area of any polygon with N points:

$$A = \frac{1}{2}\sum_{i=1}^{N}(y_i \times x_{i+1} - x_i \times y_{i+1}) \tag{13.7}$$

This formula requires the $(N + 1)$th point. However, it does not exist in the point list of the polygon. As a result, the first point is used as the $(N + 1)$th point so as to make this polygon closed.

Similarly, as shown in Figure 13.2, the area covered by a profile (or a section) consisting of N points can be computed as follows:

$$A_{\text{profile}} = \sum_{i=1}^{n-1} \frac{z_i + z_{i+1}}{2} \times D_{i,i+1} \tag{13.8}$$

where $D_{i,i+1}$ is the horizontal distance between the ith and $(i + 1)$th points.

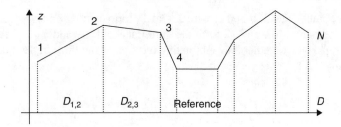

Figure 13.2 Area covered by a profile.

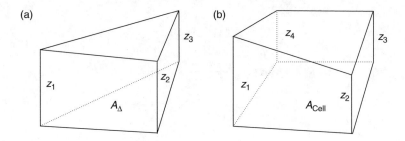

Figure 13.3 Volume calculation-based TIN and grid DTM.

13.2.2 Volume

After the horizontal area A_Δ covered by a triangular facet is computed, the volume of the triangular prism covered by this triangular facet (see Figure 13.3a) can be computed as follows:

$$V_3 = \frac{z_1 + z_2 + z_3}{3} \times A_\Delta \qquad (13.9)$$

If the DTM is in a grid form, the volume covered by a cell (Figure 13.3b) can be computed as follows:

$$V_4 = \frac{z_1 + z_2 + z_3 + z_4}{4} \times A_{\text{Cell}} \qquad (13.10)$$

where A_{Cell} is the horizontal area covered by the cell.

By using either of these two formulae, the volume required for cutoff or fill-up for an engineering design on the DTM can then be computed as follows:

$$V = V_{\text{originalDEM}} - V_{\text{newDEM}} \qquad (13.11)$$

The result of V can be interpreted as follows:

1. $V > 0$, cutting off
2. $V > 0$, filling up
3. $V = 0$, no need to do either.

13.3 MORPHOLOGICAL TERRAIN PARAMETERS

Morphometric terrain parameters are those that can be derived directly from the DTM using some local operations, such as slope and aspect, complexity index, and so on.

13.3.1 Slope and Aspect

Although slope was discussed in Chapter 2 and the use of slope information presented in Chapters 4 and 7, yet no rigorous definition has been given so far. Slope is the first derivative of a surface and has both magnitude and direction (i.e., aspect). That is, slope is a vector consisting of gradient and aspect. The term *slope* used in the previous chapters is called *gradient* in geomorphological literature. The term *aspect* is defined as the direction of the biggest slope vector on the tangent plane projected onto the horizontal plane. Aspect is the bearing (or azimuth) of the slope direction (Figure 13.4), and its angle ranges from 0 to 360°. (Note that in some literature, east is used as the reference direction for aspect instead of north.) In this context, the term slope is still used to refer to the gradient.

Suppose the surface function is

$$z = f(x, y) \tag{13.12}$$

Then, the slope is defined as

$$\text{Slope}_x = \frac{df}{dx} = f_x$$
$$\text{Slope}_y = \frac{df}{dy} = f_y \tag{13.13}$$

Slope can be derived from the TIN or grid DTM using simple local operations. Suppose the three vertices of a 3-D triangular facet are points 1, 2, and 3. The normal

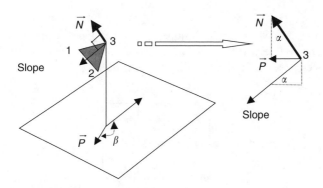

Figure 13.4 Definitions of slope and aspect.

(i.e., a vector) of this triangular facet at point 3 can be computed as follows:

$$\vec{N} = \begin{vmatrix} i & j & k \\ x_1 & y_1 & z_1 \\ x_2 & y_2 & z_2 \end{vmatrix} \tag{13.14}$$

$$= i(y_1 z_2 - y_2 z_1) - j(x_1 z_2 - x_2 z_1) + k(x_1 y_2 - x_2 y_1)$$

where i, j, and k are the unit vectors in the x, y, and z directions.

The projection of the $\vec{N}$ onto the horizontal plane $\vec{P}$ is computed as follows:

$$\vec{P} = i(y_1 z_2 - y_2 z_1) - j(x_1 z_2 - x_2 z_1) \tag{13.15}$$

The slope angle of the triangle, α, is then computed as follows:

$$\sin \alpha = \frac{|\vec{P}|}{|\vec{N}|} \tag{13.16}$$

The aspect of this slope direction, β, is computed as follows:

$$\tan \beta = \left(-\frac{x_1 z_2 - x_2 z_1}{y_1 z_2 - y_2 z_1} \right) \tag{13.17}$$

Many approaches are available to compute slope and aspect from a grid DTM. However, no attempt is made to introduce all of them. Instead, only some simple methods are presented. Figure 13.5 is a window with nine cells from a grid DTM. From this window, the slope and aspect values of the central cell, that is, with height z_0, can be estimated as follows:

$$\text{Slope} = \tan \alpha = \sqrt{\text{Slope}_{\text{Row}}^2 + \text{Slope}_{\text{Col}}^2} \tag{13.18}$$

$$\text{Aspect} = \tan \beta = \frac{\text{Slope}_{\text{Col}}}{\text{Slope}_{\text{Row}}} \tag{13.19}$$

In these formulae, $\text{Slope}_{\text{Row}}$ and $\text{Slope}_{\text{Col}}$ are the slopes in the row and column directions, respectively. If the row is west to east, then Slope_{we} is normally used to denote $\text{Slope}_{\text{Row}}$, and likewise Slope_{sn} to denote $\text{Slope}_{\text{Col}}$.

z_5	z_2	z_6
z_1	z_0	z_3
z_8	z_4	z_7

Figure 13.5 A window for the computation of slope and aspect value.

INTERPRETATION OF DIGITAL TERRAIN MODELS

Methods for the computation of the slopes in these two directions are listed in Table 13.1. In this table, the variable d is as usual the grid interval. Figure 13.6 shows an example of slope and aspect maps of an area: the contours and gray image are shown in Figure 12.9. Comparative analysis has also been made by Skidmore (1989) and Liu (2002). It has been revealed (Liu 2002) that method 1 has the highest accuracy and computational efficiency, and method 2 comes second. However, method 1 has not yet been implemented in popular commercial GIS software.

Table 13.1 Methods for the Computation of Slopes in Row and Column Directions

No.	References	Equations for Slope in Row and Column Directions	Equation No.
1	Ritter 1987; Zevenbergen and Thorne 1987	$\text{Slope}_{we} = \dfrac{z_3 - z_1}{2 \times d}, \quad \text{Slope}_{sn} = \dfrac{z_2 - z_4}{2 \times d}$	(13.20)
2	Horn 1981	$\text{Slope}_{we} = \dfrac{(z_7 + 2z_3 + z_6) - (z_8 + 2z_1 + e_5)}{8 \times d}$ $\text{Slope}_{sn} = \dfrac{(z_6 + 2z_2 + z_5) - (z_7 + 2z_4 + z_8)}{8 \times d}$	(13.21)
3	Unwin 1981	$\text{Slope}_{we} = \dfrac{(z_7 + \sqrt{2}z_3 + z_6) - (z_8 + \sqrt{2}z_1 + z_5)}{(4 + 2\sqrt{2})d}$ $\text{Slope}_{sn} = \dfrac{(z_6 + \sqrt{2}z_2 + z_5) - (z_7 + \sqrt{2}z_4 + z_8)}{(4 + 2\sqrt{2})d}$	(13.22)
4	Sharpnack and Akin 1969; Hengl et al. 2003	$G = \text{Slope}_{we} = \dfrac{(z_7 + z_3 + z_6) - (z_8 + z_1 + z_5)}{6 \times d}$ $H = \text{Slope}_{sn} = \dfrac{(z_6 + z_2 + z_5) - (z_7 + z_4 + z_8)}{6 \times d}$	(13.23)

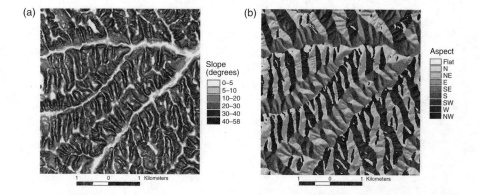

Figure 13.6 An example of slope and aspect maps of an area (as shown in Figure 12.9): (a) slope map and (b) aspect map.

13.3.2 Plan and Profile Curvatures

Hengl et al. (2003) regarded Equation (13.23) as the Evens–Young method. By this method, the three second derivatives of the terrain surface can also be derived as follows:

$$D = \frac{d^2 f}{dx^2} = \frac{(z_1 + z_3 + z_5 + z_6 + z_7 + z_8) - 2(z_0 + z_2 + z_4)}{3 \times d^2}$$

$$E = \frac{d^2 f}{dy^2} = \frac{(z_2 + z_4 + z_5 + z_6 + z_7 + z_8) - 2(z_0 + z_1 + z_3)}{3 \times d^2} \qquad (13.24)$$

$$F = \frac{d^2 f}{dx\,dy} = \frac{z_6 + z_8 - (z_5 + z_7)}{4 \times d^2}$$

Using Equations (13.23) and (13.24), the curvature can then be computed as shown in Table 13.2 (extracted from Hengl et al. 2003). The signs of the curvatures are defined in Figure 13.7. It can be seen that for plan curvature, a positive value indicates the divergence of the flow and a negative value the concentration of the flow and for profile curvature, a positive value indicates the convex profile and a negative value the concave profile. The mean curvature is the average of the plan curvature. Figure 13.8 shows an example of curvature maps of the area whose slope and aspect maps are shown in Figure 13.6.

Table 13.2 Methods for the Computation of Curvatures

Name	Equations	Equation No.
Plan curvature	$\mathrm{PlanC} = -\dfrac{H^2 \times D - 2 \times G \times H \times F + G^2 \times E}{(G^2 + H^2)^{1.5}}$	(13.25)
Profile curvature	$\mathrm{ProfC} = -\dfrac{G^2 \times D + 2 \times G \times H \times F + H^2 \times E}{(G^2 + H^2) \times (1 + G^2 + H^2)^{1.5}}$	(13.26)
Mean curvature	$\mathrm{MeanC} = -\dfrac{(1 + H^2) \times D - 2 \times G \times H \times F + (1 + G)^2 \times E}{(G^2 + H^2) \times (1 + G^2 + H^2)^{1.5}}$	(13.27)

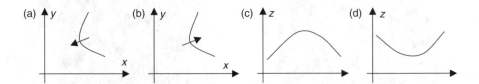

Figure 13.7 The sign of plan curvature (PlanC) and profile curvature (ProfC): (a) positive PlanC; (b) negative PlanC; (c) positive ProfC; and (d) negative ProfC.

INTERPRETATION OF DIGITAL TERRAIN MODELS

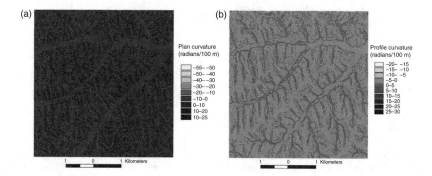

Figure 13.8 Maps of plan curvature and profile of the area as shown in Figure 12.9: (a) plan curvature map and (b) profile curvature map.

13.3.3 Rate of Change in Slope and Aspect

In Figure 13.5, suppose the slope of grid point 0 is Slope_0, and the slope of grid point j is Slope_j, $j = 1, 2, \ldots, 7, 8$, then the rates of change in slope in grid cell 0 are as follows:

$$\text{SR}_{0,j} = \begin{cases} \dfrac{\text{Slope}_j - \text{Slope}_0}{d}, & \text{for } j = 1, 2, 3, 4 \\ \dfrac{\text{Slope}_j - \text{Slope}_0}{\sqrt{2}d}, & \text{for } j = 5, 6, 7, 8 \end{cases} \quad (13.28)$$

where d is the grid interval. There are eight values for the rate of slope change. The one with the maximum magnitude is taken as the rate of slope change, that is,

$$\text{SR}_0 = \text{SGN}_{S_{\max}} |\text{SR}_{\max}| \quad (13.29)$$

where $|S_{\max}| = \text{MAX}(|\text{SR}_{0,1}|, |\text{SR}_{0,2}|, |\text{SR}_{0,3}|, |\text{SR}_{0,4}|, |\text{SR}_{0,5}|, |\text{SR}_{0,6}|, |\text{SR}_{0,7}|, |\text{SR}_{0,8}|)$ and $\text{SGN}_{S_{\max}}$ represents the sign of $S_{\max}$. For example, if $\text{SR}_{0,4}$ has the largest absolute value, then $\text{SR}_0 = \text{SR}_{0,4}$. The computation of the rate of aspect change is done exactly the same way.

13.3.4 Roughness Parameters

The *roughness* of a DTM surface is defined as the ratio of the surface area S and its projection onto the horizontal plane (i.e., the horizontal area A):

$$\text{Roughness}_A = \frac{S}{A} \quad (13.30)$$

When $\text{Roughness}_A = 1$, which is the smallest possible value, it means that the DTM surface is a horizontal surface.

It can be noted that the roughness values of two inclined planes will be different if the angles are different, although both are planes. This is a serious deficiency. Another commonly used method is to make use of the two average heights along the diagonal (see Figure 13.5):

$$\text{Roughness}_z = \left| \frac{z_5 + z_7}{2} - \frac{z_6 + z_8}{2} \right| \quad (13.31)$$

Another interesting parameter is the convexo-concave coefficient. It is defined as

$$CC = \frac{(z_{max} + z_{max}^o)/2}{z_{mean}} \quad (13.32)$$

where z_{max} is the height point of the four nodes of a grid cell; z_{max}^o is the height of the node opposite the highest node along the diagonal; and z_{mean} is the mean value of the four heights. The result of CC can be interpreted as follows:

1. $CC > 0$: convex shape
2. $CC < 0$: concave shape
3. $CC = 0$: level.

13.4 HYDROLOGICAL TERRAIN PARAMETERS

One of the major tasks in digital terrain analysis is the computation of hydrological parameters, which are used to model the mass (e.g., water, sediments, and nutrient) transportation and flow between land units. A number of important parameters have been proposed, for example, *total contributing area*, *specific catchment area*, *compound topographic index*, and *stream power index*. The results from the models form important input to, for example, the development of soil erosion models, land use and land evaluation, landslide prediction, and catchment and drainage network analysis (Zhou and Liu 2002). However, all these are the secondary terrain parameters and they are commonly derived from a more fundamental element — the flow model. A detailed discussion of these secondary parameters can be found elsewhere (e.g., Wilson and Gallant 2000; Hengl et al. 2003). In this section, only flow models are discussed, including flow direction, flow accumulation and lines, as well as catchments and drainage networks.

13.4.1 Flow Direction

The fundamental principle behind the determination of flow direction is that water will flow downhill (from a higher place to a lower place). On a terrain surface, peaks are the maxima and pits are the minima. Ridge lines connect local maxima and valleys (or ravines) lines connect local minima. Therefore, water will flow from peaks and ridge lines to valleys and pits. The direction of flow can also be determined using a DTM.

There are two general approaches:

1. *Single-flow direction (SFD)*: The total amount of flow should be received by a single neighboring cell that has the maximum downhill slope to the current cell, as shown

INTERPRETATION OF DIGITAL TERRAIN MODELS

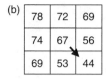

Figure 13.9 Approaches for the determination of flow direction: (a) SFD, D4; (b) SFD, D8; and (c) MFD.

in Figure 13.9(a) (only four possible directions) and Figure 13.9(b) (all eight possible directions).

2. *Multiple-flow direction (MFD)*: The flow from the current cell is distributed to all lower neighboring cells according to some criteria, slope and flow width (i.e., contour length), as shown in Figure 13.9(c) and expressed by Equation (13.33) (Quinn et al. 1991).

$$F_i = \frac{L_i \times \tan \alpha_i}{\sum_{i=1}^{n} L_i \times \tan \alpha} \qquad (13.33)$$

where F_i is the proportional flow to the ith neighboring cell; L_i is the flow width, which is equal to $(\sqrt{2}/4)d$ for the direction along the diagonal and $(1/2)d$ for four side neighbors (where d is the grid interval); and α_i is the slope angle of the ith neighboring cell.

A systematic classification of algorithms for the determination of flow direction based on these two approaches has been given by Zhou and Liu (2002). In this section, only the basic principles are introduced through simple algorithms. More precisely, only the *deterministic eight-node* (D8) is introduced because of its simplicity and wide implementation in GIS. However, it has been found from experimental testing results that D8 may produce unacceptable errors and the warning is that "care must be taken if they are used in real-world applications where accuracy is of concern" (Zhou and Liu 2002).

The principles of D8 (O'Callaghan and Mark 1984) are

1. Water can flow in only one of the eight directions (i.e., left, right, up, down, lower-left, upper-left, lower-right, and upper-right).
2. The direction must have the largest down slope.

In some literature, slope is measured by a *distance-weighted drop*, which is the height difference (between a given point and the next point) divided by the horizontal distance. In raster space, the distance is given in a unit of pixels. Therefore, the distance between two side neighbor grid cells is 1 and that between two diagonal grid cells is $\sqrt{2}$. Therefore, in the case of a 3 × 3 window, the *distance-weighted drop* is

1. the height difference in row or column
2. the height difference divided by $\sqrt{2}$ in diagonal.

(a)				(b)				(c)			
6	7	8		64	128	1		32	64	128	
5	0	1		32	0	2		16	0	1	
4	3	2		16	8	4		8	4	2	

Figure 13.10 Coding systems for the flow direction: (a) simple coding; (b) by Jenson and Domingue (1988); and (c) by Arc/Info.

(a)
78	72	68	73	60	48
75	68	56	50	46	50
70	55	45	40	39	47
65	57	53	26	30	26
67	60	48	23	18	20
75	55	45	12	10	12

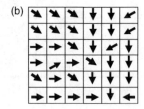

(c)
2	2	2	4	4	8
2	2	2	4	4	8
1	1	2	4	8	4
1	128	1	2	4	4
2	1	2	4	4	4
1	1	1	1	4	16

Figure 13.11 Flow directions and their coding: (a) a 6 × 6 grid DTM; (b) flow directions; and (c) flow coding.

In some literature, a coding system is assigned to each direction. The simplest coding system is as shown in Figure 13.10(a). Some researchers use numbers with a power of 2. For example, the coding used by Jenson and Domingue (1988) is shown in Figure 13.10(b) and the coding used in Arc/Info GIS is shown in Figure 13.10(c).

Figure 13.11 illustrates the flow directions of a grid DTM with 6 × 6 cells and Figure 13.12 is the flow direction map of the area as shown in Figure 12.9. In this figure, the coding system shown in Figure 13.10(c) is employed.

It must be noted here that usually a preprocessing is needed to remove depressions (Jenson and Domingue 1988) and compound depressions (Zhu et al. 2003). Detailed discussion of this lies outside this section.

13.4.2 Flow Accumulation and Flow Line

After flow directions have been determined, a flow accumulation matrix can be computed. Figure 13.13 is the flow accumulation matrix of the area shown in Figure 13.11. In this matrix, each cell is assigned a value equal to the number of cells that flow to it. Water flows to the lowest area accumulatively. Therefore, the lowest area will collect the water flow from all cells in the area. In Figure 13.13(a), the highest number is 35 because water from all 35 cells will flow to the last pixel. Those pixels with large numbers in the accumulation matrix form the flow lines. If different colors are used for the different numbers, one can see the flow lines clearly. Figure 13.13(b) is an example of the shaded flow accumulation map of Figure 13.13(a). Figure 13.14(a) is a flow accumulation map.

If a cell has a zero in the flow accumulation matrix, it means that no water from other cells flows to it, thus this cell must be a local maxima, corresponding to points at peaks and ridge lines.

INTERPRETATION OF DIGITAL TERRAIN MODELS

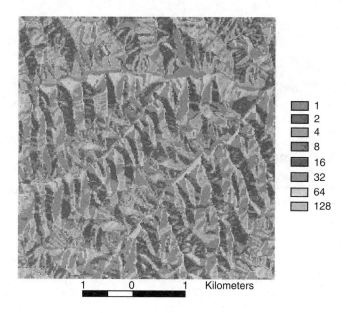

Figure 13.12 Flow direction map of the area as shown in Figure 12.9.

(a)
0	0	0	0	0	0
0	1	1	2	2	0
0	2	7	5	4	0
0	1	0	20	0	1
0	0	1	0	22	2
0	2	3	7	35	3

(b)
0	0	0	0	0	0
0	1	1	2	2	0
0	2	7	5	4	0
0	1	0	20	0	1
0	0	1	0	22	2
0	2	3	7	35	3

Figure 13.13 Flow directions and their coding: (a) flow accumulation and (b) flow accumulation with shading

13.4.3 Drainage Network and Catchments

After the flow accumulation matrix is produced, the extraction of flow lines becomes easy. From Figure 13.13, it can be seen that if one sets a threshold for flow accumulation values, then flow lines can be delineated easily. In Figure 13.13(b), a threshold value of 3 is set and the flow lines are then highlighted. Those cells with water accumulation greater than this threshold will be linked together and vectorized to become flow lines. All the flow lines together form a drainage network, which can be represented by a tree structure (Figure 13.15). Figure 13.14(b) is a drainage network map.

The level of detail of the extracted drainage network is in inverse proportion to the threshold used for flow accumulation. Experience shows that the results would be ideal if the threshold is set to the mean value of accumulation of all the cells (Tang 2000).

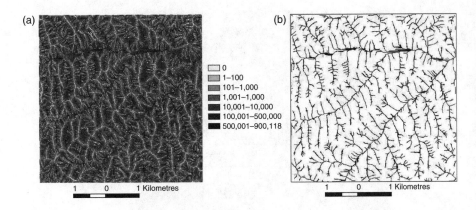

Figure 13.14 Flow accumulation and lines of the area as shown in Figure 12.9. (a) Flow accumulation map. (b) Drainage network map.

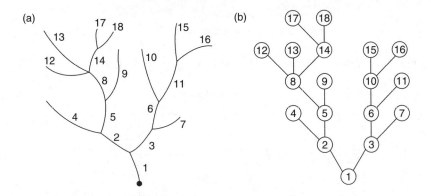

Figure 13.15 Tree structure for the representation of drainage network.

Another important hydrological parameter is the catchments (or watersheds). A watershed is the land area that drains precipitation into a particular stream or river system and is a polygon formed by the ridge lines. That is, those cells with zero water accumulation form the watershed boundary. The normal practice for delineating a watershed is to start from the outlet of the stream concerned and trace through the tree structure until the upstream limits of the basin are defined. More detailed discussion for catchment area calculation can be found in other literature (e.g., Freeman 1991).

13.4.4 Multiple Direction Flow Modeling: A Discussion

As previously stated, the SFD approach, although simple and easy to implement, has serious deficiencies as follows:

1. Because it is based on a grid and the D8 model, small-scale flow directions are biased to these eight directions, giving an implausible set of river directions.

INTERPRETATION OF DIGITAL TERRAIN MODELS

2. Modeling of physical processes (e.g., groundwater flow or heat flow) is not usually done on a "winner takes all" approach: if two exit paths from a cell are approximately equal in influence (based on the local gradient), then finite-difference flow models allocate the flow accordingly.
3. The fixed flow direction model (Figure 13.11b) does not allow for the effects of accumulation in a cell — hence the major difficulties with apparent "pits" in the flow model, where local (and usually minor) minima are found. A great deal of effort must then be expended on eliminating these artifacts that are due to the expected variability in a grid-based DEM and whose influences on the flow model are out of all proportion to their importance.

A better solution, that saves effort in the long run, is to implement a proper finite-difference scheme that resolves these issues. As in groundwater modeling, the volume of water transferred from a higher to a lower cell in one time step is equal to a conductivity factor, times the cross-sectional area of the face between the cells, times the tangent of the gradient between the cell centers. The conductivity factor is based on ground control (being lower for vegetation and higher for bare rock) and on the time step used. The cross-sectional area, for surface runoff, is based on the length of the common 2-D boundary (e.g., the grid cell size) times a putative unit height. The tangent of the gradient equals the height difference between the cells divided by the distance between the cell centers. This is described in Narasimhan and Witherspoon (1976) as the integrated finite-difference (IFD) technique.

To reduce the bias of a regular grid, random Voronoi cells may be introduced, with an estimated height at each center. (In most cases the original data points will be insufficiently dense, and additional interpolated points will be inserted at random locations.) This gives local flow directions that vary more plausibly in orientation. As the available water is allocated on the basis of the demand calculated for the neighboring cells, some flow will be allocated to all (lower) adjacent cells. If the height of a cell is considered to be the sum of the ground elevation plus the currently accumulated water, the problem of pits is eliminated — they fill up rapidly, as in the real world. The result of a simple simulation using this approach is given in Figure 13.16.

13.5 VISIBILITY TERRAIN PARAMETERS

Visibility analysis is a basic terrain analysis function used in a wide variety of applications such as resource management, urban planning, crime mapping, and military operations analysis. There are two fundamental parameters in visibility analysis:

1. intervisibility of line-of-sight (LoS), that is, point-to-point visibility
2. viewshed, that is, point-to-area visibility.

Recently, other terms such as visibility surface and visualscape have been introduced for more complex terrain analysis (Caldwell et al. 2003; Llobera 2003), although intervisibility and the viewshed are still the most important concepts for DTM-based visibility analysis.

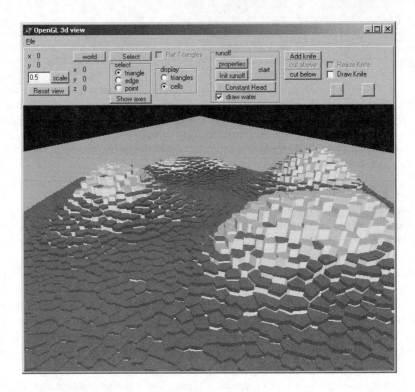

Figure 13.16 Surface runoff simulation using Voronoi cells and IFD modeling.

13.5.1 Line-of-Sight: Point-to-Point Visibility

The simplest method of visibility analysis is the LoS profiles model, as shown in Figure 13.17(a). The LoS is defined by the viewpoint V and the target T.

The basic idea of intervisibility analysis is to compute from the DEM the profile section between the viewpoint and the target (defining a LoS), as shown in Figure 13.17(a), to see whether or not any point within the profile will block the LoS (i.e., between viewer and the target). If yes, then the target is not visible. In this figure, if the height of the obstacle is lower then a critical value, then the target T is visible. Let the height of the target be h_T, the height of the viewer be h_V, let the distance between the target and the obstacle be D_{TO}, and let the distance between the viewer and the obstacle be D_{VO}; then, the critical value for the height of the obstacle is

$$h_{\text{Crit}} = \frac{D_{TO}h_T + D_{VO}h_V}{D_{TO} + D_{VO}} \qquad (13.34)$$

This means, if the height of the obstacle is lower than h_{Crit}, then the target is visible.

Alternatively, if the height of the obstacle is known, then the area blocked by this object can also be computed. This is shown in Figure 13.17(b). The length of the

INTERPRETATION OF DIGITAL TERRAIN MODELS

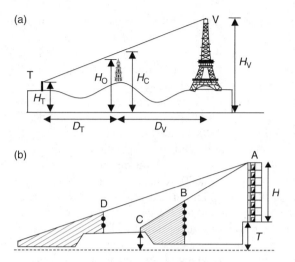

Figure 13.17 Intervisibility between two points: LoS. (a) Critical height of the obstacle between two points. (b) Shaded areas blocked by obstacles.

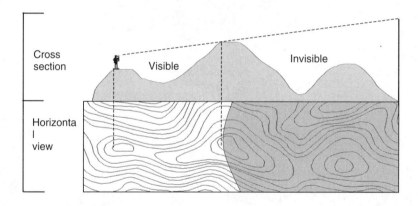

Figure 13.18 The viewshed concept.

invisible area is computed as follows:

$$D_{TO} = \frac{h_O}{h_V - h_O} \times D_{VO} \tag{13.35}$$

13.5.2 Viewshed: Point-to-Area Visibility

A viewshed is all the regions that are visible from a viewpoint. It is formed by a set of points. Figure 13.18 shows an example. In fact, the viewshed is a 2-D extension of the LoS. Viewshed analysis is a key component of visual impact assessment study.

There are more grid-based algorithms for the computation of viewshed although TIN-based algorithms are also available (e.g., De Floriani and Magillo 1994). The results of grid-based algorithms are usually represented in a discrete form, that is, each grid point is recorded as being either visible or invisible. This representation is called *visibility matrix* in some literature.

To obtain the viewsheds in a regular grid DEM from a given viewpoint, a simple method is to judge the visibility between each grid point and the viewpoint using Equations (13.34) and (13.35). However, as can be seen from Figure 13.18, there would be a huge amount of redundant computation in using such a simple method. To improve computation efficiency, several specially designed algorithms have been developed (e.g., Wang et al. 2000; Rana and Morley 2002), apart from the parallel processing algorithms (De Floriani and Magillo 1994). For example, topographic features on the terrain surface have been used to reduce the number of observer–target pair comparisons (Rana and Morley 2002); and reference planes rather than sightlines are used to save a considerable amount of computation (Wang et al. 2000). Detailed discussion of this will not be presented here in this chapter.

CHAPTER 14

Applications of Digital Terrain Models

DTMs have found wide applications since their origin in the late 1950s, in various disciplines such as mapping, remote sensing, civil engineering, mining engineering, geology, geomorphology, military engineering, land planning, and communications (Catlow 1986; Petrie and Kennie 1990; Li and Zhu 2000; Maune et al. 2001). In this chapter, brief descriptions of various applications will be given.

14.1 APPLICATIONS IN CIVIL ENGINEERING

The first application of DTM is in civil engineering, more precisely, highway engineering. In 1957, Roberts (1957) proposed the use of DTMs for highway design. One year later, Miller and Laflamme (1958) used the data to set up a cross-section (profile) model and coined the concept of DTMs for the first time. Thereafter, Roberts and his colleagues at MIT developed the first terrain modeling system. This system could not only interpolate in the sections (profiles), but also calculate the cut-and-fill between sections and provide useful data for engineering design. By 1966, they had been able to provide programs for road design using DTMs, most of which were based on the cut-and-fill calculation. Many techniques originally developed for road design have been applied to construction engineering, such as the design of reservoirs and dams. DTMs have also been widely applied to other related engineering such as mining. In this section, applications in road engineering and water conservancy will be described briefly.

14.1.1 Highway and Railway Design

The development of a transportation network is complicated, aiming to provide a network to satisfy the needs of society. The design process can be split into steps such as site investigation, route planning and design, earthwork calculation, pavement

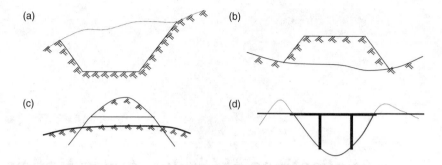

Figure 14.1 Cases involved in the design of roads and railways: (a) excavation in the form of cuts (cross section); (b) embankments in the form of fills (cross section); (c) digging in the form of tunnel (profile); and (d) construction in the form of bridge (profile).

design, bridge and tunnel design, and so on. DTMs help in route planning and design and earthwork calculation.

Due to variations in the terrain, it is unlikely that a road or railway can be constructed without any earthwork. In most cases, tunnels and bridges need to be constructed, hills and lowlands modified (Figure 14.1). Earthwork can take the form of excavation or the construction of embankments, to carry an elevated highway or railway. Normally in a road or railway design, both cuts and fills will be necessary.

Designers make every effort to select a route passing through areas with stable geological conditions, with gentle slopes and small curves to minimize earthwork. Traditionally, such work was done on contour maps. Nowadays, DTMs are widely used for drawing plans, profiles (along the designed central line), and cross sections, for computing the volume of earthwork, for generating perspective views, and even for producing 3-D animation. As various routes are possible for a given project, the aim of the design is to obtain an optimal route.

The basic requirements for landscape modification with a TIN are the ability to insert and delete points in a triangulation (see Chapter 5) and to assign to them a particular height. Although it is desirable to have additional tools that are specific to the particular application, terrain manipulation can be done with these alone — other techniques are described in Chapter 15. Figure 14.2 shows a triangulated surface where roadside elevation values have been added to the original terrain, and some of the original data points have been modified. Figure 14.3 shows a 3-D view of this model, which may then be used to evaluate the feasibility of the proposed route. Figure 14.4 shows a highway on the TIN-based DTM with shading.

In mining, DTMs have also been used to compute earth volume and to simulate mining progression.

14.1.2 Water Conservancy

There are different types of water conservancy projects, such as reservoirs and canals. A canal project is similar to a road project, but there are differences. The major difference is that water cannot naturally flow uphill. Therefore, a canal is normally not allowed to have upward slopes.

APPLICATIONS OF DIGITAL TERRAIN MODELS

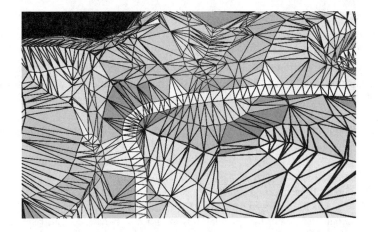

Figure 14.2 A triangulated terrain surface with road lines added.

Figure 14.3 A 3-D view of the triangulated terrain surface with road lines.

Figure 14.4 A highway designed on TIN-based DTM with shading.

In a reservoir project, the water volume needs to be estimated, the location of the dam determined, various critical water and outlet water levels designed, and drainage planned. Reservoir volume and area are two major features. DTMs can be used to replace traditional contour maps to assist in selecting a site for the dam and estimating water volume. The procedure for the computation of water volume is as follows:

1. Let points (x_A, y_A) and (x_B, y_B) be the two points on the dam axis, then the equation for this axis is
$$y = kx + b \tag{14.1}$$
where $k = (y_B - y_A)/(x_B - x_A)$ and $b = y_A + kx_A$.
2. Compute the intersection points (x_i, y_i) of the dam with the contour lines at different levels, which are derived from the DTM of the reservoir area.
3. Compute the area of the irregular polygon formed by each contour (e.g., the kth contour) and the designed dam:
$$A_k = \frac{1}{2} \sum (x_{i+1} + x_i)(y_{i+1} - y_i), \quad k = 1, 2, \ldots, m \tag{14.2}$$
4. Compute the volume between the adjacent two areas (e.g., the kth and the $(k+1)$th):
$$\Delta V = \frac{1}{2}(A_k + A_{k+1}) \times \Delta H \tag{14.3}$$
where ΔH is the height difference between the two adjacent areas.
5. Compute the volume contained by the reservoir:
$$V = \sum \Delta V \tag{14.4}$$

In the end, the curves showing the relationships between water level and reservoir volume and between water level and water area can also be produced with ease.

14.2 APPLICATIONS IN REMOTE SENSING AND MAPPING

DTMs have many applications in remote sensing and mapping, such as topographic mapping (contours), thematic mapping, orthoimage generation and image analysis, map revision, and so on. In this section, only the applications in orthoimage generation and remote sensing image analysis are discussed.

14.2.1 Orthoimage Generation

To make images useful as backdrops for other thematic information and base maps, it is desirable that the images have characteristics similar to those of maps. This means that the same scaling, orientation, and projection into a geo-referencing system. (e.g., a national geodetic system) should be adopted. To accomplish this, a number of requirements must be fulfilled:

1. All image points should be registered in a geo-referencing system such as a national geodetic (or grid) system.

APPLICATIONS OF DIGITAL TERRAIN MODELS

2. Each point (pixel) of the resulting image should have the same scale if the ground area is small, or else scale variations should follow a map projection.
3. The relative relationships between features should also be retained.

Remote sensing images, either satellite or aerial images, do not have such good characteristics due to the distortions caused by the imperfections of camera or scanner systems, the instability of platforms (tilts and flying height variations), atmospheric refraction, the earth's curvature, and terrain height variations. The two most serious factors are the instability of the platform and terrain height variations. Therefore, geometric rectification is required.

To rectify the images, the relationship between the image and ground points needs to be established. For aerial photography, relationships were discussed in Chapter 3 and expressed by Equation (3.3). For scan image, Equation (3.3) can still be used but only for each scan line. Therefore, it is not practical to use Equation (3.3) to model scan images, because there are normally thousands of scan lines in a frame, and this results in too many unknowns. In practice, a polynomial (normally second or third order), as listed in Table 4.1, is used to approximate geometric transformation models. A few control points from both the image and the ground are measured as reference points to solve the coefficients of this model. Then, all points on the image can be transformed to the ground. This is geometric rectification in which distortions are corrected, minimized, or redistributed. However, the distortion caused by terrain variation is still there. To remove this distortion, as shown in Figure 14.5, a DTM is required.

In Figure 14.5(a), the distortion caused by relief is shown. The ground point A has a height z over a reference datum. This causes a displacement "aa" on the image. In the case of rectification, if the height of each pixel on the image, for example, "a" in Figure 14.5, is found from the DTM, then a correction can be applied. The determination of the ground height is done by an iterative process (Albertz et al. 1999), as shown in Figure 14.5(b).

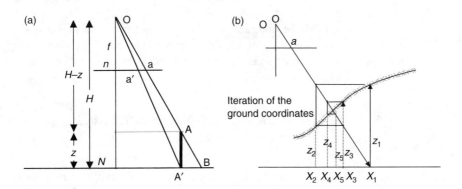

Figure 14.5 Image distortion due to relief and its correction: (a) image distortion caused by relief and (b) intersection of ground surface with light ray.

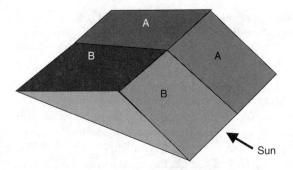

Figure 14.6 Effect of topographic variation on image brightness.

Table 14.1 Removal of Topographic Effect with Band Ratio

Unit	Image Brightness on Slope Facing Away from Sun			Image Brightness on Slope Facing Sun		
	Red Band	NIR Band	Red/NIR	Red Band	NIR Band	Red/NIR
A	20	30	0.67	40	60	0.67
B	30	60	0.50	50	100	0.50

14.2.2 Remote Sensing Image Analysis

In Chapter 12, shading was used to produce a vivid representation of terrain surface. It was argued that the surface reflectance is different if the slope and aspect are different. This means that the brightness of image pixels is also affected by the slope and aspect of the terrain surface. Figure 14.6 shows such effects.

However, the image from all the bands will suffer from the same effect for the same area. Therefore, in remote sensing, band ration is widely used to remove such effects. In Figure 14.6, there are two types of land cover, A and B. The sunlight is from the lower-right corner. Therefore, the surface on the right side will appear to be brighter on the image. Table 14.1 gives a possible example. The ratios of the red and the near infrared (NIR) bands are the same although the absolute differences are very different. With a DTM, the slope and aspect map of the terrain surface can easily be produced, as discussed in Chapter 13; thus, the topographic effect can then be removed. In the end, a more reliable analysis could be made from the image after the removal of topographic effects.

14.3 APPLICATIONS IN MILITARY ENGINEERING

14.3.1 Flight Simulation

Pilot training is difficult, costly, and sometimes dangerous. It is natural to think about simulation so that the pilot can sit down in front of a special device to learn how to control an airplane. In addition to pilot training, flight simulation can also be used for mission planning and rehearsal.

APPLICATIONS OF DIGITAL TERRAIN MODELS

Figure 14.7 Virtual battlefield environment simulation (You 1991).

In simulation, the DTM plays an important role. 3-D rendering techniques are employed to simulate the terrain, often with LOD techniques. Textures and other attributes can also be mapped onto the DTM surface to generate realistic scenery. To simulate scene changes while flying, the "fly-through" technique is used.

DTMs can also be used to guide cruise missiles. This is done by matching the DTM surface stored in the computer with the real world sensed by the detectors on board the cruise missile.

14.3.2 Virtual Battlefield

The virtual battlefield is a simulation of a potential battlefield generated in computers, which allows people to be involved. Battlefield simulation provides a dynamic and stereo environment, which can be used to recapitulate the battle, evaluate the results, and gain experience. DTM is used to simulate the battle environment. Figure 14.7 is an example of the virtual battlefield environment.

A number of parameters can be derived from the DTM for the battlefield simulation, such as intervisibility, shields of the landform, exposed distance of a moving unit, the closest shielding distance to the target, and accessibility of the battle field.

14.4 APPLICATIONS IN RESOURCES AND ENVIRONMENT

14.4.1 Wind Field Models for Environmental Study

In climatology, environment, and forestry, to predict the spread of forest fires and pollution, it is necessary to be able to predict the wind direction. To do so, enough information about the wind model of the areas of interest needs to be obtained.

A large mountain range exerts dynamic and thermodynamic effects on the atmosphere. The dynamic effect means that with rotation and gravity, mountains force the air to flow like waves at various scales, which results in changes in the wave fronts. The thermodynamic effect means that temperature differences between day and night in different parts of the mountains result in local influence of hill and valley winds. The direction and velocity of winds also vary with altitude, slope, aspect, and terrain roughness, which results in a complex and unstable wind field in mountain areas. To model this, all the essential topographic parameters can be extracted from DTMs, such as

1. the lowest, highest, and average elevations of each grid cell
2. the lowest, highest, and average elevations of certain square areas delineated with specific conditions
3. the average slope of each cell
4. the percentage of cells containing such terrain features as ridges, valleys, and flat lands
5. the standard deviation of elevations and slopes.

14.4.2 Sunlight Model for Climatology

The mountain climate is deeply influenced by terrain variations. These effects result in different climates on the two sides of the mountain range. Also, unique local climates may be formed on any part of a mountainous region because of different combinations of altitudes, slopes, and aspects, as well as the shading effect of mountain ridges. When analyzing the mountain climate, one has to consider not only the relatively invariable factors like geographic latitude and average altitude of the region, but also the micro topographic parameters such as the local height differences, aspect, and shading areas.

DTMs play an important role in sunlight modeling. The incidence direction of sunlight is defined by the functions of date, time, and the latitude and longitude of the area. The sunlight received by each cell is also dependent on the slope, aspect, and altitude of the cell. To produce a precise sunlight model, the coordinates of grid cells are transformed into latitude and longitude, the angle between the incidence direction of the sunlight and the outward normal of the grid cell is then calculated, the shade status is judged by a hidden surface algorithm of the 3-D perspective, and then the instantaneous sun radiation can be accurately calculated. The sun radiation of a grid cell in 1 day can be obtained by summarizing all instantaneous sun radiation. Similarly, the sun radiation on one grid cell during a month, a season, or a year can be calculated. Of course, the sun radiation at one time period on one slope surface can be obtained by accumulating all the radiation values of those grid cells on the surface.

14.4.3 Flood Simulation

The flat areas of river basins are often flooded after heavy rain. Therefore, it is necessary to study flood risks. To do so, potential flood levels and velocity are the two major parameters to be considered. The DTM has been used to simulate floods. In such a simulation, with a given rainfall, the amount of water from different catchments can be estimated, as described in Chapter 13. After considering the capacity of the river, the amount of water to be accumulated can be computed. Then, the area

APPLICATIONS OF DIGITAL TERRAIN MODELS

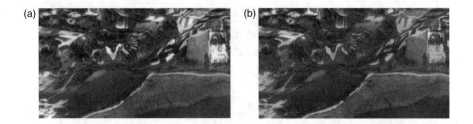

Figure 14.8 Flood areas simulated with satellite images superimposed. The color plate can be viewed at http://www.crcpress.com/e_products/downloads/download.asp?cat_no=TF1732.

to be flooded can also be estimated. Figure 14.8 shows an example of the flood areas simulated using a DTM, with satellite images superimposed.

14.4.4 Agriculture Management

Recently a popular term — precision farming — has come into use. It means that that farmers can control the quantity of water, fertilizer, and pesticides placed on different areas of the farm land, based on the attributes of the land, such as soil type and condition, slope, the condition of the crops, and so on. Slope (as well as aspect) information can be derived from a DTM.

Slope is a type of spatial information important to soil erosion. In some developing countries, areas with steep slopes are still farmed, resulting in serious soil erosion. This was also the case, for example, in China. However, in the late 1990s, the Chinese government ordered that no lands with a slope over a certain value should be farmed. In this way, the situation of soil erosion has been improved.

14.5 MARINE NAVIGATION

The topographic surface is usually observed above sea level, but clearly the sub-sea terrain surface is important in various applications. Terrain model construction is often more difficult, as the overall surface form is often not available at the time of sampling. Observations of the sea floor are often made along ships' tracks, giving a highly anisotropic distribution that requires special interpolation techniques to reconstruct plausible surfaces (Gold and Condal 1995).

However, the most important marine application of DTMs is undoubtedly in ship navigation. This is most commonly based on the use of electronic or paper marine charts, but is now being considered for 3-D representations (Gold et al. 2004). Figure 14.9 shows a navigator's view of the East Lamma Channel, Hong Kong, together with the superimposed chart symbols. Figure 14.10 shows the use of dynamic 3-D safety contours, highlighting the safe channel for a particular ship's draught. Figure 14.11 shows how the dynamic intersection of the tidal sea surface with the terrain is used for a collision-avoidance system, with the kinetic Voronoi diagram (derived from TIN) as the basis — collisions are only possible between the generators of adjacent cells.

Figure 14.9 A navigator's view of the East Lamma Channel.

Figure 14.10 Channel safety contours based on sub-sea topography.

Figure 14.11 The intersection of the terrain and sea surface: collision detection using kinetic Voronoi diagrams.

14.6 OTHER APPLICATIONS

In planning and landscape design, visual impact analysis (VIA) is applied to the new designs. That is, the designs are superimposed onto a DTM to create a virtual landscape, which is visually analyzed.

DTMs can also be used for communication network planning. Problems such as dead angles and blind areas in site selection of the radio or television transmitting station can be computed.

Indeed, as the DTM is a fundamental model of the Earth's surface, it has applications in all Earth-related sciences. However, a complete coverage of these applications lies outside the scope of this book.

CHAPTER 15

Beyond Digital Terrain Modeling

Chapter 14 discusses the more traditional development and applications of terrain models. Here, we look at some extensions of these models for specific problems.

15.1 DIGITAL TERRAIN MODELING WITH COMPLEX CONSTRUCTION

15.1.1 Manual Addition of Constructions on Terrain Surface

For simplicity it is usually assumed that terrain models are monotonic in X and Y — there is only one possible Z for each XY location. This is often true in the real world, but not always — occasionally there are caves, tunnels, overhanging cliffs, bridges, and overpasses. In the work by Tse and Gold (2002), the standard TIN model is extended by merging some aspects of terrain modeling (TINs), computational geometry (the Quad-Edge data structure), and computer aided design or CAD (Euler operators, which guarantee to preserve the connectivity of the surface after they are applied). They found it easy to combine them to give the usual operations on a 2D triangulation — as well as add an operator that generates a hole between any two nonadjacent triangles (which is really the same thing as adding a bridge or handle to the surface). Figure 15.1, Figure 15.2, and Figure 15.3 give simple examples, and Figure 15.4 and Figure 15.5 show part of a Hong Kong city model.

Thus, a simple modification of the basic triangulation algorithm allows one to interactively modify the terrain model to add complex features that are otherwise unavailable. Because one is still forming a connected surface, a variety of topological operations, such as neighborhood selection and flow modeling, may be performed. Clearly another, even higher, layer of operations would permit one to add predesigned features such as buildings, dams, tunnels, etc. to our terrain model.

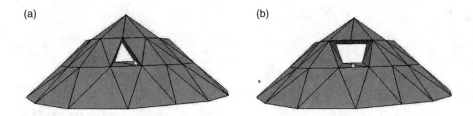

Figure 15.1 TIN model of a surface with a tunnel: (a) a tunnel on a TIN model and (b) the enlarged tunnel.

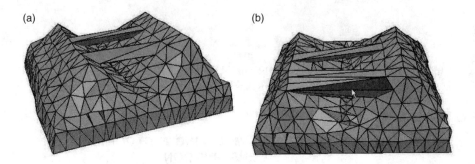

Figure 15.2 TIN model of a surface with bridges. (a) two bridges on a TIN model and (b) the enlarged bridge.

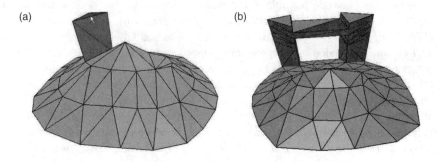

Figure 15.3 TIN model of a ground surface with buildings and bridge: (a) a building on a TIN model and (b) a bridge connecting two buildings.

15.1.2 Semiautomated Modification of the Terrain Surface

Section 15.1.1 showed simple terrain modification based on modifying the TIN by adding and deleting individual points with (X, Y, Z) coordinates. This is effective but slow to do by hand. An alternative approach is to "cut" the triangulated surface with a "knife" in order to sculpt it to the form desired. One first sets the knife size, location, and orientation and then performs the cut (or intersect) operation. One may

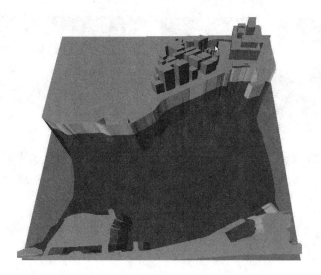

Figure 15.4 A partial view of Hong Kong harbor.

Figure 15.5 Part of the Hong Kong city model.

either lower the terrain surface to the knife position, for example, cutting into the side of a hill, or else raise the surface to the knife position, creating an embankment or dam. More points are added to the triangulation to form the intersection lines between the knife and the original terrain, and one assumes a maximum slope (less than vertical) for the edges of the cut or embankment. Figure 15.6(a) shows a simple TIN model with the knife in place. Figure 15.6(b) shows the result after the surface is lowered to the knife (with a 45° embankment specified). Figure 15.6(c) shows the knife positioned across a valley, and Figure 15.6(d) shows the result of raising the terrain surface to the knife, forming a dam structure across the valley.

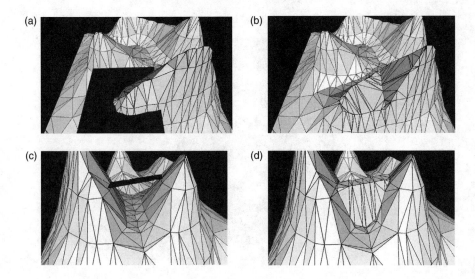

Figure 15.6 Terrain modification by "cutting" the triangulated surface with a "knife." (a) The knife positioned on the terrain. (b) Modified terrain after lowering the surface to the knife blade. (c) The knife positioned across a valley. (d) The dam created after raising the surface to the knife blade.

15.2 DIGITAL TERRAIN MODELING ON THE SPHERE

With the introduction of the concept *digital earth*, global modeling of the Earth's surface (Gold and Mostafavi 2000) has become a hot topic. The digital terrain modeling techniques described in this book can also be extended to spherical terrain modeling.

15.2.1 Generation of TIN and Voronoi Diagram on Sphere

In planimetric terrain modeling, as discussed in Chapter 4, grids and TINs have been widely used to tessellate the terrain. On the sphere, a similar tessellation model needs to be used. The concept of spherical surface tessellation was presented by Fuller, a German cartographer, for map projection in the 1940s (Dutton 1996). Since then, many researchers have approached this problem to project, analyze, and index global data. Many methods are based on *inscribed polyhedrons*, such as the *tetrahedron*, the *cube* (Snyder 1992), the *octahedron* (Dutton 1989, 1996; Goodchild et al. 1991; Goodchild and Yang 1992; Otoo and Zhu 1993; Clarke and Mulcahy 1995), the *dodecahedron* (Wickman and Elvers 1974), and the *icosahedron* (Fekete 1990; White et al. 1992; Lee and Samet 2000), as shown in Figure 15.7. The edges of the polyhedron are projected to the spherical surface and form the edges of spherical triangles.

The octahedron-based tessellation is a regular triangular mesh on the sphere, called the octahedral quaternary triangular mesh (O-QTM). Figure 15.8 shows an

BEYOND DIGITAL TERRAIN MODELING

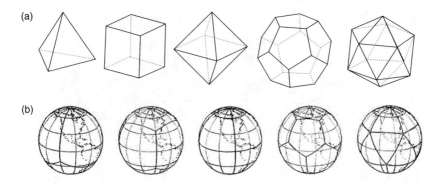

Figure 15.7 Spherical surface tessellation based on inscribed polyhedra (Reprinted with permission from White et al. 1992): (a) five polyhedra and (b) projected to the spherical surface.

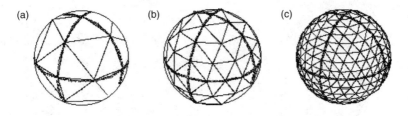

Figure 15.8 Hierarchical tessellation of the spherical facet based on octahedron (Dutton 1996): (a) level 1; (b) level 2; and (c) level 3.

example of O-QTM at three difference levels (Dutton 1996). Terrain modeling can then be applied to the QTM.

The QTM can also be used as a coordinate system on the sphere, just like the regular grid or triangular network on a 2D plane. In the QTM, a point is represented by a triangle, an arc is represented by a series of neighbor triangles, and a region is represented by a series of neighbor triangles on and within its boundary trace. From the QTM, a TIN can then be constructed. Alternatively, spherical TINs can also be derived from spherical Voronoi diagrams (Augenbaum 1985; Robert 1997; Chen et al. 2003). Figure 15.9 shows an example of spherical Voronoi diagram and its dual — the spherical TIN.

15.2.2 Voronoi Diagram for Modeling Changes in Sea Level on Sphere

Mostafavi and Gold (2004) used the dynamic Voronoi diagram on the sphere to model the continually changing height of the sea, rather than of terrain. Figure 15.10(a) shows an initial set of cells, each representing a fixed mass of water, and uses the free Lagrange method to simulate flow under lunar gravitational influence, and hence the sea height. Coastlines were modeled by a double line of fixed Voronoi cell generators. Figure 15.10(b) shows the result after simulation started: high water (HW) is indicated by smaller, and therefore higher, cells, while low water (LW) is shown by larger,

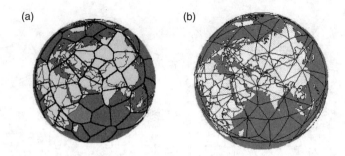

Figure 15.9 Spherical TIN formation: (a) Voronoi diagram and (b) Delaunay triangulation.

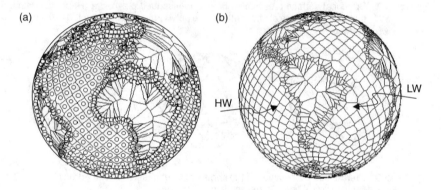

Figure 15.10 Dynamic Voronoi diagram on sphere to model the continually changing heights of sea: (a) initial configuration of Voronoi cells and (b) Voronoi cell configuration indicating lunar tides.

lower cells. Figure 15.11 is a Mercator projection showing the flow directions and velocities of each cell.

15.3 THREE-DIMENSIONAL VOLUMETRIC MODELING

Two dimensions are required in terrain modeling to generate the underlying triangulation or grid. Once elevations are added as an attribute, the result is usually known as "2.5D" modeling, although the data structures remain 2D. Once the topology (or connectedness) can no longer be represented on the plane (as in 3D objects in CAD or games), a surface representation, composed usually of triangles, is often used, as in Section 15.1.

However, for some applications a surface model is inappropriate, and a full 3D volumetric model is needed. Examples include geological, atmospheric, and oceanographic models, where attributes need to be assigned to arbitrary locations in 3D space. In some cases a 3D grid may be used, or an octree where nodes represent volumes. A more flexible approach is to replace the 2D triangulation structure

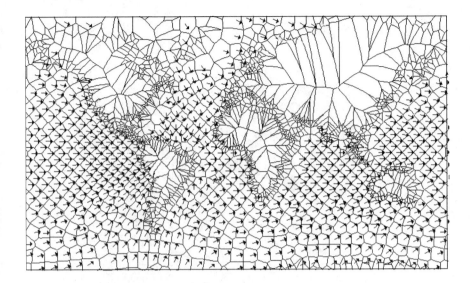

Figure 15.11 Mercator projection showing cell velocities and directions.

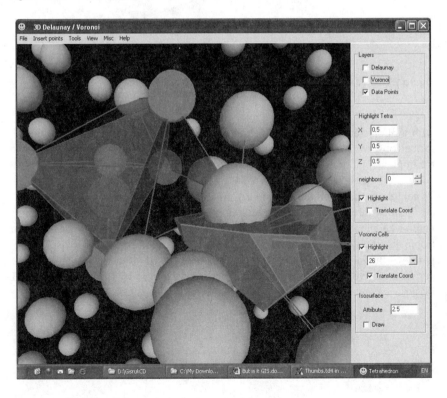

Figure 15.12 Delaunay and Voronoi cells in three dimensions.

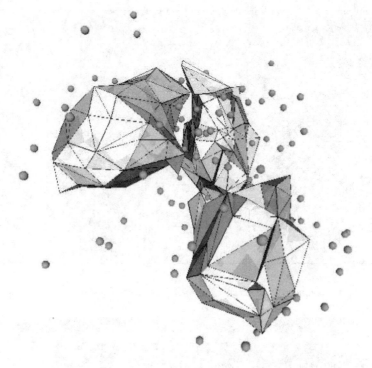

Figure 15.13 A three-dimensional isosurface. The color plate can be viewed at http://www.crcpress.com/e_products/downloads/download.asp?cat_no=TF1732.

with a 3D Delaunay tetrahedral model, thus allowing the connection of arbitrarily located observations, and then to add the 3D Voronoi cells. While conceptually simple, the implementation is often difficult, due to the large number of degenerate (coplanar, cocircular) cases. Figure 15.12 shows a 3D data set with one Delaunay cell and its neighbors shown on the left and a Voronoi cell on the right.

Various interpolation techniques, such as the equivalent of the 2D Sibson or natural neighbor interpolation, may be used, and these behave well for the anisotropic distributions of data that are often found in three dimensions. For visualization purposes the individual tetrahedral may be sliced, based on the values at the corner vertices, to give the 3D equivalent of 2D contours. Figure 15.13 shows a single 3D isosurface constructed in this way.

Epilogue

It was natural that we felt relieved and excited somehow after having completed the final draft of this book and having uploaded the materials onto the ftp site of the publisher. However, soon we started to feel obliged to write this epilogue because there are a few issues confronting us.

We thought it is really a pity that no authored book in this discipline had been made after over 40 years of development although there are two edited works, *Terrain Modelling in Surveying and Civil Engineering* by Petrie and Kennie (1990) and *Digital Elevation Model Techniques and Applications: The DEM User Manual* by Maune (2001). Our aim was to write a book systematically covering a wide range of topics in digital terrain modeling so as to fill in the gap in this area. While writing, we were faced with a number of challenges.

The first challenge was related to the selective omission of materials. It was difficult to make decisions. This is because the term "digital terrain modeling" would mean different things to different groups of terrain specialists and practitioners. To the producers (including photogrammetrists and surveyors), data acquisition and terrain surface modeling are of most concern; to geographers, terrain analysis and applications are the most important; to geologists, interpolation techniques seem to be critical; .... It is really hard to satisfy all these groups. In the end, we decided that those topics are simplified if they have rich bodies of literature available. For example, we did not include many algorithms and techniques for interpolation and triangulation as there is a huge body of literature (e.g., Su 1989; Chin 1995; Sakhnovich 1997; de Berg 2000; Phillips 2003) in these areas covering the techniques developed in computational geometry and geosciences. Contouring is a traditional topic in digital terrain modeling but is only briefly discussed in this book because a book authored by Watson (1992) has been dedicated to this topic. Similarly, DTM-based terrain analysis is briefly discussed because of a recent book edited by Wilson and Gallant (2000).

The second challenge was related to the depth of discussion. We may disappoint those readers who are interested in mathematics because we present neither mathematical proofs nor technical details. Indeed, it is the main aim of this book to present a systematic accounting of stories in digital terrain modeling at the level of principles and methodology, as the title of the book suggests.

The third challenge was related to the boundary of the discipline. Because the terrain surface is of concern to all geosciences and a huge body of related modeling methodology and applications is available, we have to cut down the contents somewhere. Therefore, we did not cover much on Voronoi diagrams (e.g., Davies 2000; Okabe et al. 2000) although we are very interested in this topic. Similarly, we did not cover much on geostatistics (e.g., Olea 1999) and even omitted the famous Kriging technique. We only simply mentioned the surface modeling on sphere and with construction in Chapter 15.

All in all, we are pleased with the compilation of some materials presented to you, but also feel guilty about the imperfection. Your comments are appreciated so that we could make improvements in another edition, if possible.

References

Ackermann, F. 1979. The accuracy of digital terrain models. In: *Proceedings of 37th Photogrammetric Week*, University of Stuttgart, Germany, pp. 113–143.
Ackermann, F. 1996. Airborne laser scanning for elevation models. *GIM International*, 10(10):24–25.
Albertz, J., Li, Z., and Zhang, W. 1999. Orthoimages from air-borne scanner images using flight parameters. *The Geomatics Journal of Hong Kong*, 1(2):13–24.
Amenta, N., Bern, M., and Eppstein, D. 1998. The crust and the beta-skeleton: combinatorial curve reconstruction. *Graphical Models and Image Processing*, 60:125–135.
Augenbaum, M. 1985. On the construction of the Voronoi mesh on a sphere. *Computational Physics*, 59:177–192.
Aumann, G., Ebner, H., and Tang, L. 1991. Automatic derivation of skeleton lines from digitized contours, *ISPRS Journal of Photogrammetry and Remote Sensing*, 46:259–268.
Aurenhammer, F. 1987. Power diagrams: properties, algorithms and applications. *SIAM Journal of Computing*, 16:78–96.
Aurenhammer, F. 1991. Voronoi diagrams — a survey of a fundamental geometric data structure. *ACM Computing Surveys*, 23(3):345–405.
Axelsson, P. 1999. Processing of laser scanner data: algorithms and applications. *ISPRS Journal of Photogrammetry and Remote Sensing*, 54(1):138–147.
Baffisfore, A. 1957. A do-it-yourself terrain model. *Photogrammetric Engineering*, 23:712–720.
Balce, A.E. 1987. Determination of optimum sampling interval in grid digital elevation models (DEM) data acquisition. *Photogrammetric Engineering and Remote Sensing*, 53:323–330.
Baltsavias, E.P. 1999a. Airborne laser scanning: existing systems and firms and other resources. *ISPRS Journal of Photogrammetry & Remote Sensing*, 54(1):164–198.
Baltsavias, E.P. 1999b. Airborne laser scanning: basic relations and formulas. *ISPRS Journal of Photogrammetry & Remote Sensing*, 54(1):199–214.
Blum, H. 1967. A transformation for extracting new descriptors of shape. In: Whaten Dunn, W. (Ed.), *Models for the Perception of Speech and Visual Form*, MIT Press, Cambridge, MA, pp. 153–171.
Borgefors, G. 1986. Distance transformations in digital images. *Computer Vision, Graphics and Image Processing*, 34:344–371.
Bowyer, A. 1981. Computing Dirichlet tessellations. *The Computer Journal*, 24(2):162–166.

Brassel, K. and Reif, D. 1979. Procedure to generate Thiessen polygon. *Geographical Analysis*, 11(3):289–303.
Burington, R. and May, D. 1970. *Handbook of Probability and Statistics with Tables*, 2nd ed. McGraw-Hill Book Company, New York.
Caldwell, D.R., Mineter, M.J., Dowers, S., and Gittings, B.M. 2003. Analysis and Visualization of Visibility Surfaces. www.geocomputation.org/2003/Abstracts/Caldwell_Abs.pdf.
Catlow, D.R. 1986. The multi-disciplinary applications of DEMs. *Auto-Carto London*, 1:447–454.
Chen, J., Zhao, X.S., and Li, Z.L. 2003. An algorithm for the generation of Voronoi diagrams on the sphere based on QTM. *Photogrammetric Engineering and Remote Sensing*, 69(1):79–90.
Chen, X. 1991. *Mathematic Morphology and Images Analysis*. Publishing House of Surveying and Mapping, Beijing (in Chinese).
Chen, Y.Q., Zhang, G., Ding, X.L., and Li, Z.L. 2000. Monitoring earth surface deformations with InSAR technology: principle and some critical issues. *Journal of Geospatial Engineering*, 2(1):3–21.
Chen, Z. and Guevara, J. 1987. Systematic selection of very important points (VIP) from digital terrain model for constructing triangular irregular networks. *Auto Carto*, 8: 50–56.
Cheng, G. 2000. *Hierarchy Representation of Virtual Terrain Environment and Research into the Real Time Shading Technology*. Ph.D. thesis, Zhengzhou Institute for Mapping and Surveying, Zhengzhou. 133 pages (in Chinese).
Chin, F. 1995. *Finding the Constrained Delaunay Triangulation and Constrained Voronoi Diagram of a Simple Polygon in Linear Time*. University of Hong Kong, Hong Kong.
Christensen, A. 1987. Fitting a triangulation to contour lines. In: *Proceedings of Auto Carto 8*, pp. 57–67.
Clarke, A., Gruen, A., and Loon, J. 1982. A contour-specific interpolation algorithm for DEM generation. *International Archive of Photogrammetry and Remote Sensing*, 14(III):68–81.
Clarke, K.C. and Mulcahy, K.A. 1995. Distortion on the interrupted modified Collignon projection. In: *Proceedings of GIS/LIS 95*, Nashville, TN, pp. 175–181.
Collins, S.H. 1968. Stereoscopic orthophoto maps. *Canadian Surveyor*, 22(1):167–176.
Cryer, J. 1986. *Time Series Analysis with Minitab*. Duxbury Press, Boston.
Curlander, J.C. and Mcdonough, R.N. 1991. *Synthetic Aperture Radar: Systems and Signal Processing*. John Wiley & Sons, New York.
Dakowicz, M. and Gold, C.M. 2002. Extracting meaningful slopes from terrain contours. In: Sloot, P.M.A., Tan, C.J.K., Dongarra, J.J., and Hoekstra, A.G. (Eds.), *Proceedings: Computational Science — ICCS 2002*, Lecture Notes in Computer Science, Vol. 2331, Springer-Verlag, Berlin, pp. 144–153.
Davies, G.A. 2000. *Voronoi Diagrams: The Modelling of Natural and Synthetic Structures*. World Scientific, Singapore.
de Berg, M. 2000. *Computational Geometry: Algorithms and Applications*, 2nd ed. Springer, Berlin.
de Berg, M. and Dobrindt, K. 1998. On levels of detail in terrains. *Graphical Models and Image Processing*, 60(1):1–12.
de Floriani, L. 1989. A pyramidal data structure for triangle-based surface description. *IEEE Computer Graphics and Applications*, 9(2):67–78.
De Floriani, L. and Magillo, P. 1994. Visibility algorithms on triangulated digital terrain models. *International Journal of Geographical Information Systems*, 8(1): 13–42.

REFERENCES

Delaunay, B. 1934. Sur la sphere vide. *Bulletin of the Academy of Sciences of the USSR, Classe des Sciences Mathématiques et Naturelles*, 7:793–800.

Demek, J. 1972. *Manual of Detailed Geomorphological Mapping*, Academica, Prague, Czech Republic.

Devillers, O. 1999. On deletion in Delaunay triangulations. In: *15th Annual ACM Symposium on Computational Geometry*, June 13–16, 1999, Miami Beach, Florida, pp. 181–188.

DiBiase, D. 1990. Visualization in earth sciences. *Earth and Mineral Sciences, Bulletin of the College of Earth and Mineral Sciences, PSU*, 56(2):13–18.

DiBiase, D.W., MacEachren, A.M., Krygier, J.B., and Reeves, C. 1992. Animation and the role of map design in scientific visualisation. *Cartography and Geographic Information Systems*, 19:201–214, 265–266.

Dikau, R. 1989. The application of a digital relief model to landform analysis in geomorphology. In: Raper, J. (Ed.), *Three Dimensional Applications in Geographic Information Systems*, Taylor & Francis, London, pp. 51–78.

Duchaineau, M., Wolinsky, M., Sigeti, D.E., Miller, M.C., Aldrich, C., and Mineev-Weinstein, M.B. 1997. ROAMing terrain: real-time optimally adapting meshes. In: Roni Yagel and Roni Yagel (Eds.), *Proceedings of the 8th Conference on Visualization '97*, October 18–24, 1997, Phoenix, AZ, pp. 81–88.

Dutton, G. 1989. Modelling locational uncertainty via hierarchical tessellation. In: Goodchild, M. and Gopal, S. (Eds.), *Accuracy of Spatial Databases*. Taylor and Francis, London, pp. 125–140.

Dutton, G. 1996. Encoding and handling geospatial data with hierarchical triangular meshes. In: Kraak, M.J. and Molenaar, M. (Eds.), *Proceeding of 7th International Symposium on Spatial Data Handling*. Delft, the Netherlands, August 12–16, 1996, pp. 34–43.

Ebisch, K. 1984. Effect of digital elevation resolution on the properties of contours. *Technical Paper, ASP-ACSM Fall Convention*, San Antonio, Texas, September 9–14, pp. 424–434.

Ebner, H., Hofmann-Wellenhof, B., Reiss, P., and Steidler, F. 1980. HIFI — A minicomputer program package for height interpolation by finite elements. *International Archives of Photogrammetry and Remote Sensing*, 23(IV):202–241.

Elfick, M.H. 1979. Contouring by use of a triangular mesh. *The Cartographic Journal*, 16:24–29.

Environmental Systems Research Institute (ESRI) 1992. *Cell-Based Modeling with GRID 6.1.* ARC/INFO User's Guide.

Evans, I. 1981. General geomorphimetry. In: Goudie, A. (Ed.), *Geomorphological Techniques*, Geoge Allen & Unwin, Boston, MA, pp. 31–37.

Evans, I. 1972. General geomorphometry: Derivatives of altitude, and the descriptive statistics. In: Chorley, R. (ed.), *Spatial Analysis in Geomorphology*, London: Methuen & Co. Ltd, 17–90.

Eyton, J.R. 1984. Raster contour. *Geo-Processing*, 2:221–242

Falby, J.S., Zyda, M.J., Pratt, D.R., and Mackey, R.L. 1993. NPSNET: hierarchical data structure for real-time three-dimensional visual simulation. *Computer & Graphics*, 17(1):65–69.

Fekete, G. 1990. Rendering and managing spherical data with sphere quadtree. In: Kaufman, A. (Ed.), *Proceedings of Visualization '90*, IEEE Computer Society, Los Alamitos, CA, pp. 176–186.

Felicísimo, A. 1994. Parametric statistical method for error detection in digital elevation models. *ISPRS Journal of Photogrammetry and Remote Sensing*, 49:29–33.

Flood, M. 2001. Laser altimetry: from science to commercial Lidar mapping. *Photogrammetric Engineering & Remote Sensing*, 67(11):1209–1218.

Forture, S. 1975. A sweep line method for Voronoi diagram. *Methodica*, 2:153–174.
Frederiksen, P. 1980. Terrain analysis and accuracy prediction by means of the Fourier transformation. *International Archives of Photogrammetry and Remote Sensing*, 23(4):284–293. (Also *Photogrammetria*, 36(1981):145–157.)
Frederiksen, P. 1981. Terrain analysis and accuracy prediction by means of the Fourier transformation. *Photogrammetria*, 36:145–157.
Frederiksen, P., Jacobi, O., and Justesen, J. 1978. Fourier transformation von hohenbeobachtungen. *ZFV*, 103:64–79.
Frederiksen, P., Jacobi, O., and Kubik, K. 1983. Measuring terrain roughness by topographic dimension. In: *Proceedings of International Colloquium on Mathematical Aspects of DEM*, Stockholm, Sweden.
Frederiksen, P., Jacobi, O., and Kubik, K. 1986. Optimum sampling spacing in digital elevation models. *International Archives of Photogrammetry and Remote Sensing*, 26(3/1):252–259.
Freeman, T.G. 1991. Calculating catchment area with divergent flow based on a regular grid. *Computers & Geosciences*, 17(3):413–422.
Fritsch, D. and Spiller, R. (Eds.) 1999. *Photogrammetric Week'99*. Wichmann, Heidelberg, Germany.
Gannapathy, S. and Dennehy, T. 1982. A new general triangulation method for planar contours. *Computer Graphics*, 16(3):69–72.
Garmin. 2003. What is GPS? http://www.garmin.com/aboutGPS/ (last accessed August 3, 2003).
Gold, C.M. 1989. Surface interpolation, spatial adjacency and GIS. In: Raper, J. (Ed.), *Three Dimensional Applications in GIS*, Taylor and Francis, London, pp. 21–36.
Gold, C.M. and Condal, A.R. 1995. A spatial data structure integrating GIS and simulation in a marine environment. *Marine Geodesy*, 18:213–228.
Gold, C.M. and Cormack, S. 1987. Spatially ordered networks and topographic reconstructions. *International Journal of Geographical Information Systems*, 1:137–148.
Gold, C.M. and Mostafavi, M. 2000. Towards the global GIS. *ISPRS Journal of Photogrammetry and Remote Sensing*, 55(3):150–163.
Gold, C.M. and Snoeyink, J. 2001. A one-step crust and skeleton extraction algorithm, *Algorithmica*, 30:144–163.
Gold, C.M., Charters, T.D., and Ramsden, J. 1977. Automated contour mapping using triangular element data structures and an interpolant over each triangular domain. In: George, J. (Ed.), *Proceedings: Siggraph'77. Computer Graphics*, 11(2): 170–175.
Gold, C.M., Chau, M., Dzieszko, M., and Goralski, R. 2004. A window into the real world. In: *Proceeding Conference on Spatial Data Handling*, Leicester, U.K., August 2004.
Goldstein, R.M., Zebker, H.A., and Werner, C.L. 1988. Satellite radar interferometry: two-dimensional phase unwrapping. *Radio Science*, 23:713–720.
Gong, J., Li, Z.L., Zhu, Q., Sui, H.G., and Zhou, Y. 2000. Effect of various factors on the accuracy of DEMs: an intensive experimental investigation. *Photogrammetric Engineering and Remote Sensing*, 66(9):1113–1117.
Goodchild, M.F. and Yang, S. 1992. A hierarchical data structure for global geographic information systems. *Computer Vision and Geographic Image Processing*, 54(1): 31–44.
Goodchild, M.F., Yang, S., and Dutton, G. 1991. *Spatial Data Representation and Basic Operations for a Triangular Hierarchical Data Structure*. NCGIA Report, pp. 91–98.
Gosper, J.J. 1998. 2D Convex Hulls. http://www.brunel.ac.uk/~castjjg/java/mscthesis/convexhull/ (last accessed August 2003).

Gouraud, H. 1971. Illumination of computer-generated pictures. *Communication of ACM*, 18(60):311–317.
Graham, L.C. 1974. Synthetic interferometer radar for topographic mapping. *Proceedings of the IEEE*, 62(2):763–768.
Green, P.J. and Sibson, R. 1977. Computing Dirichlet tessellations in the plane. *The Computer Journal*, 21(2):168–173.
Guibas, L., Mitchell, J.S.B., and Roos, T. 1991. Voronoi diagrams of moving points in the plane. In: G. Schmidt (Ed.), *The 17th International Workshop in Graph-Theoretic Concepts in Computer Science*, Lecture notes in computer science 570; 113–125. Berlin; New York, Springer-Verlag, 1992.
Haala, N., Brenner C., and Anders, K.H. 1998. 3D urban GIS from laser altimeter and 2D map data. *International Archives of Photogrammetric & Remote Sensing*, 32(Part 3/1):339–346.
Hannah, M. 1981. Error detection and correction in digital terrain models. *Photogrammetric Engineering and Remote Sensing*, 47(1):63–69.
Haralick, R., Sternberg, S., and Zhuang, X. 1987. Image analysis using mathematical morphology. *IEEE Transactions of Pattern Analysis and Machine Intelligence*, 9:532–550.
Hardy, R. 1971. Multiquadratic equations of topography and other irregular surfaces. *Journal of Geophysical Research*, 76:1905–1915.
Helava, U.V. 1958. New principles of photogrammetric plotters. *Photogrametria*, 14(2):89–96.
Helava, U.V. and Chapelle, W.E. 1972. Epipolar-scan correlation. *Bendix Technical Journal*, Spring Issue, 5:19–23.
Heller, M. 1990. Triangulation algorithms for adaptive terrain modeling. In: Brassel, K. and Kishimoto, H. (Eds.), *Proceedings, Fourth International Symposium on Spatial Data Handling*, Zurich, pp. 163–174.
Hengl, T., Gruber, S., and Shrestha, D.P. 2003. Digital Terrain Analysis in ILWIS. http://www.itc.nl/personal/shrestha/dta/ (last accessed on April 2, 2004).
Hogg, R. and Tanis, E. 1977. *Probability and Statistical Inference*. Macmillan Publishing Co., New York.
Horn, B.K.P. 1981. Hill shading and the reflectance map. *Proceedings of IEEE*, 69(1):14–47.
Imhof, E. 1965. *Kartographische Gelandedarstellung*. Walter de Gruyter & Co., Berlin.
Jensen, J.R. 1980. Stereoscopic statistical maps. *The American Cartography*, 7(1):25–37.
Jenson, S.K. and Domingue, J.O. 1988. Extraction topographic structure from digital elevation data for geographic information system analysis. *Photogrammetric Engineering and Remote Sensing*, 54(11):1593–1600.
Jiang, B. 1996. Cartographic visualization: analytical and communication tools. *Cartography*, 25(2):1–11.
Keating, T. and Wolf, P. 1976. Analytical photogrammetry from digitised image densities. Presented paper, *XIII ISP Congress*, Helsinki, Finland, 1976.
Kidner, D.B. and Smith, D.H. 2003. Advances in the data compression of digital elevation models. *Computers & Geosciences*, 29:985–1002.
Klein, R. 1988. *Abstract Voronoi Diagrams and Their Applications*. Lecture Notes in Computer Science, Vol. 333, Springer-Verlag, Berlin.
Konecny, G., Bahr, H., Reil, W., and Schreiber, H. 1979. *Use of Spaceborne Metric Camera for Cartographic Applications*. Report to the Ministry of Research and Technology of FRG.

Kraak, M.J. and Brown, A. 2001. *Web Cartography: Developments and Prospectus*. Taylor & Francis, London.

Kraus, K. and Pfeifer, N. 1998. Determination of terrain models in wooded areas with airborne laser scanner data. *ISPRS Journal of Photogrammetry and Remote Sensing*, 53(4):193–203.

Kubik, K. and Botman, A. 1976. Interpolation accuracy for topographic and geological surfaces. *ITC Journal*, 1976-2:236–274.

Kubik, K. and Roy, R. 1986. *Digital Terrain Model Workshop Proceedings*, Columbus, OH.

Leberl, F. 1990. *Radargrammetric Image Processing*. Artech House, Norwood, MA.

Leberl, F. and Olson, D. 1982. Raster scanning for operational digitizing of graphical data. *Photogrammetric Engineering and Remote Sensing*, 48(4):615–627.

Leberl, F., Domik, G., Cimino, J., and Kobrick, M. 1986a. Radar stereomapping techniques and application to SIR-B images of Mt. Shasta. *IEEE Transactions on Geoscience and Remote Sensing*, 24(4):473–481.

Leberl, F., Domik, G., Cimino, J., Raggam, J., and Kobrick, M. 1986b. Multiple incidence angle SIR-B experiment over Argentina: stereo-radargrammetric analysis. *IEEE Transactions on Geoscience and Remote Sensing*, 24(4):482–491.

Lee, D.T. and Drysdale, R.L. 1981. Generalization of Voronoi diagram in the plane. *SIAM Journal of Computing*, 10:73–87.

Lee, M. and Samet, H. 2000. Navigating through triangle meshes implemented as linear quadtree. *ACM Transactions on Graphics*, 19(2):79–121.

Ley, R. 1986. Accuracy assessment of digital terrain models. *Auto-Carto London*, 1:455–464.

Li, C., Chen, J., and Li, Z.L. 1999. A raster-based algorithm for computing Voronoi diagrams of spatial objects using dynamic distance transformation. *International Journal of Geographical Information Science*, 13(3):209–225.

Li, D. 1998. *Lecture on New Technology of Photography Measurement*. Press of Wuhan Technical University of Surveying and Mapping, Wuhan, China (in Chinese).

Li, Z.L. 1988. On the measure of digital terrain model accuracy. *Photogrammetric Record*, 12(72):873–877.

Li, Z.L. 1990. Sampling Strategy and Accuracy Assessment for Digital Terrain Modelling. Ph.D. thesis, The University of Glasgow.

Li, Z.L. 1991. Effects of check points on the reliability of DTM accuracy estimates obtained from experimental tests. *Photogrammetric Engineering and Remote Sensing*, 57(10):1333–1340.

Li, Z.L. 1992a. Variation of the accuracy of digital terrain models with sampling interval. *Photogrammetric Record*, 14(79):113–128.

Li, Z.L. 1992b. Variation of the accuracy of digital terrain models with sampling interval. *Photogrammetric Record*, 14(82):651–660.

Li, Z.L. 1993a. Theoretical models of the accuracy of digital terrain models: an evaluation and some observations. *Photogrammetric Record*, 14(82):651–660.

Li, Z.L. 1993b. Mathematical models of the accuracy of digital terrain model surfaces linearly constructed from grid data. *Photogrammetric Record*, 14(82):661–674.

Li, Z.L. 1994. A comparative study of the accuracy of digital terrain models based on various data models. *ISPRS Journal of Photogrammetry and Remote Sensing*, 49(1):2–11.

Li, Z.L. 1996. Transformation of spatial representation in scale dimension: a new paradigm for digital generalization of spatial data. *International Archives for Photogrammetry and Remote Sensing*, XXXI(B3):453–458.

Li, Z.L. 1999. Scale: a fundamental scale in spatial information science. In: Xu, G. and Chen, Y. (Eds.), *Towards Digital Earth — Proceedings of the International Symposium on The Digital Earth*, November 29 to December 2, 1999, Beijing, pp. 533–538.

REFERENCES

Li, Z.L. 2003. Scale in data integration and multi-scale representation. In: Chen, J. and Wu, L. (Eds.), *Framework for Digital China*, Science Press, Beijing, pp. 57–67 (in Chinese).

Li, Z.L. and Li, C.M. 1999. Objective generalization of DEM based on a natural principle. In: Chen, J., Zhou, Q.M., Li, Z.L., and Jiang, J. (Eds.), *Proceedings of 2nd International Workshop on Dynamic and Multi-Dimensional GIS*, October 4–6, 1999, Beijing, pp. 17–22.

Li, Z.L. and Openshaw, S. 1992. Algorithms for objective generalization of line features based on the natural principle. *International Journal of Geographical Information Systems*, 6(5):373–389.

Li, Z.L. and Openshaw, S. 1993. A natural principle for objective generalisation of digital map data. *Cartography and Geographic Information System*, 20(1):19–29.

Li, Z.L. and Zhu, Q. 2000. *Digital Elevation Models*. Wuhan University Press, Wuhan, China (in Chinese).

Li, Z.L., Lam, K., and Li, C.M. 1998. Effect of compression on the accuracy of DTM. *Geographic Information Science*, 4(1–2):37–43.

Li, Z.L., Hill, C., Azizi, A., and Clark, M.J. 1993. Exploring the potential benefits of digital photogrammetry: some practical examples. *Photogrammetric Record*, 14(81): 469–475.

Lindstrom, P., Koller, D., Ribarsky, W., Hodges, L.F., Faust, N., and Turner, G. 1996. Real-time continuous level of detail rendering of height fields. In: Rushmeier, H. (Ed.), *Proceedings of SIGGRAPH 96*, New Orleans, LA, August 4–9, 1996, pp. 109–118.

Liu, X.J. 2002. Analysis and Evaluation of Error of Interpreting Algorithm Based on Regular Grid Digital Elevation Model. Ph.D. thesis, Wuhan University, Wuhan, China (in Chinese).

Llobera, M. 2003. Extending GIS-based visual analysis: the concept of visualscapes. *International Journal of Geographical Information Science*, 17(1):25–48.

Loon, J.C. 1978. Cartographic Generalization of Digital Terrain Models. Dissertation, University Microfilm International.

Luebke, D., Reddy, M., Cohen, J., Varshney, A., Watson, B., and Huebner, R. 2003. *Level of Detail for 3D Graphics*. Morgan Kaufmann Publisher, San Franscisco, CA.

Makarovic, B. 1972. Information transfer on construction of data from sampled points. *Photogrammetria*, 28(4):111–130.

Makarovic, B. 1973. Progressive sampling for DTMs. *ITC Journal*, 1973-4:397–416.

Makarovic, B. 1975. Amended strategy for progressive sampling. *ITC Journal*, 1975-1:117–128.

Makarovic, B. 1977. Regressive rejection — a digital data compression technique. In: *Proceedings of ASP/ACSM Fall Convention*, Little Rock, October 1977.

Makarovic, B. 1979. From progressive sampling to composite sampling. *Geo-processing*, 1:145–166.

Makarovic, B. 1984. A test on compression of digital terrain model data. *ITC Journal*, 1983-2:133–138.

Mandelbort, B. 1967. How long is the coast of Britian? Statistical self-similarity and fractional dimension. *Science*, 155:636–638.

Mandelbrot, B. 1981. *The Fractal Geometry of Nature*. W.H. Freeman and Company, San Francisco, CA.

Mark, D. 1975. Geomorphometric parameters: a review and evaluation. *Geografiska Annaler*, 57A:165–177.

Masry, S.E. 1974. Digital correlation principles. *Photogrammetric Engineering*, 40(3): 303–308.

Masser, I. 1988. The Regional Research Laboratory Initiative: a progress report. *International Journal of Geographical Information System*, 2:11–22.

Maune, D.F. (Ed.) 2000. *Digital Elevation Model Techniques and Applications: The DEM User Manual*. American Society for Photogrammetry and Remote Sensing, Bethesda, MD.

Maune, D.F., Huff, L.C., and Guenther, G.C. 2001. DEM user applications. In: Maune, D.F. (Ed.), *Digital Elevation Model Techniques and Applications: The DEM User Manual*, American Society for Photogrammetry and Remote Sensing, Bethesda, MD, pp. 367–394.

McCullagh, M. and Ross, R. 1980. Delaunay triangulation of a random data set for isarithmic mapping. *The Cartographic Journal*, 17(2):178–181.

McElroy, S. 1992. *Getting Started with GPS Surveying*. The Global Positioning System Consortium (GPSCO), Australia.

McLain, D.H. 1976. Two dimensional interpolation from random data. *Computer Journal*, 19:178–181.

Meyer, W. 1985. *Concepts of Mathematical Modeling*. McGraw-Hill Book Company, New York.

Meyer, W. 1995. *Concepts of Mathematical Modelling*. McGraw-Hill Book Company, New York.

Miles, R.E. and Maillardet, R.J. 1982. The basic structures of Voronoi and generalized Voronoi polygons. *Journal of Applied Probability*, 19A:97–111.

Miller, C. and Laflamme, R. 1958. The digital terrain model — theory and applications, *Photogrammetric Engineering*, 24:433–442.

Moore, I.D., Grayson, R.B., and Ladson, A.R. 1994. Digital terrain modelling: a review of hydrological, geomorphological, and biological applications. In: Beven, K.J. and Moore, I.D. (Eds.), *Terrain Analysis and Distributed Modelling in Hydrology*, John Wiley & Sons, Chichester, U.K., pp. 7–34.

Moritz, H. 1980. *Advanced Physical Geodesy*. Wichmann, Heidelberg, Germany.

Mostafavi, M.A., Gold, C., and Dakowicz, M. 2003. Delete and insert operations in Voronoi/Delaunay methods and applications. *Computers & Geoscience*, 29:523–530.

Mostafavi, M.A. and Gold, C.M. 2004. A global kinetic spatial data structure for a marine simulation. *International Journal of Geographical Information Science*, 18(3):211–227.

Narasimhan, T.N. and Witherspoon, P.A. 1976. An integrated finite difference method for analyzing fluid flow in porous media. *Water Resources Research*, 12:57–64.

O'Callaghan, J.F. and Mark, D.M. 1984. The extraction of drainage networks from digital elevation data. *Computer Vision, Graphics, and Image Processing*, 28(4):323–344.

Ohya, T., Iri, M., and Murota, K. 1984a. A fast Voronoi-diagram method with quaternary tree bucketing. *Information Processing Letters*, 18:227–231.

Ohya, T., Iri, M., and Murota, K. 1984b. Improvement of the incremental method for the Voronoi diagram with computational comparison of various methods. *Journal of the Operations Research Society of Japan*, 27:306–336.

Okabe, A., Boots, B., Sugihara, K., and Chiu, S.N. 2000. *Spatial Tessellations: Concepts and Applications of Voronoi Diagrams*, 2nd ed. John Wiley, Chichester, U.K.

Olea, R.A. 1999. *Geostatistics for Engineers and Earth Scientists*. Kluwer Academic Publishers, Boston, MA.

Openshaw, S. 1994. *The Modifiable Areal Unit Issue*. CATMOG #38. Geo Books, Norwick, England.

O'Rourke, J. 1993. *Computational Geometry in C*. Cambridge Press, Cambridge.

REFERENCES

Ostman, A. 1986. A graphic editor for digital elevation models. *Geo-processing*, 3: 143–154.

Otoo, E. and Zhu, H. 1993. Indexing on spherical surfaces using semi-quadcodes. In: Abel, J. and Beng, C.O. (Eds.), *Advances in Spatial Databases*, Lecture Notes in Computer Science 692, Springer, Singapore, pp. 509–529.

Petrie, G. 1990a. Modelling, interpolation and contouring procedures. In: Petrie, G. and Kennie, T. (Eds.), *Terrain Modelling in Surveying and Civil Engineering*, Whittles Publishing, Caithness, England, pp. 112–137.

Petrie, G. 1990b. Terrain data acquisition and modelling from existing maps. In: Petrie, G. and Kennie, T. (Eds.), *Terrain Modelling in Surveying and Civil Engineering*. Whittles Publishing, Caithness, England, pp. 85–111.

Petrie, G. and Kennie T. (Eds.) 1987. An introduction to terrain modeling: applications and terminology. In: *Terrain Modelling in Surveying and Civil Engineering: A Short Course*. University of Glasgow.

Petrie, G. and Kennie, T. (Eds.) 1990. *Terrain Modelling in Surveying and Civil Engineering*. Whittles Publishing, Caithness, England.

Peucker, T. 1972. *Computer Cartography*. Association of American Geographer, Commission on College Geography, Washington, DC.

Phillips, G.M. 2003. *Interpolation and Approximation by Polynomials*. Springer, New York.

Phong, B.-T. 1975. Illumination for computer-generated pictures. *Communication of ACM*, 18(6):311–317.

Polidori, L. 1991. Digital terrain models from radar images: a review. In: Guyenne, T.D. and Hunt, J.J. (Eds.), *Proceedings of the International Symposium on Radars and Lidars in Earth and Planetary Science*, France, Cannes, pp. 141–146.

Quattrochi, D.A. and Goodchild, M.F. (Eds.) 1997. *Scale in Remote Sensing and GIS*. CRC Press, Boca Raton, FL.

Quinn, P.F., Beven, K., Chevallier, P., and Planchon, O. 1991. The prediction of hillslope flow paths for distributed hydrological modelling using digital terrain models. In: Beven, K.J. and Moore, I.D. (Eds.), *Terrain Analysis and Distributed Modelling in Hydrology*, John Wiley & Sons, Chichester, U.K., pp. 63–83.

Rana, S. and Morley, J. 2002. *Optimising Visibility Analyses Using Topographic Features on the Terrain*. Working Paper 44, CASA of University College London.

Ritter, P. 1987. A vector-based slope and aspect generation algorithm. *Photogrammetric Engineering and Remote Sensing*, 53:1109–1111.

Robert, J.R. 1997. Delaunay triangulation and Voronoi diagram on the surface of a sphere. *ACM Transactions on Mathematical Software*, 23(3):416–434.

Roberts, R. 1957. Using new methods in highway location. *Photogrammetric Engineering*, 23:563–569.

Saaty, T. and Alexander, J. 1981. *Thinking with Models*. Pergamon Press, Oxford.

Sakhnovich, L.A. 1997. *Interpolation Theory and its Applications*. Kluwer Academic Publishers, Boston, MA.

Sarjakoski, T. 1981. Concept of a completely digital stereo plotter. *The Photogrammetric Journal of Finland*, 8(2):95–100.

Schuts, G. 1976. Review of interpolation methods for digital terrain models. *International Archives of Photogrammetry*, 21(3).

Schwarz, P. 1982. A test for personal stereoscopic measuring precision. *Photogrammetric Engineering and Remote Sensing*, 48(3):375–381.

Serra, J. 1982. *Image Processing and Mathematical Morphology*. Academic Press, New York.

Sharpnack, D.A. and Akin, G. 1969. An algorithm for computing slope and aspect from elevations. *Photogrammetric Survey*, 35:247–248.

Sibson, R. 1980. A vector identity for the Dirichlet tesselation. *Mathematical Proceedings of the Cambridge Philosophical Society*, 87:151–155.

Sibson, R. 1981. A brief description of natural neighbor interpolations. In: Baruett, V. (Ed.), *Interpreting Multivariate Data*, John Wiley, Chichester, U.K., pp. 21–36.

Sigle, M. 1984. A digital terrain model for the state of Badenwuttemberg. *International Archives of Photogrammetry and Remote Sensing*, 25(A3/B):1016–1026.

Skidmore, A.K. 1989. A comparison of techniques for calculating gradient and aspect from a gridded digital elevation model. *International Journal of Geographical Information System*, 4:323–334.

Snyder, J.P. 1992. An equal-area map projection for polyhedral globes. *Cartographica*, 29(1):10–12.

State Bureau of Quality and Technical Supervision (SBQTS) 1999. *Geo-Spatial Data Transfer Format (GB/T 17798-1999)*. Standard Press of China, Beijing.

Strahler, A. 1956. Quantitative slope analysis. *Bulletin of the Geological Society of America*, 67:571–596.

Su, B., Li, Z.L., and Lodwick, G. 1998. Algebraic models for collapse operation in digital map generalization using morphological operators. *Geoinformatica*, 2(4):359–382.

Su, B.Q. 1989. *Computational Geometry: Curve and Surface Modeling*. Academic Press, Boston, MA.

Sugihara, K. 1992. A simple method for avoiding numerical errors and degeneracy in Voronoi diagram computation. *IEICE Transactions Fundamentals*, E75-A(4): 468–477.

Tang, G.A. 2000. *A Research on The Accuracy of Digital Elevation Models*. Science Press, Beijing.

Tang, L. 1989. Surface modelling and visualization based upon digital image processing techniques. In: Grun, A. and Kahmen, H.K. (Eds.), *Optical 3-D Measurement Techniques*, Wichmann Verlag, Karlsruhe, Germany, pp. 317–325.

Tempfli, K. 1980. Spectral analysis of terrain relief for the accuracy estimation of digital terrain models. *ITC Journal*, 1980-3:487–510.

The Aerospace Corporation. 2003. GPS Premier. http://www.aero.org/publications/GPSPRIMER/ (last accessed August 3, 2003).

Thibault, D. and Gold, C. 2000. Terrain reconstruction from contours by skeleton construction. *Geoinformatica*, 4(4):349–374.

Thiessen, A.H. 1911. Precipitation averages for Large Areas. *Monthly Weather Review*, 39:1082–1084.

Toomey, M. 1988. The Alberta digital elevation model. *International Archives of Photogrammetry and Remote Sensing*, 27(B3):775–783.

Torlegard, K., Ostman, A., and Lindgren, R. 1986. A comparative test of photogrammetrically sampled digital elevation models. *Photogrammetria*, 41(1):1–16.

Toutin, T. 2002. Impact of terrain slope and aspect on radargrammetric DEM accuracy, *ISPRS Journal of Photogrammetry and Remote Sensing*, 57(3):228–240.

Toutin, Th. 2000. Evaluation of radargrammetric DEM from RADARSAT images in high relief areas. *IEEE Transactions on Geoscience and Remote Sensing*, 38(2): 782–789.

Toutin, Th. and Gray, L. 2000. State-of-the-art of elevation extraction from satellite SAR data. *ISPRS Journal of Photogrammetry & Remote Sensing*, 55:13–33.

Tsai, V.J.D. 1993. Delaunay triangulations in TIN creation: an overview and a linear-time algorithm. *International Journal of GIS*, 7(6):501–524.

REFERENCES

Tse, O.C.R. and Gold, C.M. 2002. TIN meets CAD — extending the TIN concept in GIS. In: Sloot, P.M.A., Tan, C.J.K., Dongarra, J.J., and Hoekstra, A.G. (Eds.), *Computational Science — ICCS 2002*, Vol. 2331, Springer-Verlag, Berlin, pp. 135–143.

Turner, H. 1997. A comparison of some methods for slope measurement from large-scale air photos. *Photogrammetria*, 32:209–237.

Unwin, D. 1981. *Introductory Spatial Analysis*. Methuen, London.

U.S. Geological Survey (USGS) 1998. Standards for Digital Elevation Models. http://rockyweb.cr.usgs.gov/nmpstds/demstds.html (last accessed April 2004).

van Kreveld, M. 1996. Efficient methods for isoline extraction from a TIN. *International Journal of Geographical Information Systems*, 10:523–540.

Wang, J.J., Robinson, G.J., and White, K. 2000. Generating viewshed without using sightlines. *Photogrammetric Engineering & Romote Sensing*, 66(1):87–90.

Watson, D.F. 1981. Computing the N-dimensional Delaunay tessellation with applications to Voronoi diagram. *The Computer Journal*, 24(2):167–172.

Watson, D.F. 1992. *Contouring: A Guide to the Analysis and Display of Spatial Data*. Pergamon Press, New York.

Weibel, R. 1987. *An Adaptive Methodology for Automated Relief Generalization*. Auto-Carto 8, Baltimore, MD.

Wentworth, C. 1930. A simplified method for the determining the average slope of land surface. *American Journal of Science*, 20:184–194.

Wessel, P. 2003. Compression of large data grids for Internet transmission. *Computers & Geosciences*, 29:665–671.

Whitmore, G. and Thompson, M. 1966. Introduction to photogrammetry. In: Thompson, M. (Ed.), *Manual of Photogrammetry*, American Society for Photogrammetry, Fall Church, pp. 1–16.

White, D., Kimmerling, J., and Overton, W.S. 1992. Cartographic and geometric components of a global sampling design for environment monitoring. *Cartography & Geographic Information Systems*, 19(1):5–22.

Wickman, F.E. and Elvers, E. 1974. A system of domains for global sampling problems. *Geografiska Annaler*, 56(3/4):201–212.

Wilson, J.P. and Gallant, J. (Eds.) 2000. *Terrain Analysis: Principle and Applications*. John Wiley & Sons, Singapore.

Wu, H. 1997. Structured approach to implementing automated cartographic generalisation. *Proceedings of ICC'97*, 1:349–356.

Yeoli, P. 1977. Computer executed interpolation of contours into arrays of randomly distributed height points. *The Cartographic Journal*, 14:103–108.

You, X. 1991. The battlefield environment simulation based on VR. In: *The First Workshop on the Virtual Reality and Geography*, Shenzhen, China, May 29, 2001 (in Chinese).

Zaninetti, L. 1990. Dynamic Voronoi tessellation. *Astronomy and Astrophysics*, 233:293–300.

Zebker, H.A. and Villasenor, J. 1992. Decorrelation in interferometric radar echoes. *IEEE Transactions on Geoscience and Remote Sensing*, 30:950–959.

Zebker, H.A., Rosen, P.A., and Hensley, S. 1997. Atmospheric effects in interferometric systhetic aperture radar surface deformation and topographic maps. *Journal of Geophysical Research*, 102:7547–7563.

Zevenbergen, L.W. and Thorne, C.R. 1987. Quantitative analysis of land surface topography. *Earth Surface Processes and Landforms*, 12:47–56.

Zhao, R.L., Li, Z.L., Chen, J., Gold, C., and Zhang, Y. 2002. A hierarchical raster method for computing for Voronoi diagrams based on Quadtree. In: Sloot, P., Tan, C., Dongarra, J., and Hoekstra, A. (Eds.), *Computational Science — ICCS 2002*, Lecture Notes in Computer Science, Vol. 2331, Springer-Verlag, Berlin, pp. 1004–1013.

Zhou, Q. and Liu, J. 2002. Error assessment of grid-based flow routing algorithms used in hydrological models. *International Journal of Geographical Information Science*, 16(8):819–842.

Zhu, G., Wang, J., and Jiang, W. 1999. *Digital Map Analysis*. Publishing House of Wuhan Technical University of Surveying and Mapping, Wuhan, China (in Chinese).

Zhu, Q. and Chen, C.J. 1998. An algorithm for fast generation of TIN and its dynamic updating. *Journal of Wuhan Technical University of Surveying and Mapping*, 23(3):204–207 (in Chinese).

Zhu, Q., Zhao, J., Zhong, Z., and Sui, H.G. 2003. An efficient algorithm for the extraction of topographic structures from large scale grid DEMs. In: Li, Z.L., Zhou, Q., and Kainz, W. (Eds.), *Advances in Spatial Analysis and Decision Making*, A.A. Balkema Publishers, Lisse, Netherlands, pp. 99–107.

Zwicker, M. and Gross, M.H. 2000. *A Survey and Classification of Real Time Rendering Methods*. Mitsubishi Electric Research Laboratories, Cambridge Research Center. http://www.merl.com/papers/TR2000-09/ (last accessed April 2004).

Index

Accuracy, 1, 6, 7, 9, 13, 21, 26, 28, 29, 34–36, 50–56, 61–63, 78, 99, 115, 120, 137, 141, 142, 159–183, 185–190, 196, 225, 229, 241, 273, 277
Accuracy, DTM, 10, 78, 159, 161–163, 165–168, 170–177, 186–187, 189
Accuracy, empirical model, 170–172, 188–189
Accuracy, source data, 28
Accuracy, theoretical model, 173–188
Accuracy assessment, 115, 159, 186
Accuracy loss, 7, 78, 142, 163, 164, 174–181, 186, 206, 227, 241
Accuracy prediction, 170, 173, 177–178
Aerial photography, 33, 37
Aggregation, 191, 197, 198
Airborne laser scanning, 50–52
ALS. See Airborne laser scanning
Analytical plotter, 10, 37
Animation, 249, 262–266
Area subdivision algorithm, 157
Aspect map, 273, 290
Autocorrelation, 17, 18
Azimuth resolution, 41–43

Battlefield simulation, 291
Bilinear surface, 70, 76, 116, 178–180
Binary contouring, 239, 240
Bit boundary block transfer, 263
bitblt, See Bit boundary block transfer
BLOB, 218, 220, 221, 223, 225
Block, 214, 218–223, 263–265
Block diagram, 4
Block encoding, 227
Break lines, 23, 24, 26–28, 68, 70, 71, 74, 170, 184, 223
Brightness variable, 247, 248
Budget-based simplification, 207
bzip2, 227

Cartographic digitization, 31, 56, 63
Cartography, 11, 12, 191
Catchment, 10, 276, 279, 280, 292
Caves, 297
CDT. See Constrained Delaunay triangulation
Central projection, 202
Chebyshev's theorem, 163
Checkpoints, 160, 166, 168, 169, 170, 171
Chessboard distance, 112, 113
CIPA. See Contour-specific methods
Circumcircle, 79, 80, 89, 94, 97, 98, 101, 105
Circumcircle, empty, 89, 94, 97, 98
Circumcircle, smallest, 98, 99
CISS. See Contour-specific methods
City block distance, 112
Clustered data, 27
Clusters of gross errors, 153, 154
Code assignment, 228
Coding, data, 12, 226–228
Coding, flow, 278
Colinearity condition, 38
Collapse, edge, 206
Collapse, triangle, 206
Collision-avoidance system, 293
Color variables, 247
Composite sampling, 26, 28, 69, 74, 80, 216
Compound topographic index, 276
Compression, lossless, 226–229
Compression, lossy, 227
Compression ratio, 228, 229
Concave point, 23, 24, 182
Concave profile, 274
Concentration of the flow, 274
Conceptual model, 5
Conductivity factor, 281
Conic function, 121
Constrained Delaunay triangulation, 99, 101
Continuous scales, 200, 205
Continuous surface, 66, 68, 70, 72, 74, 116
Contour intervals, 35, 188–190

Contour smoothing, 238
Contouring threading, 242
Contour tracing, 236, 242
Contouring, binary, 239, 240
Contouring, edge, 239
Contouring, gray-tone, 241
Contouring, raster-based, 233, 238
Contouring, vector-based, 233, 238, 241
Contour-specific methods, 83, 84
Convex hull, 90–94
Convex point, 24
Convex profile, 274
Convex slope, 182, 183
Convexity, plane, 20
Convexity, profile, 20
Convexo-concave coefficient, 276
Convolution, 139, 140, 142, 202
Coplanarity condition, 49
Co-registration, 46
Course line, 23
Covariance, 17, 18, 175
Cross-sectional area, 281
Crust, 104
Cubic function, 82, 121
Curvature, 16, 19, 274, 275
Curve fitting, 238, 239, 242
Cut operation, 298
Cutoff frequency, 28
Cycle ambiguity, 48

Data block, 221, 265, 275
Data coding, 12, 226–228
Data exchange standard, 230–231
Data organization, 218, 225, 264
Data redundancy, 22, 57, 74, 77, 227
Delaunay tetrahedral model, 304
Delaunay triangulation, 73, 79, 80, 87–102, 104, 108, 109, 113, 114
Delaunay triangulation, constrained, 99, 101
Depth sorting algorithm, 257
Descriptors, qualitative, 13
Descriptors, quantitative, 13
DGPS, See Differential GPS
Differences in slope change, 144, 145
Differential GPS, 50
Diffuse reflection, 258
Diffusion model, 259
Digital earth, 300
DEM, see digital elevation model
Digital elevation model, 7
Digital photogrammetric workstation, 37, 39, 239
Digital terrain model, 4, 6–7, 65, 250, 293, 297
Digital terrain modeling, 9, 11, 31, 18, 26, 31, 36, 63, 83, 115, 117, 126, 127, 133, 137, 155, 164, 166, 189, 191, 192, 227, 230, 247, 297, 300

Digitization, cartographic, 31, 56, 63
Digitization, line-following, 56–57
Dilation, 112, 113
Dirichlet tessellation, 73
Direct interpolation, 82
Discontinuous surface, 68, 72, 116
Discrete scale, 196
Dispersion, 162, 181
Distance transformation, 111–113
Divergence of the flow, 274
Doppler equations, 49, 50
DPW, See Digital photogrammetric workstation
Drainage network, 276, 279, 280
Drum scanner, 58
DTM, see digital terrain model
Dynamic variables, 249, 263

Ear, 98, 99
ECS, See Eye-coordinate system
Edge collapse, 206
Edge contouring, 239
Edge detection, 239, 240
Edge swapping, 95, 97
Empty circumcircle principle, 89, 94
Enumeration unit, 192
Erosion, morphological, 113, 114
Error, DTM, 20, 78, 134–136, 159–162, 168–169
Error, gross, 135, 142–144, 146–157
Error, random, 74, 134, 136, 138, 139, 142
Error, round-off, 138
Error, source data, 134, 168
Error, systematic, 134
Euclidean space, 192–195
Euler's theorem of planar graphs, 79
Exploratory acts, 249
Exponential function, 17, 18, 121, 175
Extreme Error, 181, 185
Eye-coordinate system, 254, 255

Feature-specific data, 23, 172–174
Fidelity-based simplification, 207
File structure, 213–216
Filter, high-pass, 138
Filter, low-pass, 138, 139, 141
Finite elements, 126, 127
Flat triangle, 80, 81, 102
Flat-bed scanner, 58
Flight simulation, 12, 290
Flood simulation, 292
Flow accumulation, 276, 278–280
Flow direction, 276–278, 281
Flow line, 278, 279
Flow model, 276, 281
Fly-through, 263, 265, 266, 291
Format standards, 231

INDEX

Fourier series, 4, 120, 174
Fractal dimension, 15, 16
Frame animation, 263
Function, Arthur, 121
Function, conic, 121
Function, cubic, 82, 121
Function, exponential, 17, 18, 121, 175
Function, Gaussian, 139, 140, 175
Function, geometric, 121
Function, hyperbolical, 121
Function, Lu, 122
Function, weighting, 129, 130, 139
Function, Wild, 122, 123
Functional model, 5
F-S data, See Feature-specific data

Gaussian function, 139, 140, 175
General polynomial, 66, 67, 70
Geographical space, 8, 192–195
Geometric function, 121
Geometric variables, 247
Geo-scale, 192, 193
Gift wrapping algorithm, 91, 92
Global interpolation, 115
Global surface, 72, 74, 148
Gouraud shading model, 259
Gradient, 19, 20, 271, 281
Grain, 19
Gray-tone contouring, 241
Gross error detection, 142–148, 156
Ground truth, 166
gzip, 227, 229

Heighting with InSAR, 46
Highway design, 6, 8, 285
High-pass filter, 138
Hill shading, 251
Huffman coding, 227–229
Hydrological parameters, 267, 276
Hyperbolical function, 121
Hypermetric tints, 2, 251, 252, 260

IKONOS, 3, 34
Illumination model, 250, 253, 254, 257–259
Image matching, 10, 39, 47, 50, 134, 135, 140, 142, 152
Incremental method, 110
Indirect interpolation, 82
InSAR, See Interferometric SAR
Inscribed polyhedrons, 300
Interferogram, 43–48
Interferometric SAR, 43, 48
International standardization agreements, 231
Interpolation, global, 115
Interpolation, bicubic spline, 119, 120, 126
Interpolation, bilinear, 82, 117–119
Interpolation, Hardy method, 120

Interpolation, linear, 78, 82–84, 117, 170, 174, 178, 179, 235, 259
Interpolation, local, 115
Interpolation, Lu function, 122
Interpolation, nearest point, 82
Interpolation, patchwise, 115
Interpolation, pointwise, 84, 116
Interpolation, random-to-grid, 68, 71, 81–83, 87, 163, 166, 241
Interpretation, DTM, 267
Intervisibility, 281–283, 291

Kernel function, 120, 121
Koch line, 15, 16
Kriging, 9, 117

Lambert cosine law, 258
Least-squares collocation, 126
Level of abstraction, 194
Level of detail, 191, 194, 279
LIDAR, See Light detection and ranging
Light detection and ranging, 50–53, 62, 227
Linear depression, 252
Linear stretching, 252
Line-following, 56–58
Line-of-sight, 281, 282
Local interpolation, 115
Local optimization procedure, 89
Local surface, 72, 73, 123
LOD, See Level of detail
LOP, See local optimization procedure
Lossless compression, 226–229

Mate contour map, 246
Mate DTM, 244
Mathematical morphology, 112, 113
Maximum unambiguous range, 54
Medial axis, 104
Medial axis transform, 103
Metadata, 213–216, 225, 226
Mirror reflection, 258, 259
Model, conceptual, 5
Model, functional, 5
Model, mathematical, 5, 6, 9, 10, 131, 173, 178, 186, 187, 250
Model, physical, 5, 6
Model, qualitative, 5
Model, quantitative, 5
Model, stochastic, 5, 116, 131
Model, sunlight, 292
Model, wind, 291
Modifiable areal unit, 192
Moving averaging, 127–131
Moving surface, 130–132
Multiple-flow direction, 277
Multiple-scale representation, 191, 196, 197, 199, 200

Multi-scale representation, metric, 196, 200
Multi-scale representation, visual, 196, 197, 205

National spatial data infrastructure, 208, 225, 229
Natural Principle, 194, 200–204
Nearest point interpolation, 82
Network, drainage, 276, 279
Network, DTM, 66, 216
Network, grid, 66, 71, 75, 79, 80, 87, 166
Network, triangular, 66, 71, 75, 77, 79, 80, 87, 89, 99, 129, 199, 206, 216, 264, 301
Normal distribution, 134, 160, 163, 167, 181

Object-relational database, 218
Operational strategy, 212
Optimal linear predictor, 228
Optimum sampling intervals, 10
Orthoimage, 1, 135, 243, 244, 288
Overpasses, 297

Page flipping technique, 263
Parsimony, principle of, 6
Pass, 74, 116
Patchwise interpolation, 115
Perspective contour diagram, 4
Phase unwrapping, 48
Phong model, 258, 259
Physical model, 5, 6
Plan curvature, 274, 275
Point cloud, 55
Point deletion, 98, 99
Point insertion, 95, 98, 110, 111
Pointwise interpolation, 84, 116, 148
Polynomial, general function, 66, 67, 70
Polynomial surface, 119, 123, 131, 136, 148, 238
Precision farming, 293
Prediction by area, 160
Prediction by production, 159
Profile curvature, 16, 274, 275
Profiling, 23, 25, 27, 69, 140, 164
Progressive rejection process, 78
Progressive sampling, 10, 25–27, 70
Projection area, 267, 268
Projective interpolation, 9
Pyramid representations, 197, 198, 200

QTM, See Quaternary triangular mesh
Quadtree structure, 111, 198, 199, 204, 227
Qualitative descriptors, 13
Qualitative model, 5
Quality control, 9, 115, 133, 135
Quantitative descriptors, 13
Quantitative model, 5
Quaternary triangular mesh, 300

Radarclinometry, 40
RADARSAT, 50
Random data, 27
Random noise, 134, 137–139, 142
Random sampling, 22
Random-to-grid interpolation, 68, 71, 81–83, 87, 163, 166, 241
Range equations, 49
Range resolution, 41, 42, 53
Rate of change, 20, 249, 263, 275
Ratio of distances, 194
Real-time optimally adapting meshes, 207
Rectification, 11, 289
Region, 218–222
Regional surface, 72, 148
Regular grid sampling, 25, 26, 69, 70, 77, 80, 171, 189
Removal, triangle, 206
Removal, vertex, 206, 207
Rendering, 249–250, 253–254, 259–260, 264, 291
Resampling, 46, 81, 166, 171 177
Resolution, azimuth, 41–43
Resolution, range, 41, 42, 53
Ridge line, 23, 99, 166, 223, 276, 278, 280
ROAM. See Real time optimally adapting meshes
Robustness, 6, 188
Roughness parameters, 19, 267, 275
Roughness vector, 18, 19
Round-off error, 138
Run-length encoding, 227, 231

Sampling strategy, 21, 24, 25, 137
Sampling theorem, 21, 22
Sampling, composite, 26, 28, 69, 74, 80, 216
Sampling, progressive, 10, 25–27, 70
Sampling, random, 22
Sampling, systematic, 22
SAR, See Synthetic aperture radar
Scale spectrum, 192
Scanner, drum, 58
Scanner, flat-bed, 58
Scanning lines algorithm, 257
Seamless pan-view, 264
Selective sampling, 24, 26, 69, 182
Semivariogram, 17, 18, 175
Shading, 2, 241, 250–251, 254, 257, 259, 260, 263, 264
Shading, hill, 251
Shading, slope, 251
Simple incremental algorithm, 94
Simplicity, 6, 53, 100, 139, 188, 277, 297
Simplification, 191, 194, 206, 207
Simplification, budget-based, 207
Simplification, fidelity-based, 207

INDEX

Single-flow direction, 276
Skeleton circle, 105
Skeleton lines, 102–104, 106, 114
Skeleton points, 104
Slicing, 239, 240
Slope consistency, 143
Slope constraining test, 143
Slope shading, 251
Smallest circumcircle, 98
Smallest visible object, 201
Smallest visible size, 201
Sobel operator, 240
Soil erosion, 11, 276, 293
Source reduction, 227, 228
Specific catchment area, 276
Specification, DTM accuracy, 162
Specification, map, 20, 35, 174, 188
Specular reflection, 258
Spherical surface tessellation, 300
Spherical terrain modeling, 300
Spherical TINs, 301
Spherical Voronoi diagrams, 301
SPOT, 3, 15, 34, 62, 200, 201
STANAG 3809, 231
Stereo contour map, 243, 246
Stereo pair, 9, 31, 36, 38, 135
Stereocomparator, 36
Stereomate, 233, 243–246
Stochastic model, 5, 116, 131
Stream power index, 276
String data, 27, 28
Structuring element, 113
Sunlight model, 292
Superimposition, 136
Surface area, 267, 268, 275
Surface reconstruction, 65, 66, 70, 104, 115, 166, 174, 177
SVS, See Smallest visible size
Synthetic aperture radar, 39, 40, 42
Systematic error, 134
Systematic sampling, 22

Tangent of gradient, 281
Terrain classification, 20
Terrain descriptor, 13, 14, 18, 20, 175
Terrain roughness, 18, 19, 140, 175, 292
Terrain roughness vector, 18, 19
Texture mapping, 250, 260, 261
Thiessen diagram, 73
Thiessen polygon, 73, 109
Threading, 9, 237, 242
Three-point predictor, 229
Tie points, 47
Tile, 212, 218–221

TIN, see triangular irregular network
Total contributing area, 276
Transfer function, 175
Transformation in scale dimension, 194, 201
Trend, 136
Trend analysis, 136
Triangle collapse, 206
Triangle removal, 206
Triangular irregular network, 66, 77, 79, 80, 87, 88
Triangulation, Bowyer–Watson algorithm, 94
Triangulation, manual, 109
Tunnel, 286, 297

Uniform distribution, 169
Unimodal distribution, 167

Valley, 23, 106, 166, 170, 276, 292, 299
Variables, color, 247
Variables, dynamic, 249, 263
Variables, geometric, 247
Variables, primary, 247
Variables, secondary, 248
Variables for visualization, 247
Variogram, 17, 175–177
Vertex removal, 206, 207
Viewpoint animation, 263
Viewshed, 267, 281, 283, 284
Visibility matrix, 284
Visible surface identification, 256
Visual impact analysis, 296
Visual inspection, 136
Visual landscape, 251
Voronoi diagram, 69, 73, 79, 88, 103, 104, 107–113, 129, 130, 293, 300, 301, 306
Voronoi diagram, raster-based algorithms, 111
Voronoi diagram, vector-based algorithms, 108
Voronoi region, 73, 79, 109, 110, 112, 114, 129, 130

Walk-through, 263, 266
Walk-through algorithm, 95
Water conservancy, 285, 286
Watershed, 280
Weighting function, 129, 130, 139
Wigner–Seitz cells, 73
Wild function, 122, 123
Wind model, 291
Window size, 140, 149, 150, 152–154

Z-buffer algorithm, 257
Zero stereo model, 135

**GA 139 .L5 2005
Li, Zhilin, 1960-
Digital terrain modeling**